MW01470675

Designing and Applying Treatment Technologies

Remediation of Chlorinated and Recalcitrant Compounds

Editors

Godage B. Wickramanayake
Battelle

Robert E. Hinchee
Parsons Engineering Science, Inc.

The First International Conference
on Remediation of Chlorinated and
Recalcitrant Compounds

Monterey, California, May 18–21, 1998

 BATTELLE PRESS

Columbus • Richland

Library of Congress Cataloging-in-Publication Data

International Conference on Remediation of Chlorinated and Recalcitrant Compounds
(1st : 1998 : Monterey, Calif.)
    Designing and applying treatment technologies : remediation of chlorinated and
recalcitrant compounds / editors, Godage B. Wickramanayake, Robert E. Hinchee.
        p.   cm.
    "First International Conference on Remediation of Chlorinated and Recalcitrant
Compounds, Monterey, California, May 18–21, 1998."
    Includes bibliographical references and index.
    ISBN 1-57477-061-6 (hardcover : alk. paper)
    1. Organochlorine compounds--Biodegradation--Congresses. 2. Hazardous
waste site remediation--Congresses. 3. Hazardous waste sites--Design and
construction--Congresses. I. Wickramanayake, Godage B., 1953–   .
II. Hinchee, Robert E. III. Title.

TD1066.073I58   1998b
628.5'2—dc21   98-25126
                                                                                CIP

Printed in the United States of America

Battelle Press
505 King Avenue
Columbus, Ohio 43201, USA
614-424-6393 or 1-800-451-3543
Fax: 1-614-424-3819
Internet: press@battelle.org
Website: www.battelle.org/bookstore

For information on future symposia and conference programs, write to:
    Bioremediation Symposium
    Battelle
    505 King Avenue
    Columbus, Ohio 43201-2693
    Fax: 614-424-3667

CONTENTS

Barrier Design and Construction

Performance Evaluation of Permeable Barriers

Remediation of Pesticides/Herbicides

Regulation and Remediation of PCBs/Dioxins

Remediation of Explosives/Nitroaromatics

Applying Multiple Remediation Technologies

FOREWORD

Remediation practitioners and site managers often face complex environmental contamination challenges that prove to be either resistant to conventional remediation approaches or too costly to treat by such methods. Fortunately, there are new technologies waiting in the wings, and sometimes a combination of existing and/or emerging technologies proves to be the most cost-effective solution. *Designing and Applying Treatment Technologies: Remediation of Chlorinated and Recalcitrant Compounds* combines technical guidance on some of the newer remediation technologies with case studies on remediation strategies that have worked effectively at real-world sites. Chapters cover bench-scale testing, modeling, and performance evaluation for permeable barriers; barrier design and construction; remediation of explosives and nitroaromatics; remediation of pesticides and herbicides; regulation and remediation of PCBs/dioxins; emerging technologies; and applying multiple remediation technologies.

This is one of six volumes published in connection with the First International Conference on Remediation of Chlorinated and Recalcitrant Compounds, held in May 1998 in Monterey, California. The 1998 Conference was the first in a series of biennial conferences focusing on the more problematic substances—chlorinated solvents, pesticides/herbicides, PCBs/dioxins, MTBE, DNAPLs, and explosives residues—in all environmental media. Physical, chemical, biological, thermal, and combined technologies for dealing with these compounds were discussed. Several sessions dealt with natural attenuation, site characterization, and monitoring technologies. Pilot- and field-scale studies were presented, plus the latest research data from the laboratory. Other sessions focused on human health and ecological risk assessment, regulatory issues, technology acceptance, and resource allocation and cost issues. The conference was attended by scientists, engineers, managers, consultants, and other environmental professionals representing universities, government, site management and regulatory agencies, remediation companies, and research and development firms from around the world.

The inspiration for this Conference first came to Karl Nehring of Battelle, who recognized the opportunity to organize an international meeting that would focus on chlorinated and recalcitrant compounds and cover the range of remediation technologies to encompass physical, chemical, thermal, and biological approaches. The Conference would complement Battelle's other biennial remediation meeting, the In Situ and On-Site Bioremediation Symposium. Jeff Means of Battelle championed the idea of the conference and made available the resources to help turn the idea into reality. As plans progressed, a Conference Steering Committee was formed at Battelle to help plan the technical program. Committee members Abe Chen, Tad Fox, Arun Gavaskar, Neeraj Gupta, Phil Jagucki, Dan Janke, Mark Kelley, Victor Magar, Bob Olfenbuttel, and Bruce Sass communicated with potential session chairs to begin the process of soliciting papers and organizing the technical sessions that eventually were presented in Monterey. Throughout the process of organizing the Conference, Carol Young of Battelle worked tirelessly to keep track of the stream of details, documents, and deadlines involved in an undertaking of this magnitude.

Each section in this and the other five volumes corresponds to a technical session at the Conference. The author of each presentation accepted for the Conference was invited to prepare a short paper formatted according to the specifications provided. Papers were submitted for approximately 60% of the presentations accepted for the conference program. To complete publication shortly after the Conference, no peer review, copy-editing, or typesetting was performed. Thus, the papers within these volumes are printed as submitted by the authors. Because the papers were published as received, differences in national convention and personal style led to variations in such matters as word usage, spelling, abbreviation, the manner in which numbers and measurements are presented, and type style and size.

We would like to thank the Battelle staff who assembled this book and its companion volumes and prepared them for printing. Carol Young, Christina Peterson, Janetta Place, Loretta Bahn, Lynn Copley-Graves, Timothy Lundgren, and Gina Melaragno spent many hours on production tasks. They developed the detailed format specifications sent to each author, tracked papers as received, and examined each to ensure that it met basic page layout requirements, making adjustments when necessary. Then they assembled the volumes, applied headers and page numbers, compiled tables of contents and author and keyword indices, and performed a final page check before submitting the volumes to the publisher. Joseph Sheldrick, manager of Battelle Press, provided valuable production-planning advice and coordinated with the printer; he and Gar Dingess designed the volume covers.

Neither Battelle nor the Conference co-sponsors or supporting organizations reviewed the materials published in these volumes, and their support for the Conference should not be construed as an endorsement of the content.

Godage B. Wickramanayake and Robert E. Hinchee
Conference Chairman and Co-Chairman

Sequential Anaerobic/Aerobic Biodegradation of Chlorinated Solvents: Pilot-Scale Field Demonstration

Ronald F. Lewis (U.S. EPA, Cincinnati, Ohio)
Maureen A. Dooley, Jaret C. Johnson and Willard A. Murray (ABB Environmental
Services, Wakefield, Massachusetts)

INTRODUCTION

ABB Environmental Services, Inc. (ABB-ES) has developed a Two-Zone Plume-Interception Treatment Technology designed for enhanced, in-situ biodegradation of chlorinated and non-chlorinated organic solvents in ground water and saturated soils. The technology is designed to enhance the natural biodegradation process for a more rapid and complete degradation of the organic contaminants to non-toxic compounds (e.g., ethylene) or to complete mineralization. This paper presents the results from a field demonstration which has been operated and partially funded as an Emerging Technology under the Superfund Innovative Technology Evaluation (SITE) Program by the U.S. Environmental Protection Agency (USEPA).

Technical Approach. The technology consists of a two-zone process for enhanced bioremediation of PCE and TCE using sequential anaerobic/aerobic biological degradation. The two zones - first anaerobic followed by aerobic - can either be sequential in space or in time. The first zone is designed to operate under highly reducing conditions to stimulate anaerobic bacteria (e.g. methanogens, sulfate reducers) to dechlorinate the solvents. Complete dechlorination can occur - that is, transformation of PCE to TCE to dichloroethylene (DCE), vinyl chloride (VC) and ultimately ethylene. But in some cases the dechlorination of DCE to VC is slow, and VC to ethylene even slower. Aerobic biodegradation of DCE and VC, on the other hand, is relatively fast. Therefore a second zone is designed to stimulate methanotrophic bacteria to complete the degradation by oxidizing the DCE and VC after the methanogens have reduced the PCE and TCE.

CASE STUDY: WATERTOWN, MA

Site Description. The site is situated in a historically industrial section of Watertown, MA. The general soil profile consists of approximately 13 feet of sand and gravel over approximately 7 feet of silty sand; then glacial till (an aquitard) is encountered. Groundwater occurs at approximately 8 feet below land surface and is contaminated with chlorinated solvents, including PCE, TCE and degradation products characteristic of natural biological reductive dechlorination.

Feasibility of In-Situ Bioremediation. Treatability studies, consisting of bench-scale laboratory tests of both anaerobic and aerobic biodegradation, were conducted using the site groundwater. In summary, the treatability study findings indicated 1) complete biodegradation to ethylene was occurring naturally at the site, 2) no toxic conditions to inhibit bacterial growth, 3) biodegradation could be enhanced best using lactic acid but methanol also worked, and 4) aerobic biodegradation of both TCE and DCE was rapid. Therefore, the in-situ enhancement of natural biological degradation activity appeared to be a feasible treatment technology.

Design, Construction and Operation. In the field demonstration, groundwater was extracted from three downgradient wells, nutrients and a carbon source were added to the water stream, and the amended water injected into three wells 17 feet upgradient. This recirculating treatment cell was a "temporal" two zone design in which the system was first maintained under anaerobic conditions and monitored for progress in degrading the contaminants. When most of the PCE and TCE had been anaerobically transformed to DCE and VC, the cell was converted to aerobic conditions through the addition of oxygen (and methane as needed).

Five 2-inch PVC monitoring wells were positioned between the injection and extraction wells to monitor the progress of the biodegradation process; and a groundwater recirculation rate of 0.25 gallons per minute (gpm) established a single recirculation cell of about 30 feet in diameter.

Operation. The system was in full operation from November 1996 to November 1997 with a relatively constant recirculating flow rate of 0.25 gpm and an amendment injection rate of 10 milliliters/minute (or about 4 gallons per day which is about 1% of the recirculating flow).

Establishment of Anaerobic Conditions. Amendments initially added to the recirculating flow to enhance anaerobic biological dechlorination, included ammonia chloride (25 milligrams per liter [mg/L]), potassium tripolyphosphate (25 mg/L), lactic acid (100 mg/L initially), yeast extract (5 mg/L), and sodium hydroxide (sufficient to neutralize the pH of the amendment batch).

Beginning in April 1997, amendments were pulsed into the system rather than metered in continuously and the lactic acid concentration was increased from approximately 100 mg/L to 250 mg/L to lower redox conditions and attempt to enhance reductive dechlorination processes. These conditions were maintained until June 1997, when the lactic acid concentration was increased to approximately 350 mg/L. The anaerobic phase was maintained for approximately 8 months until late July, 1997, when it was switched to an aerobic system by introducing oxygen into the injection wells.

Establishment of Aerobic Conditions. Aerobic conditions were established by suspending oxygen release compound (ORC) contained in "socks" inside the three injection wells (July 1997). This system provided a continuous release of dissolved oxygen into the recirculating groundwater flow. Approximately 1 month was required to degrade residual carbon in the system and create aerobic conditions in the injection wells. At the time the system was converted to aerobic conditions, most of the TCE had been transformed to cis-DCE and VC. DCE and VC levels were approximately 2 mg/L each when aerobic conditions were established.

Methane levels were approximately 150 to 200 micrograms per liter (ug/L) in groundwater at the beginning of the field demonstration but had decreased to approximately 50 to 100 ug/L when the aerobic phase of treatment began. Since aerobic/methanotrophic bacteria (which known to biodegrade VOCs) require methane for their activity, methane-saturated water was added to the system on a weekly basis to increase the methane concentration in groundwater. This methane addition was initiated approximately two months after the ORC "socks" were placed in the wells.

Results. Results of the enhanced, in-situ bioremediation indicate that under anaerobic conditions significant reductions in the concentration of TCE were observed (12 mg/L to less than 1 mg/L). There was no substantial degradation over the first 4 months, but after that, TCE concentrations were reduced and transformation products cis-DCE, VC and ethylene were generated. Bar graphs showing VOC concentrations in wells along the flow path from the injection wells to the extraction wells are presented in Figures 1a, 1b and 1c. These figures show the spatial distribution of VOCs at three points in time - February 11, April 23 and July 21, 1997. It can be seen that in February 1997 (as it was since December 1996) the concentrations of ethenes were fairly uniform throughout the recirculation cell (PCE = 1500 ug/L, TCE = 12000 ug/L, DCE = 3500 ug/L, and VC = 100 ug/L). By April 1997, a little more than 2 months later, some significant transformations had taken place with the ratio of TCE to DCE much smaller, and some increases in VC. By July 1997, some 3 months later, it can be seen that the PCE had been reduced to about 100 ug/L or less, TCE was now less than 1000 ug/L, DCE was generally much greater than TCE (a complete reversal from February), and VC had increased in all wells. All of this is an indication of a significant amount of reductive dechlorination.

Figure 2A shows concentrations of VOCs versus time in well EPA-2. It can be seen that significant reductive dechlorination was not apparent until March or April 1997, 4 to 5 months after the initiation of amendment addition to the treatment cell. After this lag period, the ratio of DCE to TCE began to increase significantly, however, substantial increases in VC did not occur until July 1997. At this point in time, PCE and TCE levels had decreased to less than 10% of their initial concentrations, DCE had increased marginally, and VC had increased by an order of magnitude. Figure 2C shows the time variation of redox for one extraction well, one injection well and all five monitoring wells. It can be seen that after March or April 1997 when most of the reductive dechlorination occurred, the redox had been lowered to -100 or less. Methane levels continuously declined from about 0.2 to 0.3 mg/L at the beginning of the field demonstration, to about 0.05 to 0.1 mg/L in July. It is therefore apparent that methanogenic conditions were not achieved during the anaerobic phase, and most of the dechlorination had probably occurred due to the action of sulfate-reducing bacteria. There was evidence of sulfate reduction based upon the detection of sulfide in wells during the last two months of the anaerobic study.

There had been an overall reduction (about 80%) in total VOC mass (as moles) in well IN-2. Initial mass of VOCs (February 1997) was 200 umoles, and approximately 40umoles of VOCs remained in July 1997, which was the end of the anaerobic phase of treatment. Results from volatile acid analysis indicated that lactic acid was transformed to acetic acid and propionic acid.

Aerobic conditions were established about one month after placing ORC in the injection wells. There was no apparent change in the concentration of VOCs over that period, but over the next month VOC levels (primarily DCE and VC) started to decrease. VOC results are presented for well EPA-3 (Figure 2B) and there is an apparent reduction in DCE and VC. DCE epoxide, which is a transient biodegradation product formed when DCE is aerobically biodegraded by methanotropic bacteria, was detected during the last two sampling events confirming aerobic VOC degrading bacteria were stimulated.

CONCLUSIONS
Anaerobic biodegradation of PCE and TCE was enhanced through the addition of organic and inorganic nutrients at the field demonstration site in Watertown, MA. Initial results from the aerobic phase indicated no immediate reduction in the concentration of VOCs when oxygen was added, but this was likely due to the presence of residual carbon from the anaerobic phase (lactic acid). Data (VOC, methane, DO, redox) will continue to be collected now that aerobic conditions have been established.

Specific conclusions that can be drawn from the results at the Watertown field demonstration are:

- The reductive dechlorination of PCE and TCE to DCE, VC and ethene has been accomplished primarily by sulfate-reducing bacteria.
- A time lag of about 4 months was required before significant reductive dechlorination occurred. This corresponded to the time and lactic acid dosing required to reduce the redox to about -100 throughout the treatment cell.
- If the lactic acid dosage at the beginning had been higher (close to the concentration used in June and July of 1997), the lag period may have been shortened.
- Sequential anaerobic-aerobic (Two-Zone) biodegradation of PCE and its degradation products appears to be a viable and cost-effective treatment technology for the enhancement of the natural reductive dechlorination process.

Designing and Applying Treatment Technologies

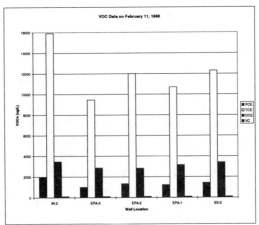

Figure 1A. VOC Results From February 11, 1997 Sampling Event

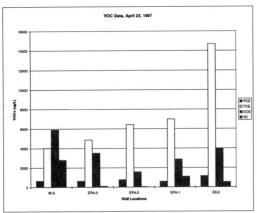

Figure 1B. VOC Results From April 23, 1997 Sampling Event

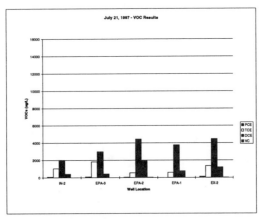

Figure 1C. VOC Results From July 21, 1997 Sampling Event

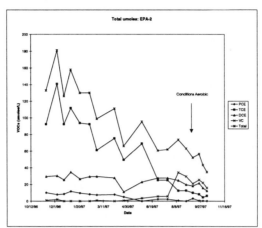

Figure 2A. VOC Results Presented as umoles from Well EPA-2

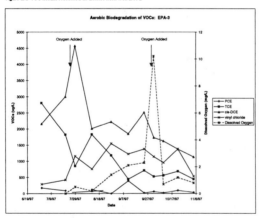

Figure 2B. VOC Results from Well EPA-3 During Aerobic Phase

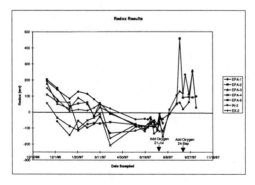

Figure 2C. Results from Redox Analysis

DESIGN AND OPERATION OF A HORIZONTAL WELL, *INSITU* BIOREMEDIATION SYSTEM

Brent Weesner, ConsuTec, Inc., Tampa, FL, USA
Steve Acree, U.S. Environmental Protection Agency, Ada, OK, USA
Todd McAlary, Beak International Incorporated, Guelph, Ontario, Canada
Joseph J. Salvo, General Electric Corp. (GE) R & D, Schenectady, NY, USA

ABSTRACT: A large field demonstration using nutrient addition to stimulate insitu anaerobic bioremediation of chlorinated solvent contaminated soil and ground water was performed at the former U.S. Department of Energy Pinellas Plant in Largo, Florida from January through June, 1997. Insitu anaerobic bioremediation was evaluated after recommendation by participants in the Innovative Treatment Remediation Demonstration (ITRD) Program, as an innovative technology that could accelerate the removal and destruction of chlorinated volatile organic compounds (VOCs) in a sandy, shallow, anaerobic aquifer at the Northeast Site.

Laboratory batch and column studies conducted through the ITRD Program using site soil and ground water showed that a population of anaerobic microorganisms existed capable of remediating the chlorinated contaminants of concern. These studies suggested that contaminant degradation could be significantly enhanced by the addition of nutrients such as benzoate, lactate, and methanol. Therefore, the challenge of the design and operation of the field bioremediation demonstration was to insure proper nutrient delivery to the treatment area.

System design required a relatively accurate characterization of site contamination and ground water flow directions and rates since the study area included about four soil layers with higher VOC concentrations in layers of lower hydraulic conductivity. Field and laboratory-scale tests were conducted to characterize the hydraulic properties. Computer modeling was used to evaluate various nutrient delivery systems. A vertical flow system with two horizontal wells and a series of surface infiltration galleries was selected to facilitate nutrient delivery vertically across the low conductivity layers.

Construction and operation of the system required groundwater extraction and infiltration capabilities, a control scheme to allow continuous operation and nutrient injection, and aggressive construction techniques for proper location of the system components. These activities, including numerical modeling, laboratory studies, construction techniques, and operational results are summarized in this paper.

INTRODUCTION

The USDOE Office of Environmental Restoration (EM-40) established the Innovative Treatment Remediation Demonstration (ITRD) Program to help accelerate the adoption and implementation of new and innovative soil and groundwater remediation technologies (Hightower et al, 1998). In September 1993, the ITRD Program chose the Pinellas Plant Northeast Site as a project location. After evaluating site conditions and reviewing a host of technologies, anaerobic bioremediation was chosen as one of the technologies to be demonstrated. Field activities associated with this demonstration were completed in July 1997 (USDOE, 1998). In the site remediation strategy, other technologies were intended to complete initial mass removal (Rice, 1998) and bioremediation was intended to be the technology that would address widespread, low level contamination. Evaluation of the site data suggested that dechlorination was already occurring but the rate of that dechlorination was unknown and there was insufficient evidence whether compounds were being completely or partially dechlorinated. Laboratory studies were performed using site soil and groundwater to evaluate the ability of the indigenous microbial population to provide complete dechlorination and further to attempt to accelerate the dechlorination through addition of electron donors (Flanagan et al, 1995). In addition to this study, EPA performed a concurrent study having similar results at the RS Kerr Environmental Research Laboratory in Ada, OK. Several combinations of electron donors were evaluated in these studies. The main chlorinated compounds of concern included methylene chloride, trichloroethene, cis-1,2-dichloroethene, and vinyl chloride. In an attempt to ensure success and because calculations showed that nutrients would not be a substantial element of the overall cost (approx. 1 percent of total project), the ITRD Pinellas Advisory Group decided to use a combination of nutrients: methanol, benzoate, and lactate. Of these, lactate was 90% of the cost.

In mid-1996, the ITRD committee agreed on an installation configuration and construction began in August. Construction of this system was completed by the end of October and system commissioning was performed in November and December. Upon satisfactory understanding of the system's operational capabilities, startup including groundwater recirculation and nutrient additions began in February 1997. A tracer study using both bromide and iodide was performed as well. Tracer additions began approximately three weeks after nutrient additions were started and were used to evaluate the distribution of nutrient rich groundwater. The elaborate monitoring scheme employed in this study makes this project one of the most well documented field trials of anaerobic cometabolic bioremediation to date.

PROJECT DETAILS
Groundwater Extraction and Infiltration. The goal in designing this system was to create as near as possible a closed-loop groundwater recirculation system. One initial design included a single horizontal extraction well installed at the base of the aquifer (approximately 28 feet bls, just above the Hawthorn Group) and three parallel infiltration trenches at the surface. Hydraulic properties were evaluated using a pumping test, a borehole flowmeter study (Acree et al, 1997), and laboratory tests of soil cores. This data was used to refine the site conceptual model and allow more refined groundwater flow predictions. At this point an external constraint on the pilot study came to the forefront. Site use and funding constraints dictated that field activities had to be completed by July 1997. This was a major limitation in that layers discovered by the borehole flowmeter had an order of magnitude lower conductivity than the bulk of the aquifer. The impact of the low

conductivity layers was that nutrient breakthrough was no longer predicted to occur within the life of the project. Soil core analysis showed that the low conductivity layers contained the highest concentrations of contaminants. The existence of these layers further supported the use of a horizontal well/vertical flow system. If vertical wells were used which relied on lateral flow from extraction to injection points, the path of least resistance would carry the nutrients through the higher conductivity layers and would leave the low conductivity layers and most of the site contamination untreated. The vertical flow system was chosen for implementation so that the path from injection to extraction had to pass the low conductivity/high concentration layers. Even though flow through these low conductivity layers would be difficult, it had to be addressed in order to obtain complete nutrient distribution. To address the time limitation, the distance from injection to extraction was cut relatively in half by locating the extraction well in the center of the aquifer (approx. 16 feet bls), deepening the surface trenches (approx. 8 feet bls), and adding a second horizontal well for injection (approx. 26 feet bls).

To monitor the performance of this system, a three dimensional monitoring network was installed consisting of 64 developed, 3/4 inch drive points. In addition, a field GC was located in the equipment shed that continuously sampled the extracted groundwater.

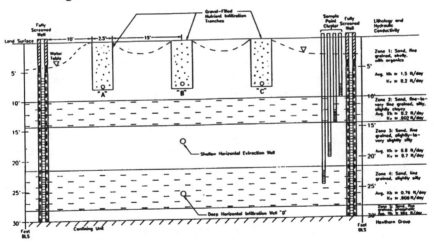

Cross—Section View Facing East
Toward Control Pad / Shed

Figure 1.

Laboratory Studies. In the first study, the methods employed both continuous recycle soil columns and serum bottles. The three nutrient treatments included: (1) complex nutrients (methanol, lactate, sulfate); (2) benzoate/sulfate; and (3) methanol alone. In the soil column studies, only the complex nutrients drove TCE to complete dechlorination (the other terminated at vinyl chloride). In the serum bottle studies, both the complex nutrients and the methanol alone treatments stimulated biodegradation of methylene chloride and drove TCE dechlorination all the way to ethene. All of the nutrient amendments included RAMM (a nutrient supplement containing inorganic nitrogen, phosphorus, calcium, magnesium, iron, and micronutrients). In the above studies, there was a significant time delay from the introduction of nutrients to the onset of dechlorination (83 to 120 days).

Modeling. Visual MODFLOW® was used to develop a quantitative understanding of the nutrient delivery rates. The modeling was performed iteratively, beginning with a simple model and relatively sparse hydrologic data and progressing to a multi-layer model input extensively with measured parameters. At each stage, simulations were performed using reasonable upper and lower estimates of key parameters. These simulations were used to guide d ata collection. This evaluation suggested that nutrient breakthrough to the extraction well would be questionable based on predicted half lives in the system. This concern initiated the idea of multiple electron donors with the consideration that the relative degradation rates would allow for the delivery of reducing power to spread throughout the treatment system. At the conclusion of the 5 month field test, the measured pumping rates, tracer data, and head distributions were used to generate a final calibrated model simulation. The final model provided numerous useful insights to assist in understanding the system (Sewell et al, 1998).

Construction Techniques. The horizontal wells were installed using the river crossing technique resulting in a surface to surface completion. One end of each of the horizontal wells was used for plumbing connections. The other end was capped so that the well could be pressurized. During installation of the shallower well, the walkover method was used for tracking the drill bit. During installation of the deeper well, the wireline method was used. The wells were constructed out of 4" HDPE pipe slotted laterally (0.020"). The slotted section included a 20/30 mesh sand-filled, 4"x2" prepack. This method of construction allows for accurately installed filter pack while maintaining a uniform outer diameter for easy pullback. This method was first used at the Pinellas Plant West Fenceline Site for installing dual-purpose air sparging/groundwater extraction wells.

    Installation of the surface trenches was challenging because the shallow sands at this site had no cohesion and would not allow a trench 8 feet deep to remain only 2 feet wide. To construct the trenches as designed, the required 8 feet was excavated and vertical sheeting was inserted to form the trench walls. Gravel was placed inside the sheets and the preexisting soils were replaced simultaneously outside the sheets until the area was built back up to original grade. At this time the sheets were removed. No surface subsidence occurred. After installation of all recirculation hardware, the monitoring network was installed.

Control Scheme. To create a constant driving force for flow, the extraction well pump (2" submersible) was operated continuously without automatic control thus setting this as the independent variable. A pressure transducer was mounted on the pump to indicate if the well was not keeping up with the pump flowrate. Because the surface trenches were the most limited flow element, these were configured for on-demand flow. Float mechanisms were constructed to control the level in the trenches within several inches. Tighter control was initially installed but the automatic valves operated too rapidly. Whenever a float in a trench dropped below the set point, a designated valve on the pump discharge line opened and allowed flow to that trench. Whenever all three trenches were full (all three valves closed), a fourth automatic valve opened and directed the extracted groundwater into the deeper horizontal well.

While recirculating the groundwater, the nutrients had to be added. This was accomplished in-line. Immediately downstream of each automatic valve was a mixing tee to which tubing from a nutrient injection pump (peristaltic) was connected. Because the injection goal was a fixed concentration for each of the electron donors (60 mg/L methanol, 120 mg/L benzoate, 180 mg/L lactate), nutrient concentrate (1%, 2%, and 3% by weight of each of these components) was volumetricly proportioned into the groundwater entering each injection line. Because the flow in each line would be different based on pressure in the line (i.e., one or more than one trench demanding water at a time), a nutrient concentrate drum was designated for each injection point so that daily groundwater flow totals could be compared to daily volume of nutrient concentrate added and pump speeds could be adjusted accordingly. After the first several weeks, the nutrient pumps required very little adjustment.

Operational Results. The system achieved steady state flow characteristics in less than a week and total groundwater flow distribution through the system was extremely consistent from one day to the next following startup. Over the course of the roughly five month operational period, several brief shutdowns of the system occurred because of extraction well redevelopments, flowmeter cleanings, and electrical outages. Upon initial startup and all subsequent startups, the surface trenches quickly filled and the system became dependent on the deeper horizontal well for recirculating the majority of the groundwater. The typical flow profile was 1.5 gpm from the extraction well, 0.9 gpm into the deep well, and 0.2 gpm to each of the trenches.

A number of conditions existed which negatively impacted the groundwater recirculation in this system. First, the low conductivity layer existing between 10 and 14 feet below land surface caused groundwater flow downward from the trenches to be diverted laterally which increased its travel time. Second, the natural groundwater elevation during the operational period was only about three feet below land surface and thus the amount of groundwater mounding (driving force) that could be created was limited. And third, the drilling fluid used to install the horizontal extraction well did not break down (decrease viscosity) as effectively as anticipated. This resulted in borehole skin effects that significantly reduced the efficiency of the well. In fact, the extraction well had to be redeveloped (through surging) approximately once a month to keep it operating at a constant rate. Because of these impediments, the shallow portion of the aquifer was not as effectively recirculated as the deeper portion.

The deeper horizontal well used for reinjection was installed in a zone of relatively low conductivity yet it was able to be operated continuously. The key to its effectiveness was that the wellhead was sealed to allow pressurization. At the operating flowrate, the pressure experienced at the wellhead was only several pounds per square inch above atmospheric. Without this ability to force water into one of the injection points, the system would not have been able to operate in a closed loop mode. Designing the system to operate in a closed loop fashion is a key element to truly performing remediation insitu.

Despite these obstacles, in less than five months, nutrients were distributed throughout a substantial portion of the study zone and given more time, complete recirculation would have been achieved. In the monitoring points with nutrient breakthrough, significant dechlorination was observed. This dechlorination was distinguished from groundwater movement in that toluene concentrations remained relatively unchanged while the chlorinated compounds showed significant reduction. Further, at a number of these points, significant increases in ethene concentrations were measured. This pilot study (1) verifies that the successes experienced in the laboratory are achievable in the field, (2) suggests that the time delay prior to onset of dechlorination experienced in laboratory may not be significant factor in the field, and (3) provides useful insights to operational hydraulics for full scale system design and operation.

REFERENCES

Acree, S., M. Hightower, R. Ross, G. Sewell, B. Weesner, 1997. *Site Characterization Methods for the Design of In-Situ Electron Donor Delivery Systems.* '97 Battelle Conference, New Orleans, LA. Battelle Press. 4:261-266.

Flanagan, W., M. Brenan, K. Deweerd, J. Principe, J. Spivack, M. Stephens, 1995. *Anaerobic Microbial Transformation of Trichloroethylene and Methylene Chloride in Pinellas Soil and Ground Water.* GE Corporate Research and Development, Schenectady, NY. Prepared for Sandia National Laboratories.

Hightower, M., P. Beam, D. Ingle, R. Steimle, J. Armstrong, M. Trizinsky, 1998. *Cooperative Approach in Implementing Innovative Technologies at the Pinellas STAR Center.* The First International Conference on Remediation of Chlorinated and Recalcitrant Compounds, Monterey, CA. Battelle Press.

Rice, B., 1998. *Remediation Demonstration of Dual-Auger Rotary Steam Stripping,* The First International Conference on Remediation of Chlorinated and Recalcitrant Compounds, Monterey, CA. Battelle Press.

Sewell, G., B. Weesner, E. Lutz, M. deFlaun, N. Baek, B. Mahaffey, 1998. *Performance Evaluation of an Insitu Anaerobic Bioremediation System for Chlorinated Solvents.* The First International Conference on Remediation of Chlorinated and Recalcitrant Compounds, Monterey, CA. Battelle Press.

USDOE, 1998. *Cost and Performance Report, In Situ Anaerobic Bioremediation, Pinellas Northeast Site,* Innovative Treatment Remediation Demonstration, Sandia National Laboratories, Albuquerque, NM.

PERFORMANCE EVALUATION OF AN IN SITU ANAEROBIC BIOTREATMENT SYSTEM FOR CHLORINATED SOLVENTS.

Guy W. Sewell (US-EPA, Ada, OK)
Mary F. DeFlaun (Envirogen, Inc., Lawrenceville, NJ)
Nam H. Baek (Occidental Chem. Corp., Dallas, TX)
Ed Lutz (DuPont, Wilmington, DE)
Brent Weesner (ConsuTec, Inc. Tampa, FL)
Bill Mahaffey (Pelorus Environ. & Biotechnol. Corp., Evergreen, CO).

ABSTRACT: A pilot scale demonstration of nutrient injection to stimulate in situ bioremediation of chlorinated solvents was performed at the Pinellas Science, Technology and Research (STAR) Center, formerly the U.S. Department of Energy Pinellas Plant in Largo, Florida. This project was implemented (January through June, 1997) for the Innovative Treatment Remediation Demonstration (ITRD) program to evaluate reductive anaerobic biological in situ treatment technologies (RABITT) as an innovative remedy. Based on laboratory studies and additional site characterizations, a vertical flow system with two horizontal wells and a series of infiltration galleries was constructed, allowing development of an effective ground water recirculation pattern to enable continuous nutrient (benzoate, lactate, and methanol) addition and enhance system performance. The performance of this system was closely monitored by tracking ground water flow rates, measuring hydraulic head changes in the aquifer, analyzing flow paths and biotransformation pathways. A three dimensional aquifer monitoring network was used to measure nutrient movement and associated contaminant degradation. Extracted ground water was also continuously monitoring for contaminants before recirculation. Significant dechlorination was observed in areas where nutrients were effectively delivered. Evaluation of the results suggests the process is applicable and cost effective with appropriate implementation, and site characterization.

INTRODUCTION

Reductive anaerobic biological in situ treatment technologies (RABITT) for the remediation of ground water contaminated with chloroethenes is a promising approach to an all too common environmental problem. The design, implementation and evaluation of these treatment systems is a complex environmental engineering challenge requiring a clear understanding of the contaminant distribution, the hydrogeologic setting and the geochemistry. To a greater extent than with pump-and-treat systems or aerobic bioremediation, the application of microbially mediated reductive dechlorination requires that the site conceptual model be a detailed three-dimensional representation, incorporating flow/time dynamics to ensure interaction of electron donor, contaminants and active microorganisms under appropriate conditions. The U.S. Department of Energy (DOE), the U.S. Environmental Protection Agency (EPA), the State of Florida, and industry partners have been involved in the design and implementation of a pilot-scale field demonstration of RABITT to support the DOE's Innovative Treatment Remediation Demonstration (ITRD) Program (Hightower et al., 1998). The study involved the creation of an in-situ circulation cell and the injection of electron donor/nutrients to stimulate biological transformation processes mediated by indigenous microorganisms. Design of an effective and efficient system under the projected test constraints, such as electron donor half life, maximum travel time and mixing efficiency, required a three-dimensional characterization of contaminant distribution, hydrology, and geochemistry (Acree et al., 1997. Weesner et al., 1998).

SITE BACKGROUND

The site of the pilot study, at the Pinellas Science, Technology and Research (STAR) Center, is located in west central Florida and was used in the 1960's for disposal of drums of waste and construction debris. Subsurface contamination, as indicated by contaminant concentrations in ground water samples from monitoring wells, is heterogeneously distributed in the shallow aquifer. The geology of the upper 30 ft (9.1 m) of the saturated zone is predominantly fine sands with varying fractions of silt and clay. Fill material, construction debris, and lagoon sediments are present in the first few feet below land surface. The top of the Hawthorn formation is encountered at a depth of about 30 ft in this area and consists of clay and limestone. The water table at the site varies seasonally from 3 to 5 ft. below land surface. Ground water flow at the site is locally influenced by a ground water extraction system that is currently in operation. Bulk hydraulic conductivity of aquifer materials was estimated using data from a 72-hour multi-well pumping test. Estimates of horizontal hydraulic conductivity clustered in a relatively narrow range from approximately 1 ft/d to 3 ft/d (0.3-0.9 m/d). Estimates of vertical hydraulic conductivity were less certain. The most reliable data set indicated that a horizontal to vertical anisotropy ratio of about 10:1 may be representative of bulk conditions in the shallow saturated zone of interest. The design chosen for the pilot study incorporated infiltration galleries and horizontal wells for fluid circulation that could be scaled up in a cost-effective manner.

The two primary objectives of this pilot study were to evaluate the use of nutrient injection to enhance *in situ* anaerobic biological degradation of chlorinated VOCs in areas of moderate contaminant concentrations and to obtain operating and performance data to optimize the design and operation of a full-scale system. During the short operational period of this pilot study, the reduction of contaminants to specific regulatory levels was not the primary emphasis.

PILOT TEST DESIGN

The pilot system was located in an area of the site that was projected to have chlorinated contaminant concentrations in ground water compatible with the targeted biotreatment range (10 - 100 ppm). The bioremediation pilot system consisted of three parallel, 8-ft deep, infiltration galleries and two horizontal wells (16 and 26 ft BLS) with 30 ft screened intervals (US-DOE. 1998). The study area was approximately 45 feet by 45 feet and extended from the surface down to a thick clay confining layer 30 feet below the surface. Ground water was extracted from the upper horizontal well (16 ft BLS) and recirculated via the surface trenches and the lower horizontal well (26 ft BLS). Benzoate, lactate, and methanol were added as electron donors, bromide (26 ft horizontal well) and iodide (surface infiltration galleries) were added as conservative tracers. The nutrient concentrations were selected based on a laboratory treatment study conducted through the ITRD Program (Flanagan et al., 1995) and earlier studies by the US-EPA (Sewell and Gibson, 1991, Gibson and Sewell, 1992). VOCs, tracers, organic acids and metabolic products, such as ethene, were monitored at 16 well clusters distributed throughout the test area. Each cluster consisted of 24 in. screened sample points at 4 vertically discrete intervals,(A 8-10 ft BLS, B 12-14 ft, C 18-20 ft, D 22-24 ft). VOC concentrations were also monitored in the extracted ground water from the upper horizontal well (16 ft BLS).

RESULTS AND DISCUSSION

The system operated from February 7, 1997 to June 30,1997. During this period, ground water was extracted and recirculated at a rate of approximately 1.5 gpm. Roughly 250,000 gallons of water, or about two pore volumes were circulated during the pilot study period (5 months). Tracer and nutrient monitoring data indicated that electron donors were delivered to 90% of the central treatment area during operations (US-DOE. 1998).

Figure 1 shows contaminant concentration data for the 16 ft BLS horizontal extraction well. After 50 days of operation, parent contaminants such as trichloroethene (TCE) and methylene chloride were decreasing in concentration. After 70 days of operation, the aqueous concentrations of these contaminants had decreased to minimum quantification levels. The daughter products *Cis* 1, 2-dichloroethene (*cis*-DCE) and vinyl chloride, showed a general decline in concentration, but not as significant a reduction as seen with TCE. This would seem to fit the conceptual model of the RABITT process, in that, while these contaminants are degraded they are also transiently produced as intermediates. Toluene concentrations seemed to show the smallest relative decrease. Again this was consistent with our conceptual model of the microbial transformation processes with RABITT application. Addition of electron donors is not predicted to significantly augment the oxidative catabolism of toluene. While there was a general decrease in ground water contaminant concentrations (Figure 1) which is often seen with extraction processes, it is important to note that only site water was recirculated, that uncontaminated water was not injected and that the pilot test site resided within a larger contaminated area at the site.

Extraction Well

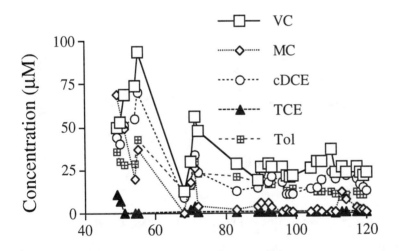

FIGURE 1. Concentration of contamiants vs time in produced ground water from the 16 ft BLS horizontal extraction well. VC-vinyl chloride, MC-methylene chloride, cDCE-*cis* 1,2-dichloroethene, TCE-trichloroethene, Tol-toluene.

While tracer breakthrough and nutrient delivery was eventually detected for most monitoring locations in the test area, only level D monitoring points were seen to have significant exposure of electron donor in terms of both time and concentration. As such, evaluation of the in situ process was limited to the D level. Figure 2 is representative concentration data from a level D monitoring point. To an even greater extent than the extraction well results, the data in Figure 2 supported our conceptual model of the RABITT process. After 30 days of operation, TCE, methylene chloride, *cis*-DCE and vinyl chloride concentrations had decreased to minimum quantification levels. The decreases in contamiant concentrations coincided with the arrival of added electron donor as indicated by the arival of bromide at the 8D monitoring location and direct measurement of benzoate and lactate (data not shown). Toluene concentrations again showed some decrease, but much less than the reductively transformed chlorinated compounds.

Monitoring Point 8D

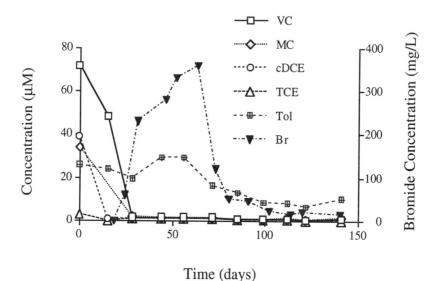

Time (days)

FIGURE 2. Concentration of contamiants vs time in monitoring well 8D. VC-vinyl chloride, MC-methylene chloride, cDCE-*cis* 1,2-dichloroethene, TCE-trichloroethene, Tol-toluene, Br-bromide.

Figure 3 represents the comparable data set for the 8B location. No significant decrease in concentrations were observed for the contaminants, with the possible exception of vinyl chloride. During the test, level B proved to be the most difficult level to treat due to the low hydraulic conductivities associated with the layer. No tracer (iodide) breakthrough was seen at monitoring point 8B (Figure 3). Direct analysis of ground water sample from the 8B location did not yield significant levels of electron donors (benzoate, lactate, methanol or acetate) or of ethene (data not shown). Thus while level B and location 8B were under the same gradient, if not the same flow conditions as level D and location 8D (located 10 ft directly below 8B), we did not have evidence of significant nutrient delivery or chemical augmentation of the dechlorination process.

CONCLUSIONS

At monitoring points where nutrient breakthrough was observed, significant declines in total chlorinated VOC concentrations (70-99%) were generally observed. These values correlate well with the results from the extraction well. For those wells where electron donor arrival was not observed, generally in areas of lower permeability or in perimeter wells, only modest contaminant reductions were recorded. Degradation rates as high as 1-2 ppm per day were observed in higher concentration areas (US-DOE, 1998), while in areas with lower concentrations degradation rates ranging from 0.05 to 0.10 ppm per day were observed. There was little evidence of significant daughter product buildup at monitoring wells after tracer breakthrough.

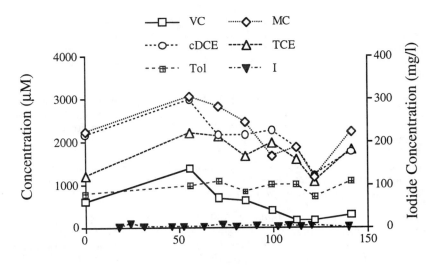

Monitoring Point 8B

Time (days)

FIGURE 3. Concentration of contamiants vs time in monitoring well 8B. VC-vinyl chloride, MC-methylene chloride, cDCE-*cis* 1,2-dichloroethene, TCE-trichloroethene, Tol-toluene, I-Iodide.

The cost of the pilot system totaled approximately $400,000, with over half the costs associated with sampling and analyses. Most of the sampling and analyses were discretionary and were used to verify the system concept and design. This level of sampling would not be needed during a full-scale bioremediation project. System construction costs were about $90,000, while operating costs were about $30,000 or $0.12 per gallon of water treated. The extensive modeling, hydrogeologic, nutrient transport, and operating cost data (Weesner et al 1998) developed during this pilot system operation suggest that the Northeast Site could be remediated using nutrient injection in approximately 2-3 years at a cost of about $4-6M. From the results of the pilot study, nutrient addition to stimulate *in* situ anaerobic biological degradation of chlorinated solvent contaminated soil and ground water appears to be a feasible and cost-effective remediation approach at the Pinellas Northeast Site for areas with moderate contaminantion levels. The limiting factors for successful, cost-effective implementation include the ability to deliver appropriate nutrients to all contaminated areas, and hydraulic travel times.

ACKNOWLEDGMENTS

We would like to thank the other ITRD members who have contributed to this effort. We would also like to thank Mike Cook and Frank Beck of the US-EPA for their support of the field research efforts. Although the research described in this paper is supported in part by the US-Environmental Protection Agency through an in-house research program, it has not been subjected to Agency review and therefore does not necessarily reflect the views of the Agency, and no official endorsement should be inferred.

REFERENCES

Acree, S.D., M. Hightower, R.R. Ross, G.W. Sewell, and B. Weesner. 1997. "Site Characterization Methods for the Design of In-Situ Electron Donor Delivery Systems." In B. C. Alleman and A. Leeson (Eds.), *Proceeding of the Battelle Conference on In-Situ and On-site Bioremediation.* April 27-May 1. New Orleans, LA. Battelle Press. 4:261-266.

Flanagan,W., M. Breenan, K. Deweerd, J. Principa, J. Spivak, and M. Stephans. 1995. Anaerobic Microbial Transformation of Trichloroethylene and Methylene Chloride in Pinellas Soil and Ground Water. General Electric Research and Development Center, Schenectady, NY

Gibson, S.A., and G.W. Sewell. 1992. "Stimulation of Reductive Dechlorination of Tetrachloroethene in Anaerobic Aquifer Microcosms by Addition of Short-Chain Organic Acids or Alcohols." *Appl. Environ. Microbiol.* 58:(4) 1392-1393.

Hightower, M., P. Beam, D. Ingle, R. Steimle, J. Armstrong and M. Trizinsky. 1998. "Cooperative Approach in Implementing Innovative Technologies at the Pinellas Plant." In *Proceeding of the First International Conference on Remediation of Chlorinated and Recalcitrant Compounds.* May 18-21. Monterey, CA. Battelle Press.

Sewell, G.W., and S.A. Gibson. 1991. "Stimulation of the Reductive Dechlorination of Tetrachloroethene in Anaerobic Aquifer Microcosms by the Addition of Toluene." *Environ. Sci. Technol.* 25: (5) 982-984.

Weesner, B., S.Acree, T. McAlary and J. Salvo. 1998. "Design and Operation of a Horizontal Well In Situ Bioremediation System." In *Proceeding of the First International Conference on Remediation of Chlorinated and Recalcitrant Compounds.* May 18-21. Monterey, CA. Battelle Press.

US-DOE. 1998. Cost and Performance Report-In Situ Anaerobic Bioremediation Pinellas Northeast Site Largo, FL. Innovative Treatment Remediation Demonstration U.S. Department of Energy.

SOIL RESTORATION WITH SUPERHEATED STEAM REMEDIATION, RECYCLE, AND RECOVERY (SUPERSTR3)

David H. Holcomb (Focus Environmental, Inc., Knoxville, Tennessee)
David M. Pitts, P.E. (Focus Environmental, Inc., Knoxville, Tennessee)
William L. Troxler, P.E. (Focus Environmental, Inc., Knoxville, Tennessee)
Primo Marchesi, P.E., (Consultant, Surfside Beach, South Carolina)
John P. Cleary, P.E., (TH Agriculture & Nutrition Company, Inc., Lenexa, Kansas)

ABSTRACT: SuperSTR3 uses superheated steam to remove contaminants from the soil and has been successfully tested and patented. Recent bench-scale testing has been successfully completed on soils contaminated with volatile organics, semivolatile organics (2,4,5-Trichlorophenol), organochlorine pesticides, herbicides (2,4,5-T), and dioxins/furans using the SuperSTR3 technology. The results of the testing confirmed that SuperSTR3 has technical viability.

A batch process using the SuperSTR3 concept has been conceptually designed to an appropriate level where capital and operating costs can be reasonably estimated. The process has been developed to operate in a batch mode because it offers the advantages of smaller size and ease of operation over the large continuous indirect thermal desorption processes offered by thermal remediation contractors. The SuperSTR3 batch process has an economic advantage for remediating small sites in that it is easier and less expensive to move from site to site. Likewise, SuperSTR3 is expected to require less time, effort, and expense for testing on each site. Because of the simplicity of operation, SuperSTR3 can be operated by existing plant personnel.

TREATABILITY TEST RESULTS

In October 1996, treatability tests were conducted using the SuperSTR3 technology on soils from two confidential sites designated as Site A and Site B. The tests were conducted following the guidelines established by the U.S. EPA's "Guide for Conducting Treatability Studies under CERCLA" (U.S. EPA, 1989). A summary treatability test soil conditions are described in Table 1.

The tests were designed to assess the viability of SuperSTR3 technology and project capital and operating costs. In addition, the test criteria included scale-up design considerations.

Treatability tests typically focus on the ability of the technology being tested to remediate soil to a target set of treatment standards. For the purpose of this testing, the RCRA Universal Treatment Standards were the target treatment standards for the selected contaminants of concern.

The initial superheated steam test conditions were fixed at 1,000 °F (538 °C), and a mass flow rate of steam based on heat transfer and velocity considerations.

Heat transfer calculations indicated that a higher gas mass flow relative to the mass of solids in the reactor would increase the heat transfer and reduce the batch time necessary to reach the operating temperature. With a smaller batch time to reach operating temperature, a smaller design diameter for the reactor could be considered. However, higher steam flow could yield higher particulate matter carryover. Based on these considerations, acceptable superficial steam velocities considered for the treatability tests ranged from 2 to 8 ft/sec (0.6 to 2.4 m/sec). Because of test equipment limitations, the treatability tests were conducted at approximately 2 ft/sec (0.6 m/sec). Subsequent pilot testing for pressure drop and bed fluidization indicated an optimum superficial bed velocity of approximately 5 ft/sec (1.5 m/sec).

TABLE 1. Treatability test soil conditions

Site A	
Atterberg Limit:	plastic limit - 16.40% moisture (point that soil becomes brittle)
Soil Classification:	sandy lean clay
Moisture Content:	16.41% (as received)
Site B	
Atterberg Limit:	plastic limit - non plastic
Soil Classification:	silty sand
Moisture Content:	11.47% (as received)

Site A test soil was treated to target temperatures of 350, 500, and 650°F (177, 260, and 343°C). Site B test soil was treated to target temperatures of 500, 650, and 800°F (260, 343, and 427°C). The target temperatures for treated soil were established based on the types of contaminants of concern (COC) for each site soil matrix, experience with thermal processes and the associated COCs, and best engineering judgment. The COCs, treatment standards, and test results are presented in Table 2. The overall results indicate that SuperSTR3 is a viable technology for the subject COCs.

Another notable difference included the amount of total elemental carbon removed from the soil during treatment. The SuperSTR3 showed 12 to 15% lower overall removal of carbon from the soil than is typically experienced with other technologies. This may present advantages in full scale operation, were recovery type air pollution control equipment is utilized. The amount of organic residue and vapor phase activated carbon usage could be lower with application of the SuperSTR3 technology than other available thermal technologies.

PROCESS MODELING

A heat and mass transfer model was developed to aid in the conceptual design of the batch process. The model sizes the reactor and calculates the required batch time and steam usage based on heat transfer and soil carryover considerations. A minimum steam flow to soil ratio (lb/hr steam to lb of soil in the reactor) of approximately 1.15 was calculated for the required heat transfer.

TABLE 2. Superheated steam treatability test results.

Contaminant of Concern (COC)	Soil Feed (µg/kg)	Target [a] UTS (µg/kg)	Treated Soil (µg/kg)	Treatment Temperature (°F,°C)
SITE A				
Volatile organics				500, 260
Tetrachloroethene	495,000	6,000	2,900	
Trichloroethene	11,500	6,000	1,400	
Xylenes (total)	40,500	30,000	1,100	
Organo chlorinated pesticides				650, 343
Aldrin	490,000	66	650	
alpha-chlordane	38,500	260	<MDL	
gamma-chlordane	125,000	260	<MDL	
Dieldrin	53,000	130	<MDL	
SITE B				
Volatile organics				500, 260
Toluene	110,000	6,000	<MDL	
Trichloroethene	29,000	6,000	<MDL	
Xylenes (total)	45,500	30,000	<MDL	
Semi volatile organics				650, 343
2,4,5-Trichlorophenol	50,500	7,400	<MDL	
Organo chlorinated pesticides				650, 343
Aldrin	140	66	<MDL	
4,4'-DDD	6,550	87	10	
Herbicides				650, 343
2,4,5-T	147,000	7,900	24,000	
2,4,5-TP (Silvex)	26,600	7,900	<MDL	
Dioxins and Furans				800, 427
TCDDs (total)	68	1	0.8	
PeCDDs (total)	1.9	1	1.7	
HxCDDs (total)	<MDL	1	3.8	
TCDFs (total)	2.2	1	0.5	
PeCDFs (total)	15	1	1.8	
HxCDFs (total)	<MDL	1	2.1	

[a] UTS = RCRA Universal Treatment Standards
MDL = Method detection limit

The model also estimated the composition of the steam recycle and purge gas streams at discrete increments of time. A significant aspect of the viability of the SuperSTR3 process is linked to the superheated steam recycle concept. Based on the results of the model calculations, a steam recycle ratio of 10 to 1 (10 lbs/hr offgas from the reactor recycled back through the superheater for every 1 lb/hr of reactor offgas purged to the water recovery and offgas polishing system) was viable based on consideration of the buildup of organics in the recycle stream.

CONCEPTUAL DESIGN

An overview of the SuperSTR3 technology is shown in Figure 1. SuperSTR3 involves the utilization of superheated steam to remove contaminants from a soil matrix. SuperSTR3 is a semi-fluidized batch process where a continuous flow of superheated steam is induced through the bed of soil to remove organic contaminants. The SuperSTR3 technology uses a process gas (superheated steam) recycle stream to build and maintain a sufficient mass flow of superheated steam, and reduce the makeup steam requirements; thus reducing the energy required to boil water.

FIGURE 1. Conceptual design of the SuperSTR3 technology

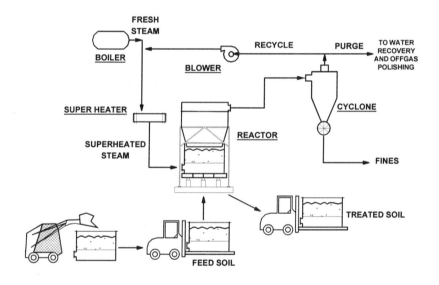

Saturated steam is continuously produced by a trailer mounted packaged boiler at a fixed mass flow rate. The saturated steam is superheated in a process heat exchanger (superheater) to a temperature of 1,000°F. Soil is processed in semi-fluidized batch reactor designed to minimize air leakage and particulate carryover. Superheated steam is delivered to a bed of soil in the reactor. The reactor contains up to 7.5 tons of contaminated soil. Mass and heat transfer occurs as the superheated steam contacts the soil. Moisture and volatile organic contaminants are removed from the soil and enter the gas stream. The gas stream exits the reactor at a reduced temperature and passes through a particulate removal device (cyclone) and through a high temperature blower. The blower is the prime mover for the process gas stream throughout the SuperSTR3 system. The blower recycles the majority of the process gas stream back through the superheater. The recycle gas stream combines with fresh makeup steam prior to superheating. The portion of the process gas stream not recycled is purged from the system and is directed to a water recovery system. The influx of fresh steam to the system and purge from the process gas stream reduces the concentration of contaminants in the soil bed and recycled process gas stream.

The purged process gas stream enters the water recovery system. Non-condensible purge gas, primarily inert gas and light hydrocarbons, will exit the condenser and be directed through vapor phase carbon beds. The polished gas stream will be vented to the atmosphere.

ECONOMIC ANALYSIS

Most commercially available thermal desorption systems are designed for application at larger sites (>15,000 tons). These systems generally have not been cost effective at small sites (<5,000 tons) because of the size of process equipment and cost of mobilization and demobilization. The remedial technology of choice for small many sites has been offsite incineration with unit treatment costs in excess off $600 per ton. This cost does not include disposal of dioxin contaminated soil. Aptus was the only incinerator in the US that would accept dioxin contaminated soil but has suspend operation.

An economic analysis was performed for application of the SuperSTR3 technology on six sites ranging in size from 3,000 tons with a hopper to hopper cost estimate of $144 per ton to 13,500 tons with a hopper to hopper cost estimate of $117 per ton (± 30%).

The option of offsite incineration when compared to the SuperSTR3 is cost prohibitive. If a company is facing remediation of dioxin and other recalcitrant contaminated soils, SuperSTR3 appears to be an extremely viable alternative for site remediation.

REFERENCES

Holcomb, D. H., Troxler, W. L., Pitts, D. M., December 16, 1996, *SuperSTR3 Technology Bench-Scale Treatability Test Summary Report, Site A*. Prepared for TH Agriculture & Nutrition Company, Inc and American Color & Chemical Corporation.

Holcomb, D. H., Troxler, W. L., Pitts, D. M., February 3, 1997, *SuperSTR3 Technology Bench-Scale Treatability Test Summary Report, Site B*. Prepared for TH Agriculture & Nutrition Company, Inc and American Color & Chemical Corporation.

Pitts, D. M., Holcomb, D. H., May 1997, *SuperSTR3 Technology Conceptual Design Manual*. Prepared for TH Agriculture & Nutrition Company, Inc and American Color & Chemical Corporation.

Holcomb, D. H., Troxler, W. L., Pitts, D. M., December 16, 1997, *SuperSTR3 Technology White Paper*. Prepared for TH Agriculture & Nutrition Company, Inc and American Color & Chemical Corporation.

DESIGN OF *IN SITU* BIOREMEDIATION SYSTEM TO TREAT GROUNDWATER CONTAMINATED BY CHLORINATED SOLVENTS

Douglas E. Jerger, IT Corporation, Findlay, Ohio
Rodney S. Skeen, Battelle Pacific Northwest Laboratories, Richland, Washington
Lewis Semprini, Oregon State University, Corvallis, Oregon
Daniel P. Leigh, IT Corporation, Findlay
Steve Granade, U.S. Navy, Point Mugu, California
Todd Margrave, U.S. Navy, Washington D.C.

ABSTRACT: *In situ* bioremediation combined with pump and treat can shorten treatment times, reduce costs, and increase destruction of chlorinated solvents. Only minor modifications to the pump and treat system are necessary to enhance the activity of the naturally occurring microorganisms. Scaling laboratory tests to a viable *in situ* bioremediation field process for the treatment of chlorinated solvents has been challenging since the underlying phenomena are complex. To aid in the scale-up process a rigorous design process has been developed that combines laboratory and field characterization activities and reactive flow and transport analysis. This process is being applied to treat chloroethene-contaminated groundwater at the Naval Air Weapons Station in Pt. Mugu, California. Numerical simulators have been adapted to reflect both site-specific conditions and pertinent microbial metabolisms. The simulators were then used to design and analyze pilot scale field tests to develop estimates for microbial kinetic and hydraulic properties. This information is being compiled to formulate an overall cleanup strategy involving accelerated and intrinsic *in situ* bioremediation.

INTRODUCTION

One of the primary technical barriers to widespread application of *in situ* bioremediation is the inability to accurately account for geochemical, hydrogeologic, and microbial phenomena during the design process. Recent bioremediation efforts have focused on this problem by developing and testing simulation tools for scale-up of *in situ* processes which combine fluid mixing and transport predictions with numerical descriptions for biological transport and reaction kinetics.

Engineering design of *in situ* bioremediation technologies is conceptually similar to the design of other systems. Specifically, measured or estimated system parameters and constraints are converted into process parameters through engineering design calculations. These process parameters are then used to evaluate the applicability of a technology against performance parameters such as cost and effectiveness.

BioSolve[SM], an *in situ* bioremediation design tool developed by Battelle Pacific Northwest Laboratory, was used for development and application of *in situ* chlorinated ethene bioremediation. BioSolve^{SM} is a suite of reactive flow and transport codes, coupled with a design process, that together enable rational scale- up and operation of in situ bioremediation systems. BioSolve^{SM} allows efficient evaluation of flow, transport, and biological reactions associated with *in situ* bioremediation processes to effectively optimize the system design and operating procedures. Efficient bioremediation of high concentration areas within groundwater is accomplished by stimulating microorganisms that can use chloroethenes, such as TCE, directly in energy-producing reactions. Stimulation of this metabolism can result in significantly enhanced destruction rates with better nutrient efficiency when compared to traditional co-metabolic bioremediation processes. Additionally, this technology can be applied for concentrations of chloroethenes greater than 100 mg/L. Complete dechlorination of chloroethenes to the drinking water standard using this metabolism is possible with ethylene and CO_2 as the final products. Transport of oxygen to the aquifer is not required because the microbial metabolism occurs under anaerobic conditions.

The objective of this project is to design and demonstrate a cost-effective *in situ* bioremediation technology to remediate chloroethene contamination.

SITE CHARACTERIZATION

A soil and groundwater quality investigation was conducted at Point Mugu Naval Air Weapons Station (NAWS), California, to delineate the extent of contamination and to collect sufficient samples for bench scale tests, evaluation of natural attenuation and the pilot test design. The investigation results are summarized below

Site Description. The focus of the investigation was Installation Restoration Program (IRP) Site 24. Soil and groundwater at that site had been affected by a discharge of chloroethenes and petroleum hydrocarbons. The source of the contaminants is considered to be two underground storage tanks located at two former UST Sites.

Site Hydrogeology. Site geology consists of a coarse to fine grained sand from surface to approximately 5 feet below ground surface (bgs); a clay layer from approximately 5 to 10 feet bgs; and a coarsening upward silt, silty sand and sand from approximately 10 to 100 feet bgs. A clay unit of undetermined extent or thickness was identified at a depth of approximately 100 feet bgs. The upper clay layer, where present, acts as an aquitard.

The water table is encountered about 5 feet bgs. Three arbitrary zones in the upper aquifer were defined as the "A" Zone, extending from 5 to 20 bgs, "B" zone from 20 to 40 feet bgs and "C" zone from 40 to 60 feet bgs. The site-wide groundwater gradient is approximately 0.001 ft/ft to the south in the A Zone and 0.01 ft/ft to the south in the C Zone.

Groundwater Quality. Separate groundwater plumes were identified that were associated with each UST. Contaminants detected included tetrachloroethene (PCE,), 1,1,1-trichloroethane (TCA) trichloroethene (TCE,), 1,1-dichloroethane (1,1-DCA) 1,1-dichloroethene (1,1-DCE), 1,2-dichloroethane (1,2-DCA), 1,2-

dichloroethene (1,2-DCE), methylene chloride (DCM), vinyl chloride (VC), toluene, fuel hydrocarbons, and carbon disulfide. Chlorinated solvents detected above regulatory standards and their maximum detected concentration included PCE (18 μg/L), TCE (2700 μg/L), 1,1-DCA (35 μg/L), 1,2-DCA (1 μg/L), 1,1-DCE (2,300), 1,2-DCE (2300 μg/L), and VC (1500 μg/L). Chlorinated organics were detected only in the A and B zones. The maximum detected concentrations of TCE were in the B Zone at UST Site 23. TPH was detected at concentrations up to 2600 μg/L. Sulfate concentrations increased with depth and ranged from 10.4 mg/L in the A zone to 5460 mg/L in the C Zone.

Benchscale Testing. The benchscale test results indicate that an accelerated anaerobic *in situ* treatment process with the addition of lactate as the initial electron donor would be the most effective alternative. Lactate stimulation under anaerobic conditions appears to be the most viable approach because of the presence of a strong lactate utilizing-sulfate reducing culture with the ability to completely dechlorinate TCE to ethylene. The accelerated transformation of TCE to VC and the subsequent dechlorination to ethylene for the nutrient media amended tests indicate that nutrients have a substantial effect on the rate and extent of dechlorination. Therefore, nutrient addition may also prove beneficial for pilot and full-scale *in situ* biological treatment.

BIODEGRADATION MODELING

BioSolve[SM] was used to design a pilot test and develop an operating protocol for conducting a pilot-scale *in situ* bioremediation test. First a numerical representation of the pertinent hydraulic, chemical, and biological properties was developed from field and laboratory data. Microbial kinetic data was integrated into BioSolve[SM]. A detailed description of the computational approach, and accuracy of the design simulators have been published elsewhere (Clement et al. 1995; 1996; 1997). The source of the kinetic information was obtained from laboratory treatability tests. This computational platform was used to simulate different nutrient feeding strategies. Using a baseline injection strategy estimated from the laboratory tests, initial simulations were completed and bioremediation performance was evaluated. Evaluation criteria included the extent of the biologically active Zone and the steady state contaminant destruction rate. Further simulations were then used to maximize both destruction rate and the size of the biologically active zone through refining the nutrient injection strategy. Once an optimum nutrient injection strategy was identified, locations for additional monitoring wells were chosen based on the spatial distribution of biological activity. The predicted changes in nutrient and chloroethene concentrations are presented in Figure 1.

The final step in the design phase was to simulate the full remediation system using the optimized feeding and recirculation strategy. This evaluation focused on the amount of nutrient breakthrough occurring at extraction wells, the level of hydraulic containment, the extent of the biologically active zone, and the overall contaminant destruction rate. Based on this evaluation, the operating and monitoring strategy was refined. Results from this stage in the design process include: 1) the final layout and spacing in the well network; 2) recirculation, nutrient feeding, and monitoring strategies for each phase of operation; and 3) the approximate duration of each operating phase.

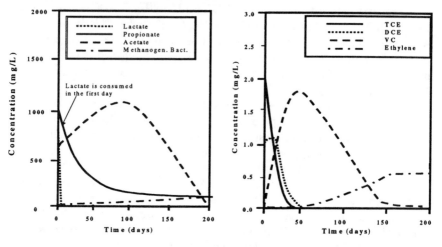

Figure 1. Model results (a) Predicted temporal changes in concentrations of lactate, propionate, acetate and methanogenic bacteria during the TCE remediation phase of pilot test. (b) Predicted temporal changes of chloroethenes during the TCE remediation phase of pilot test.

IN SITU BIOREMEDIATION DESIGN

The B Zone at UST Site 23 was selected as the location of the enhanced anaerobic bioremediation pilot test. The highest concentration of TCE was observed in the B Zone at UST Site 23 (2,700 µg/L). Lower concentrations of 1,2-DCE and vinyl chloride (1,200 µg/L and 1.8 µg/L, respectively) were observed in the B Zone indicating ongoing biodegradation of TCE by indigenous microflora. In contrast, however, TCE was not detected in the A Zone at UST Site 23 while elevated concentrations of 1,2-DCE and vinyl chloride (1,500 µg/L and 600 µg/L respectively) were detected indicating considerable intrinsic biodegradation of TCE. Therefore, acceleration of the biodegradation rate would be necessary to achieve the treatment goals.

Another consideration in the selection of the B Zone at UST Site 23 was the ability to evaluate the effect of the pilot test in the aquifer. Aquifer characteristics, including grain size distribution and hydraulic conductivity, are more uniform in the B Zone. The uniformity of these characteristics permits a more accurate determination of fate and transport of injected substrate and anticipated changes in aquifer characteristics. The accurate determination of these parameters will be required to expand the design to include the entire chloroethene plume.

System Design. The pilot system contains one injection well and one extraction well. Three monitoring locations between the injection and extraction wells are included in the design since this is the region where much of the biological activity will occur. The two off-center wells will be installed to monitor changes in flow paths caused by biological activity. This information is necessary to assess whether irreversible aquifer plugging occurs.

All wells will have a 15-foot screened interval that extends into the B zone of the aquifer. During operation, groundwater will be extracted at 10 gpm from the extraction well, amended with electron donor, and reinjected into the injection well. The gpm flow rate was chosen to provide high hydraulic control and give an appropriate residence time in the flow field for biological reaction.

Process Equipment. The primary components of the aboveground system include: substrate injection equipment, automated sampling equipment, process control equipment, and the data management system.

Dedicated in-well pumps will be used to pump water to the surface for obtaining samples from each monitoring well. Groundwater will also be sampled from ports in the surface piping that connects the extraction and injection wells (recirculation piping). Manual samples for chloroethenes, anions, and microbes will be collected from each sampling location using a syringe and specially designed sampling ports on the sample lines.

The feed pumps, the feed-line pressure and flow rate, and the feed tank liquid level will all be automatically controlled or monitored. If an adverse condition is encountered, an alarm will be activated on the computer and the feed pump will be shut down (if appropriate). The feed tank liquid level will be monitored for a low liquid level. A low liquid level condition will also activate an alarm on the computer.

System Operation. The pilot-scale test will be conducted in three phases: an abiotic recirculation control phase (Phase 1) and two biologically active phases (Phases 2 and 3, respectively).

Phase 1, Recirculation Control Phase. The purpose of this phase is to provide data on the variation in aqueous concentrations of chemical species prior to biostimulation. In addition, system trouble-shooting will take place during this period to ensure proper operation during the more critical nutrient injection phases.

Groundwater chemistry data will be monitored at the indicated wells for the duration of Phase 1. In addition, a bromide tracer test will be conducted after 2 weeks of recirculation to provide data to calibrate the flow-and-transport model to aid in process control and data evaluation in subsequent phases

Phase 2, Startup Phase. The objective of the startup phase is to lower the sulfate concentration within the flow field to allow fermentation and methanogenic process to dominate during Phase 3. This will be achieved by continuously recirculating the flow field at 10 gpm while adding periodic pulses of lactate. This feeding process will continue until the sulfate level at monitoring well MW3 is below 100 mg/L. Simulations predict that this sulfate level will be achieved after 50 days.

Phase 3, TCE Remediation Phase. This phase of the pilot-scale test will rely on a fed-batch operating strategy to achieve measurable contaminant destruction in the flow field without fouling the injection and or extraction wells. Each addition of lactate will be comprised of one high concentration pulse. Pumping will be stopped after 24 hours of recirculation to allow the injected electron donor to fully react to methane. The time between lactate injections will be determined by sampling the groundwater at each monitoring location until acetate is no longer detected (Figure 1).

CONCLUSION

Efficient bioremediation of high concentration areas within groundwater is accomplished by stimulating microorganisms that can use chloroethenes such as TCE directly in energy-producing reactions. Stimulation of this metabolism can result in significantly enhanced destruction rates with better nutrient efficiency when compared to traditional co-metabolic bioremediation processes. Additionally, concentrations of chloroethenes greater than 100 mg/L can be treated. Complete dechlorination of chloroethenes using this metabolism to the drinking water standard is possible with ethylene and CO_2 as the primary products. Transport of oxygen to the aquifer is not required because the microbial metabolism occurs under anaerobic conditions.

BioSolve[SM] allows efficient evaluation of flow, transport, and biological reactions associated with *in situ* bioremediation processes to effectively optimize the system design and operating procedures. Coupling the design tool approach with use of a newly discovered, efficient microbial metabolism will streamline the field application of this technology.

REFERENCES

Clement, T. P., B. S. Hooker, and R. S. Skeen, 1995. "Modeling Biologically Reactive Transport in Porous Media," Proceedings of the International Conference on Mathematics and Computations, Reactor Physics, and Environmental Analyses, Portland, Oregon, 1, 192-201 .

Clement, T. P., B. S. Hooker, and R. S. Skeen, 1996 "Numerical Modeling of Biologically Reactive Transport Near a Nutrient Injection Well," *ASCE J. Env. Eng.*, 122, 833-839.

Clement, T. P., Y. Sun, B. S. Hooker, and J. N. Petersen, 1997 "A Modular Computer Code for Modeling Biologically-Reactive Multi-Species Transport in Three-Dimensional Saturated Aquifers," submitted for publication.

Jerger, D. E., R. S. Skeen, L. Semprini, D. P. Leigh, S. Granade and B. Harre, 1998. "Scale-Up for In Situ Bioremediation of Groundwater Contaminated by Chlorinated Solvents" *Proceedings of HW98 Conference - HLW, LLW, Mixed Wastes and Environmental Restoration*, Tulsa, Oklahoma

DESIGN OF AN *IN-SITU* INJECTION/EXTRACTION BIOREMEDIATION SYSTEM

Brett T. Kawakami (Stanford University, Stanford, California)
John Christ (Air Force Institute of Technology, Wright-Patterson AFB, Ohio)
Mark N. Goltz (Air Force Institute of Technology, Wright-Patterson AFB, Ohio)
Perry L. McCarty (Stanford University, Stanford, California)

ABSTRACT: *In-situ* bioremediation can be an effective method for the treatment of contaminated groundwater. *In-situ* treatment involves extracting contaminated groundwater from an aquifer at one location, adding amendments necessary to induce the desired biological activity, and then reinjecting the water back into the aquifer at another location, where it undergoes treatment by native microorganisms in a bioactive zone. This injection/extraction scheme will create a recirculation system within the aquifer, causing groundwater to pass through the bioactive zone many times. If there is a regional hydraulic gradient, recirculated water is mixed with untreated water from upgradient, the exact mix ratio being a function of the regional flow velocity, well spacing, pumping rates and aquifer thickness. Recycled flow is desirable since it allows contaminated water to be treated numerous times, enhancing the overall effectiveness of treatment. The unrecycled portion of the flow is what is captured from the regional flow. Design variables such as well spacing and pumping rates will affect the recycle fraction and indirectly, the width of this capture zone.
A method is presented that allows design of *in-situ* bioremediation systems which attain design objectives of 1) reducing contaminant concentrations to acceptable levels and 2) ensuring capture of the migrating contaminant plume. An analytical model is used to determine how the design variables mentioned above affect the hydraulic characteristics of the system, which in turn affect both design objectives. A dimensionless number τ, derived from the analytical model, is presented that allows the recirculation fraction and capture zone widths to be determined for any combination of pumping rates, well spacing and regional flow. Finally, the use of these relationships is presented in a hypothetical design for Edwards Air Force Base, where a full scale recirculation system was used for *in-situ* cometabolic trichloroethylene biodecomposition.

INTRODUCTION

In-situ bioremediation of TCE contaminated groundwater has proven effective under field conditions at Edwards Air Force Base (McCarty et al, 1998). It was demonstrated that a dual well injection/extraction system was practical and could attain remediation objectives. This type of system is depicted in Figure 1. Here, contaminated water is extracted from the aquifer, mixed with necessary amendments (oxygen, co-substrate), and then re-injected back into the aquifer at another location. This reintroduced groundwater passes through the bioactive zone, where the contaminant concentrations are reduced by the active degrading microorganisms. The amount of degradation that takes place during one pass through the biostimulation zone is termed here as the one pass efficiency (E_{op}). The one pass efficiency is a function of parameters such as donor concentration, and various biological parameters. As seen in Figure 1, some portion of the flow entering the extraction well is pulled in from the regional flow which moves through the aquifer due to the natural hydraulic gradient. This regional flow contains the contaminated water that we wish to treat. The remainder of the flow entering the extraction well is actually recycled flow from the injection well. This water has passed through the bioactive zone and is thus presumably at a lower contaminant concentration. Similarly, at the injection well, a portion of the injected flow will eventually flow back to the extraction well, while the other portion will leave the system - this is the treated "effluent" that is being returned downgradient to the aquifer. Due to the blending of the higher concentration regional flow water with the lower concentration recycle water, the net effect of recycle is to lower the concentrations that the bioactive zone must treat. With higher recycle, the effluent concentration decreases. It is the unrecycled flow that provides the capture and treatment of the contaminant plume, while the recycled flow increases the system's performance. Since the quantity of captured regional flow must decrease if the recycle flow increases, there is a tradeoff between quantity of the plume treated (as measured by the width of the capture zone) and overall system performance (as measured by the difference between the contaminant concentrations in incoming regional flow and the contaminant concentrations exiting downgradient (i.e. not recycled).

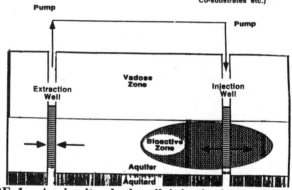

FIGURE 1.   An in-situ dual well injection/extraction system

Using a dual well system as a basis for design, the key engineering design parameters to be decided are the pumping rates, well spacing, and number of injection/extraction well pairs needed to satisfy objectives at the site. Usually, these objectives are 1) to reduce contaminant concentrations to acceptable levels and 2) to ensure capture of the contaminant plume. We will present an overview of a design method based on hydraulic principles and simple reactor theory that can be used to quickly design an injection/extraction well system to meet these objectives. The use of this method to develop a hypothetical design for an actual contamination scenario at Edwards Air Force Base is demonstrated.

THEORETICAL MODELS AND CALCULATIONS

An injection/extraction well system, in the presence of a regional hydraulic gradient and flow, produces a flow pattern as shown in Figure 2a. Both wells are assumed to operate at the same pumping rate. The streamlines entering the extraction well from the upper left indicate the captured regional flow. At the injection well, the streamlines leading off to the lower right indicate the treated water that is returned downgradient to the regional flow. The streamlines connecting the two wells indicate the recycled flow. The injection/extraction system can be represented as a reactor, as shown schematically in Figure 2b. The system is analogous to a wastewater treatment plant with recycle flow, with influent contaminated flow (Q_o) and effluent treated flow (Q_e). The recycle flow (Q_r) dilutes the influent flow concentration to C_i.

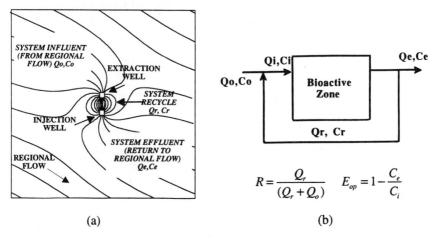

$$R = \frac{Q_r}{(Q_r + Q_o)} \qquad E_{op} = 1 - \frac{C_e}{C_i}$$

(a)   (b)

FIGURE 2.   a)   Flow pattern of injection/extraction well pair within regional flow.   b)   Schematic representation of injection/extraction well system.

At steady state, a mass balance on the system yields equation 1. Assuming the one-pass efficiency to be fixed, C_e/C_o then becomes solely a function of the recycle fraction R. The relationship between C_e/C_o and R (for $E_{op} = 0.85$) is shown graphically in Figure 3.

$$\frac{C_e}{C_o} = \frac{(1-R)(1-E_{op})}{(1-RE_{op})} \qquad (1)$$

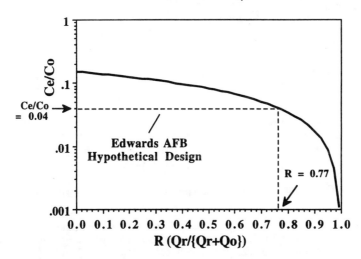

FIGURE 3. C_e/C_o vs. Recycle Percentage (R) for Injection/Extraction Well System ($E_{op} = 0.85$)

C_e/C_o is a measure of the system treatment performance. Since C_e/C_o is solely a function of R, R can be used as the design master variable. By controlling R, the designer can control the downgradient concentration from the treatment system. This can be done by adjusting the hydraulic factors that affect the fraction of flow that is recycled between the two wells. The recycle fraction for an injection/extraction well pair is a function of five variables

Q = pumping rate of each well (L^3T^{-1})
v = Darcy velocity of regional flow (LT^{-1})
B = thickness of aquifer (L)
d = distance between the wells (L)
α = angle of well line relative to regional flow (-)

DaCosta and Bennett (1960) developed an analytical method that can be used to calculate the recycle fraction with an injection/extraction well pair in regional flow knowing the values of these five system variables. For a fixed α, we define a dimensionless number τ (Equation 2).

$$\tau = \frac{Q}{vBd} \qquad (2)$$

The advantage of this number is that recycle fraction can then be plotted solely as a function of τ. Recycle fraction vs. τ (α=90 degrees) is shown in Figure 4.

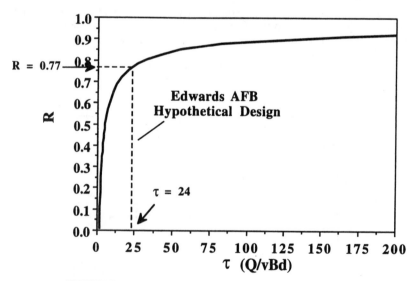

FIGURE 4.   Recycle fraction vs. τ (α = 90 degrees)

Once the target recycle fraction is known from Figure 3, the corresponding value of τ to attain this recycle fraction can be determined by using Figure 4. Once τ is known the system can be designed by choosing values for Q and d (v and B are characteristics of the aquifer) that will result in the correct value for τ. From the recycle fraction and selected pumping rate, the capture zone width for one well pair can be calculated by using equation 3.

$$W_{cz} = \frac{(1-R)Q}{vB} \qquad (3)$$

Once the capture zone width is established, the approximate number of well systems can be quickly estimated by dividing the desired length of capture by the capture width.

RESULTS AND DISCUSSION

The use of the above method is demonstrated below for a hypothetical design of an injection/extraction well system at Edwards Air Force Base, where an actual *in-situ* bioremediation demonstration was accomplished (McCarty, et al, 1998). This hypothetical design will be done using actual data from that site. Site 19 at Edwards Air Force Base has a plume of TCE contaminated groundwater as shown in figure 5. Pertinent information is summarized in Table 1.

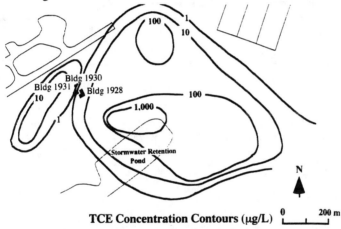

TCE Concentration Contours (µg/L)   0   200 m

FIGURE 5. TCE contamination at Edwards Air Force Base

Parameter	Value
Q (Pumped Flow rate - Each Well)	0.038 m³/min
v (Darcy Velocity of Regional Flow)	1.44e-05 m/min
B (Aquifer Thickness)	8 m
E_{op}	0.85
TCE influent (C_o)	1000 µg/L
Target effluent (C_e)	40 µg/L
Plume width	250 m

TABLE 1. System parameters for Edwards Air Force Base

The plume is approximately 250m wide (as defined by the 100 µg/L isoconcentration boundary). One remediation strategy is to place a line of wells downgradient from the contamination source to intercept the migrating plume. We will conservatively assume the upgradient concentration of TCE to be 1000 μg/L. Aquifer testing indicates 38 L/min is a reasonable pumping rate for this aquifer. Based on microcosm testing, we may assume E_{op} is 0.85. We wish to design for system performance to be 96% (C_e/C_o=0.04), which means effluent concentration should be 40 μg/L. The site at Edwards Air Force consisted of two aquifer separated by an aquitard. Here, we consider only the upper aquifer, although similar design principles would apply to the lower aquifer.

Using Figure 3 it was found that a recycle fraction of 0.77 would provide the desired effluent concentration (C_e/C_o = 0.04). Using figure 4, a τ of 24 is indicated. The well spacing and capture zone width can then be calculated:

$$d = \frac{Q}{vB\tau} = 13.7 \text{ m}; \quad W_{cz} = \frac{(1-R)Q}{vB} = 76 \text{ m}; \text{ Number of well pairs} = \frac{250}{76} = 3.33$$

The results of this hypothetical design are summarized in Table 2 below:

Design Parameter	Value
Pumping rate	0.038 m^3/min
Well Spacing	14 m
Number of well pairs	4

TABLE 2. Results of Hypothetical Design

CONCLUSIONS

An overview of a simple method to design an *in-situ* bioremediation injection/extraction well system based on groundwater hydraulic principles and reactor theory has been presented. In a hypothetical design the method produced expected TCE concentrations that were similar to those observed in an actual injection/extraction well system at Edwards Air Force Base (McCarty et al., 1998). This method can be used to make rapid assessment of the feasibility of *in-situ* bioremediation for a given site.

The simple design method shown here has some potential limitations. For example, it assumes a single-line of treatment well pairs. In reality, two or more lines in series may lead to a more efficient design. Also, the design is based on an assumed constant biodegradation efficiency, and constant values for the various aquifer parameters. Nevertheless, this simple model should be useful for quickly assessing the feasibility of implementing an in-situ injection/extraction bioremediation system for a given site.

REFERENCES

DeCosta, J.A., and R.R. Bennett. 1960. "The Pattern of Flow in the Vicinity of a Recharging and Discharging Pair of Wells in an Aquifer Having Areal Parallel Flow." International Association of Scientific Hydrology IUGG. General Assembly of Helsinki, Pub. No. 52. 524-536.

McCarty, P.L., Goltz, M.N., Hopkins, G.D., Dolan, M.E., Allan, J.P., Kawakami, B.T., and T.J. Carrothers. 1998. "Full-Scale Evaluation of In Situ Cometabolic Degradation of Trichloroethylene in Groundwater through Toluene Injection". *Environ. Sci. Technol.* **32**: 88-100.

TECHNOLOGY INNOVATION FOR SITES CONTAMINATED WITH CHLORINATED SOLVENTS, PESTICIDES, AND POLYCHLORINATED BIPHENYLS

Richard A. Brown, Fluor Daniel GTI, Trenton, New Jersey
Jacqueline A. MacDonald, National Research Council, Washington, D.C., USA
Richard G. Luthy, Carnegie Mellon University, Pittsburgh, PA, USA
Richelle Allen-King, Washington State University, Spokane, WA, USA

INTRODUCTION

The National Research Council recently completed a report, *Innovations in Ground Water and Soil Cleanup: From Concept to Commercialization*, that assesses the state of the art in technologies for ground water and soil cleanup and the strength of the market for those technologies. The report is the product of a multi disciplinary committee appointed by the National Research Council to study issues related to commercialization and development of innovative remediation technologies. The committee included experts in hydrogeology, soil science, environmental engineering, environmental policy, patent law, finance, and public opinion (see the list of committee members at the end of this paper). Among the report's findings is that the current state practice in remediation technology focuses primarily on the most readily solved contamination problem such as mobile and/or reactive contaminants and at permeable and homogeneous geologies. The greatest successes in remediation to date have been in the treatment of petroleum hydrocarbon fuels - gasoline diesel, and jet fuel - which are generally mobile (volatile) and reactive (biodegradable), and, to a lessor extent, in the treatment of chlorinated solvents which are generally mobile (volatile). Increasingly, the need for technology development is for the more recalcitrant contaminants such as chlorinated solvents, polychlorinated biphenyls (PCBs), and pesticides, and more difficult geologies. These, however present a greater challenge for remediation as their mobility and/or reactivity is less than that of hydrocarbons. Much of successful remedial technology is driven by the fate (reactivity) and transport (mobility) of the contaminants. Understanding fate and transport of a contaminant is a key part to developing a successful remediation process.

PRINCIPLES OF REMEDIATION

In general, the efficacy of treatment technology(s) is a function of the properties of the contaminant and the context (geological and hydrogeological) in which it exists. The more mobile and/or reactive a contaminant is, generally the more treatable it will be. Solubility and volatility are the primary factors that control the mobility of a contaminant in soil and groundwater. Reactivity is measure of the biodegradability or chemical stability of the contaminant. The more a treatment technology makes use of a fundamental mobility/reactivity property of the contaminant the more successful it will be. For example, the soil vapor extraction treatment of methylene chloride (highly volatile) could be highly successful in the right context while the soil vapor extraction treatment of lindane (low volatility) would not be in any geological context; methylene chloride is highly volatile, lindane is not.

The geological complexity also effects the efficacy of remediation technology. The more conducive a geological context is to fluid flow (air and/or water) the more easily contaminants may be treated. Permeability and degree of saturation are the two geological/hydrogeological factors that affect treatability. For example the soil vapor extraction treatment of methylene chloride would be highly successful in a coarse gravel, while in a silty clay it would be difficult. Gravel is permeable to air flow, silty clay is not.

Treatment options vary with both the type of contaminant to be treated. Treatment technology options for the different classes of chemicals are summarized in Table 1. A spectrum of treatment alternatives have been developed for relatively biodegradable and mobile contaminants, such as the hydrocarbon and chlorinated solvent classifications. The number of potential treatment technologies is much less for the other classes of chemical, such as chlorinated solvents, PCBs and pesticides.

The following discusses the remedial technologies available for chlorinated solvents, PCBs, and pesticides. On a comparative basis, these different types of contaminants illustrate the central importance of understanding fate and transport in developing and applying remedial technology.

Chlorinated Solvents. Chlorinated solvents are among the most common groundwater contaminants at NPL (Superfund) sites. There are a number of compounds commonly referred to as chlorinated solvents. The three most common classes encountered are halogenated methanes, chlorinated ethenes, and chlorinated ethanes.

How chlorinated solvents were lost to the environment determines the appropriate remediation technology. Chlorinated solvents are lost to the environment through two primary routes, 1.)through the use, loss, or disposal of the neat liquids or, 2.) through the use or disposal of wash/rinse waters containing residual solvents. The movement and dispersion of the chlorinated solvents in the subsurface is quite different depending on whether the release was as a neat liquid or as dissolved contaminants. If released as a solution, the contamination is almost exclusively a groundwater problem. With the release of large quantities of chlorinated solvent several distinct problems result, including gas phase solvent in the vadose zone, sorbed solvent and residual DNAPL both above and below the water table, and dissolved phase contamination that can occur both shallow and deep in the aquifer.

Once the solvents have migrated through the subsurface, their fate is quite complex. Chlorinated solvents are stable, undergoing neither rapid chemical or biological transformations. Chlorinated solvents are relatively soluble and have a high volatility. Thus, dissolution and dispersion or volatilization are significant attenuating mechanisms. A second attenuating mechanism is biological transformation. There has been considerable recent research demonstrating that chlorinated solvents do, in fact, biodegrade under certain conditions. A third mechanism of attenuation is chemical transformation. Chlorinated solvents can hydrolyze, loosing chlorine atoms and resulting in a less chlorinated "daughter" product. This hydrolysis reaction occurs more readily for chlorinated ethenes than either the chloromethanes or ethanes.

In general, the ease of treating these different aspects of chlorinated solvent sites is, in increasing order of difficulty and without regard to the geologic matrix: Residual phase (unsaturated soils)< dissolved phase < residual phase (saturated soils) < DNAPL.

Since chlorinated solvents are relatively non-reactive but are relatively volatile and soluble, the primary treatment technologies for chlorinated solvents are extractive processes. The four most widely used technologies at chlorinated solvent sites are: groundwater pump and treat, soil vapor extraction, air sparging, and dual phase extraction.

Chemical oxidation is a developing technology that has promise for the direct oxidation of chlorinated ethenes. Either ozone or Fenton's reagent (iron catalyzed hydrogen peroxide) can be used. The chlorinated ethenes are vulnerable to oxidative attack because of the presence of the double bond.

The bioremediation of chlorinated solvents has been developing along two pathways. First is the continued study of aerobic co-metabolic pathways, using toluene, natural gas, propane, etc. as the added substrate. This technology with methane was successfully demonstrated at Moffet Naval Air Station in California. A second avenue of development has been the use of sulfate reducing conditions. This technology, developed by DuPont adds sulfate and benzoate to degrade TCE and PCE.

Polychlorinated biphenyl (PCB). Polychlorinated biphenyl (PCB) compounds are a biphenyl structure with one to ten chlorine atoms. The environmental fate of PCBs is driven by the very low solubility of these compounds in water.

PCB compounds are stable, resisting both chemical or biological transformation. In anaerobic sediments higher-chlorinated PCBs undergo a slow process of microbial-mediated reductive dehalogenation whereby a chlorine on the biphenyl ring is replaced by a hydrogen atom. The less-chlorinated PCBs, e.g., mono- through tetra-PCB homologs, are amenable to aerobic microbial degradation, with the rate decreasing with the greater number of chlorine atoms.

PCBs are very sparingly soluble in water and are practically nonvolatile and, thus containment or stabilization processes is used to manage contaminated soils. The alternative, separation processes, require significant energy inputs (i.e., thermal treatments or chemical extractions).

Laboratory and field monitoring studies indicate that PCBs do biodegrade in the environment, but at a very slow rate. In the case of PCBs a combination of anaerobic and aerobic microbial treatments may result in removal of these contaminants in soils and sediments. Biological degradation under anaerobic conditions can result in reductive dechlorination of highly chlorinated PCBs. As the anaerobic processes progress, the accumulated degradation products may be destroyed aerobically. Such two-stage processes represent a promising means for treating these pollutants.

Pesticides. There are four general classes of pesticides. These are, in order of use are: complex synthetic organics, volatile organics (fumigants), naturally occurring organics, and inorganic compounds.

Pesticide contamination of soil and groundwater resulted from the manufacture, transportation, formulation and application of herbicides and insecticides. Pesticides are applied as solutions (in water or oil), dusts, or as fumigants (vapors). The pure compound or a concentrate is diluted in the field to application strength. Often contamination results from dumping wash waters and residuals from tank cleaning. There are, generally, three types of contamination scenarios encountered with pesticides - two are point source contamination, involving the use and handling of the pure compounds or concentrated mixtures, and the third is non-point source resulting from the application of the pesticide. In the first case soil and groundwater contamination results from shipping, distributing and handling the pure compound. In the second case, the formulation of the pure compound into dusts or spray could result in the release of formulated mixtures. The third contamination scenario results from the application of the pesticide in the field. At application levels the soil concentrations are generally less than a few hundred parts per billion.

With the release of pure compound or formulated products, transport occurs via the same processes as are observed with any organic compound. It is a function of the properties of the pesticide - solubility, volatility, mobility, and of the soil and groundwater dynamics. There are three primary routes of transport for pesticides after they are applied - leaching, soil particles migration, and volatilization. Since some pesticides sorb to soils, they may be transported on soil particles. With applied pesticides, volatilization can account for significant loss.

There are three attenuation pathways for pesticides - photochemical, biological, and chemical. Photochemical reactions are a significant decomposition mechanism for pesticides that are applied to the soil at the surface. With pesticides below the soil surface biological attenuation occurs primarily through both aerobic and anaerobic processes. Naturally occurring organics degrade quite readily. The primary chemical reactions for many pesticides are hydrolysis, protonation (for amine groups), and oxidation. These reactions may be catalyzed by soils, especially clay minerals.

The range of remedial alternatives available for pesticides is highly varied. It is impacted both by the nature of the compound itself and also by the type of contamination - point source or non-point source. Generally, with any pesticide, diffuse, wide- spread plumes are difficult to remediate because of the area involved. The types of remedial technologies vary with the class of pesticide and type of problem (point/non-point source). Fumigants an be remove through volatilization; Natural organics through biodegradation or oxidation; inorganics through water extraction for stabilization; and synthetic organics, depending on type, through oxidation, biodegradation or stabilization.

CONCLUSIONS

Existing technologies are inadequate for solving many types of ground water and soil contamination problems, such as those caused by PCBs and pesticides and, to a lesser extent, by chlorinated solvents. What makes it difficult to remediate these types of contaminants is that their mobility and/or reactivity are limited. Future development of remedial technology should be directed at enhancing the mobility or reactivity of these contaminants.

Acknowledgment. A complete copy of the report, *Innovations in Ground Water and Soil Cleanup: From Concept to Commercialization,* can be obtained from the National Academy Press at (800) 624-6242, http://www.nap.edu/bookstore. The report was prepared by the Committee on Innovative Remediation Technologies, chaired by P. Suresh Rao of the University of Florida. Committee members were Richard A. Brown (vice chair); Richelle M. Allen-King, Washington State University; William J. Cooper, University of North Carolina, Wilmington; Wilford R. Gardner, University of California, Berkeley; Michael A. Gollin, Spencer & Frank; Thomas M. Hellman, Bristol-Myers Squibb; Diane F. Heminway, Citizens' Environmental Coalition; Richard G. Luthy, Carnegie Mellon University; Roger L. Olsen, Camp Dresser & McKee; Philip A. Palmer, DuPont Specialty Chemicals; Frederick G. Pohland, University of Pittsburgh; Ann B. Rappaport, Tufts University; Martin N. Sara, RUST Environment & Infrastructure; Dag M. Syrrist, Vision Capital; and Brian J. Wagner, U.S. Geological Survey. Jacqueline A. MacDonald was study director.

Table 1. Treatment Technology Options for Chemical Data Classes.

	Petroleum Hydrocarbons	Chlorinated Solvents	PAHs & SVOCs	PCBs	Pesticides & Explosives
Solidification/Stabilization & Containment					
Excavation (stabilization)	X	X	X	X	X
Pump & Treat	X	X	na	na	X
Lime	X(h)	na			
Pozzolonic Agents	X(h)	na	?	?	
Asphalt batching	X	na	X		
Vitrification	na	na	na	?	
Grout walls			X	X	
Polymer wall		?			
Passive Barriers (sorption/ppt.)	?	?	?		
Biostabilization	X	na	X	?	
Slurry walls		X	X	X	X
Sheet pile walls		X	X	X	X
Reaction, Chemical and Biological					
Biopile	X	na	X	?	X
Land farming	X	na	X	?	X
Bioslurry	X	na	X	?	X
Bioventing	X(l)	na	?		na
Biosparging	X(l)	?	?		na
Intrinsic Bioremediation	X	X	?	?	X
Chemical Oxidation	X	X	X		X
Chemical Reduction		X			X
Thermal Destruction/Reduction	X	?	X	?	
Substitution	na	X	na	X	
Incineration	X	X	X	X	X
Passive-Reactive barriers					
-Fe	na	X	na	na	
-organic/					
microbiological	X	X	?		
-enhanced sorption	na	?	?		
-passive/active nutrient additions	X	X	?	?	?
Phytoremediation	X	X			X

Table 1 (continued)

	Petroleum Hydrocarbons	Chlorinated Solvents	PAHs & SVOCs	PCBs	Pesticides & Explosives
Separation & Extraction					
SVE	X(l)	X	na	na	
Thermal desorption	X	X	X	X	X
Soil Washing	X	X	X	?	?
Soil Flushing					
-acid, base, or chelatant					
-steam	?	?	?		
-foam	?	?	?		
-surfactant/cosolvent	?	?	?	?	?
Pump & Treat	X(l)	X	naX	na	
Sparging-air/steam	X(l)	X	?	na	
NAPL recovery	X	X	X	na	
Dual phase extraction	X(l)	X	na	na	
Electrokinetic	na	?	na	na	
Recycling/re-refining	X	na	na	na	
Thermally Enhanced SVE	X(h)	X	?	na	
Solvent extraction	X		X	?	

X(h): applicable primarily for heavy fuels or high molecular weight solvents
X(l): applicable to light hydrocarbons only
na: technology not applicable to this class of contaminants
?: application not commercially available or exists at an experimental stage
blank: lack of information for qualitative comparison

SUBSURFACE HYDROGEN ADDITION FOR THE
IN-SITU BIOREMEDIATION OF CHLORINATED SOLVENTS

Charles J. Newell (Groundwater Services, Inc., Houston, Texas)
Joseph B. Hughes (Rice University, Houston, Texas)
R. Todd Fisher (Groundwater Services, Inc., Houston, Texas)
Patrick E. Haas (Air Force Center for Env. Excellence, Brooks AFB, Texas)

ABSTRACT: In-situ bioremediation via direct hydrogen addition has the potential to become a simple and low-cost treatment approach for sites contaminated with chlorinated solvent compounds. Based on the results of recent research, the role of hydrogen as electron donor is now widely recognized as the key factor governing the biologically-mediated dechlorination of these common environmental contaminants. Because of hydrogen's low cost, its ability to be delivered safely and inexpensively in a variety of ways, and its ability to promote rapid dechlorination, direct hydrogen addition represents a potentially superior approach for managing and remediating chlorinated solvent groundwater plumes. A field test program is currently underway through the Technology Transfer Division, Air Force Center for Environmental Excellence (AFCEE/ERT).

BACKGROUND

Generally, organic compounds represent potential electron donors to support microbial metabolism (e.g., the oxidation of fuel hydrocarbons). However, halogenated compounds such as chlorinated solvents (PCE, TCE, DCE, etc.) can act as electron acceptors and thus become reduced in the reductive dehalogenation process. Specifically, dehalogenation by reduction is the replacement of a halogen (such as chloride, bromide, or fluoride) on an organic molecule by hydrogen, as described by the following half-reaction:

$$R\text{-}Cl + H^+ + 2e^- \rightarrow R\text{-}H + Cl^-$$

Figure 1 shows the various reductive transformation pathways for the common chlorinated solvent compounds.

Reductive dechlorination (also called halorespiration) requires a source of reducing equivalents to drive the reaction, but many contaminated sites are deficient in a suitable electron donor. Most laboratory research concerning the anaerobic degradation of chlorinated aliphatic compounds has focused on methanogenic systems. Such systems typically involve the introduction of an electron donor such as acetate, lactate, methanol, ethanol, or even a co-contaminant such as toluene, to stimulate methane producing bacteria. While chlorinated solvent compounds have been observed to be degraded in a variety of

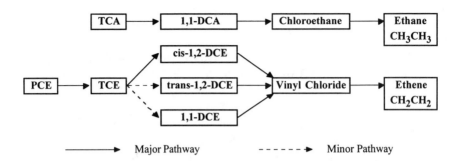

FIGURE 1. **Reductive transformation pathways for the common chlorinated solvent compounds.**

such laboratory systems (Vogel and McCarty, 1985; Freedman and Gossett, 1989; Sewell and Gibson, 1991), more recent work indicates that the methanol and other substrates used in these systems merely serve as precursors for the formation of an intermediate hydrogen pool through fermentation, and that hydrogen is the actual electron donor for dechlorination (DiStefano et al., 1992; Smatlak et al., 1996; Fennell et al., 1997; Ballapragada et al., 1997; Carr and Hughes, submitted for publication).

The role of hydrogen as the key electron donor for reductive dechlorination is not surprising given the multitude of growth substrates that have been observed to support this microbial process (e.g., acetate, lactate, benzoate, methanol, etc.) and the fact that hydrogen has been long-known as an electron donor for other microbial processes, such as methanogenesis (Brock and Madigan, 1991). However, it has only been in the recent literature that several researchers have reported unique organisms that are able to couple growth with the reductive dechlorination of PCE and TCE by using the chlorinated ethenes as electron acceptors and hydrogen as their electron donor (Holliger et al., 1993; Smatlak et al., 1996).

ADVANTAGES OF DIRECT HYDROGEN ADDITION

Direct hydrogen addition is a highly flexible process that may be implemented in a variety of process configurations for either dissolved plume management or reduction of NAPL source zones, depending upon the goals of the intended remediation effort. Example delivery systems include low-pressure biosparging, in-situ controlled-release reaction involving the placement of metal fillings or other hydrogen-releasing compound within a well or borehole, and dissolved hydrogen injection (see Hughes et al., 1997).

When compared to traditional remediation technologies (e.g., pump-and-treat, vacuum extraction, etc.), or other emerging bioremediation technologies (e.g., indirect generation of hydrogen via subsurface addition of lactate, benzoate, etc.), direct hydrogen addition represents a potentially superior method for addressing the remediation of chlorinated solvent groundwater plumes. In-situ bioremediation via direct hydrogen addition represents an extension of naturally-occurring dechlorination processes, and simply eliminates the rate-limiting biological step in natural systems (i.e., the production of hydrogen through the slow fermentation of organic electron donors).

Some of the additional advantages of the direct hydrogen addition process may be summarized as follows:

- Hydrogen addition provides highly favorable reaction stoichiometry and can tolerate process inefficiencies. For every 1 mg of hydrogen utilized by dechlorinating bacteria, 21 mg of PCE are completely converted to ethene. (Comparatively, the aerobic degradation of benzene requires 3 mg of oxygen to biodegrade just 1 mg of benzene.) Based on this stoichiometry, a dissolved groundwater plume with 2 mg/L PCE can be completely degraded through the utilization of only 0.1 mg/L hydrogen, a concentration much lower than the solubility limit for hydrogen (approximately 1.6 mg/L).

- Hydrogen is a very inexpensive substrate for chlorinated solvent remediation. Hydrogen can be delivered directly as a gas, indirectly as a dissolved solute, or liberated from a hydrogen-releasing compound. Compressed hydrogen gas in cylinders costs about $2 per cubic meter or $0.04 per mole hydrogen. Assuming 100% utilization of hydrogen, this translates to a potential treatment cost of between $0.65 and $1.00 per kg of pure chlorinated solvent in groundwater, or less than $0.000001 per liter of groundwater containing 1 mg/L PCE.

- Hydrogen produces no environmental side effects, does not leave any environmentally harmful residue in the subsurface, and is an environmentally acceptable compound to add to groundwater.

- Direct hydrogen addition is a much simpler and more flexible process than other biological treatment approaches for chlorinated solvents (e.g., methane/oxygen addition (Semprini et al., 1990) or benzoate/sulfate addition (Beeman, 1994)).

FIELD TESTING

A project to field test direct hydrogen addition is currently underway through the Technology Transfer Division, Air Force Center for Environmental Excellence (AFCEE/ERT), Brooks AFB, Texas. The test program consists of both short-term (1 week) treatability tests and long-term (1 year) pilot tests to be conducted at multiple Air Force installations as described below.

Treatability Tests. The treatability tests are designed as site screening tests, and will evaluate hydrogen utilization by indigenous microorganisms via a field test method known as "push-pull." This type of test has been described by Istok et al. (1997) for use in determining microbial activities related to degradation of petroleum hydrocarbons. The method, as adapted for the measurement of hydrogen utilization and dechlorination, consists of the following steps:

1) *Initial Groundwater Extraction:* Extraction of a known quantity of groundwater (e.g., 1000 L) from within the test area through an existing monitoring well.

2) *Amendment Addition:* Addition of known quantities of hydrogen and various volatile and non-volatile tracers (e.g., bromide, helium, sulfur-hexafluoride) to the extracted groundwater, followed by thorough mixing to create a homogeneous test solution.

3) *Initial Sampling:* Collection of a representative test solution sample which is analyzed for chlorinated organic compounds, hydrogen, tracers, and other constituents of interest (e.g., oxygen, nitrate, sulfate, etc.).

4) *Re-Injection of Groundwater Test Solution:* Pulse injection ("push") of amended groundwater into the saturated zone through the same monitoring well used for groundwater extraction.

5) *Final Groundwater Extraction:* Extraction ("pull") of the test solution/ groundwater mixture from the test well following a contact/reaction period (typically 12 to 36 hr).

6) *Final Sampling:* Collection of a final representative test solution sample which is again analyzed for chlorinated organic compounds, hydrogen, tracers, and other constituents of interest.

During the injection phase, the test solution enters the test zone through the screened area of the monitoring well. Within the test zone, biologically reactive components of the test solution (e.g., hydrogen and chlorinated organics) are utilized by the indigenous microorganisms. During the final extraction phase, the test solution is recovered and solute concentrations are measured to determine the quantities of reactants used (e.g., hydrogen, PCE, TCE) and/or products formed (e.g., DCE, vinyl chloride, ethene). The tracers are used to evaluate abiotic losses of reactants during the test process.

Pilot Tests. The year-long pilot tests, scheduled to begin in mid-1998, will each consist of a two-well extraction/injection system. In these systems, groundwater will be pumped from a location downgradient of the test area, and passed through an above-ground gas dissolution device (downflow bubble contactor or eductor/degas separator as manufactured by GDT Corporation, Phoenix, AZ). In the gas dissolution device, hydrogen will be added to the flowstream while other dissolved gases (nitrogen, carbon dioxide, methane, etc.) are removed. The resulting hydrogen-saturated groundwater will then be reinjected into the subsurface through an injection well located upgradient of the test area . A circular flow system will thereby be created wherein groundwater containing dissolved hydrogen is moved through the treatment zone stimulating biological activity throughout the zone. Sampling will be conducted at periodic intervals to evaluate hydrogen utilization and dechlorinating activity.

SUMMARY

As suggested by the results of a variety of laboratory and field studies appearing in the published literature, hydrogen is a fundamental component of the biologically-mediated reduction of chlorinated solvent compounds, and represents the key to promoting the biodegradation of these constituents in the environment. In-situ bioremediation via direct addition of hydrogen to the subsurface represents a logical extension of naturally-occurring processes, and a potentially simple and low-cost approach to the management of chlorinated solvent groundwater plumes.

Hydrogen may be delivered using a variety of process configurations, providing for a flexible approach to site remediation. Hydrogen may be used in passive barrier systems for control of dissolved-phase plumes as well as in systems designed for active source reduction. Field testing of the hydrogen addition process is currently underway through the Technology Transfer Division of the Air Force Center for Environmental Excellence. The results of these field tests, expected in 1998, will demonstrate if direct addition of hydrogen is a viable remediation/management technology for chlorinated solvents.

REFERENCES

Ballapragada, B.S., H.D. Stensel, J.A. Puhakka, and J.F. Ferguson. 1997. "Effect of Hydrogen on Reductive Dechlorination of Chlorinated Ethenes." *Environmental Science and Technology.* 31(6): 1728-1734.

Beeman, R.E. 1994. *In Situ Biodegradation of Groundwater Contaminants.* U.S. Patent No. 5,277,815, issued January 11, 1994.

Brock, T.D., and M.T. Madigan. 1991. *Biology of Microorganisms.* 6th Edition. Prentice-Hall, Englewood Cliffs, NJ.

Carr, C.S., and J.B. Hughes. Submitted for publication. "Enrichment of High Rate PCE Dechlorination and Comparative Study of Lactate, Methanol, and Hydrogen as Electron Donors to Sustain Activity." Submitted to: *Environmental Science and Technology*, November 1997.

DiStefano, T.D., J.M. Gossett, and S.H. Zinder. 1992. "Hydrogen as an Electron Donor for Dechlorination of Tetrachloroethene by an Anaerobic Mixed Culture." *Applied Environmental Microbiology.* 58(11): 3622-3629.

Fennell, D.E., J.M. Gossett, and S.H. Zinder. 1997. "Comparison of Butyric Acid, Ethanol, Lactic Acid, and Proprionic Acid as Hydrogen Donors for the Reductive Dechlorination of Tetrachloroethene." *Environmental Science and Technology.* 31(3): 918-926.

Freedman, D.L., and J.M. Gossett. 1989. "Biological Reductive Dechlorination of Tetrachloroethylene and Trichloroethylene to Ethylene Under Methanogenic Conditions." *Applied Environmental Microbiology.* 55(9): 2144-2151.

Holliger, C., G. Schraa, A.J.M. Stams, and A.J.B. Zehnder. 1993. "A Highly Purified Enrichment Culture Couples the Reductive Dechlorination of Tetrachloroethene to Growth." *Applied Environmental Microbiology.* 59(9): 2991-2997.

Hughes, J.B., C.J. Newell, and R.T. Fisher. 1997. *Process for In-Situ Biodegradation of Chlorinated Aliphatic Hydrocarbons by Subsurface Hydrogen Injection.* U.S. Patent No. 5,602,296, issued February 11, 1997.

Istok, J.D., M.D. Humphrey, M.H. Schroth, M.R. Hyman, and K.T. O'Reilly. 1997. "Single-Well 'Push-Pull' Test for In-Situ Determination of Microbial Activities." *Ground Water.* 35(4): 619-631.

Semprini, L., P.V. Roberts, G.D. Hopkins, and P.L. McCarty. 1990. "A Field Evaluation of In-Situ Biodegradation of Chlorinated Ethenes: Part 2, Results of Biostimulation and Biotransformation Experiments." *Ground Water.* 28(5): 715-727.

Sewell, G.W., and S.A. Gibson. 1991. "Stimulation of the Reductive Dechlorination of Tetrachloroethene in Anaerobic Aquifer Microcosms by the Addition of Toluene." *Environmental Science and Technology.* 25: 982-984.

Smatlak, C.R., J.M. Gossett, and S.H. Zinder. 1996. "Comparative Kinetics of Hydrogen Utilization for Reductive Dechlorination of Tetrachloroethene and Methanogenesis in an Anaerobic Enrichment Culture." *Environmental Science and Technology.* 30(9): 2850-2858.

Vogel, T.M., and P.L. McCarty. 1985. "Biotransformation of Tetrachloroethylene to Trichloroethylene, Dichloroethylene, Vinyl Chloride, and Carbon Dioxide Under Methanogenic Conditions." *Applied Environmental Microbiology.* 49(5): 1080-1083.

RAPID PAH DEGRADATION IN SOIL USING ENZYME MIXTURES

Thomas R. Stolzenburg, Ph.D., (RMT, Inc., Madison, Wisconsin)
Marianne D. Duner (RMT, Inc., Madison Wisconsin)

ABSTRACT: A bench-scale treatability study was conducted to determine if the addition of enzymes to manufactured gas plant (MGP) soil was an effective remediation treatment for polynuclear aromatic hydrocarbons (PAHs). The goal of the project was to determine if the application of enzymes applied directly to target compounds would result in much faster degradation rates than biodegradation by indigenous bacterial populations. Degradation of 4-ringed PAHs in soil was achieved in 11 days. The degradation was not due to volatilization nor *in situ* biodegradation, as shown by control microcosm results. The rate of PAH degradation by enzyme addition was dramatic, as the contaminated soil had been weathered for approximately 90 years.

INTRODUCTION

Natural attenuation is attracting more attention in the scientific and regulatory communities as a viable remediation process for addressing organic contamination. However, some organics (e.g., PAHs, petroleum oils, etc.) gain a reputation for "recalcitrance" primarily due to fundamental physico-chemical properties that subsequently slow down biodegradation by indigenous bacteria. In the case of PAHs, low water solubility and large molecular size combine to prevent rapid movement across the cell membranes of live bacteria. So, although enzymes within bacteria are capable of PAH degradation, basic molecular properties limit degradation rates.

A recent development in the degradation of recalcitrant organic compounds in soil is the use of enzymes. Enzyme mixtures are now being marketed in the United States for full-scale treatment applications. The degradation process is more accurately characterized as chemical rather than biological, since live bacteria are not used. It is attractive because enzyme degradation of recalcitrant organics in soil is faster than biodegradation.

Enzymes applied directly to target compounds should result in much faster degradation rates, because live bacteria are not used so there is no waiting period for acclimation or exponential growth. Reports of full-scale success are impressive, but sometimes observed degradation in the field cannot be attributed to the action of enzymes alone, as *in situ* biodegradation and volatilization may also contribute to observed target compound reductions.

In this study, a bench-scale treatability test, with a control, was conducted to determine the efficacy of treating recalcitrant organics in soil with commercially available mixtures of enzymes. The activity of the enzymes was separated from volatilization and biodegradation by comparison to the control.

Objective. The objective of this study was to demonstrate the efficacy of commercially available enzymes for remediating PAH-contaminated soil under controlled conditions to factor out volatilization and *in situ* biodegradation contributions. A secondary goal was to determine the rate of degradation under simulated field conditions. The PAH compounds of interest were fluoranthene, chrysene, and benzo(a)anthracene.

MATERIALS AND METHODS

A pan-scale bench test was conducted on soil from a 90-year old MGP site contaminated with PAHs. Three 5-gallon buckets of MGP-contaminated soil were homogenized by mixing thoroughly in a 55-gallon drum. One 5-gallon bucket of soil was removed and further homogenized. Subsamples were then placed into three stainless-steel pans (533.4 mm x 165.1 mm x 152.4 mm), which were filled to a depth slightly greater than 2 inches. The soil in two of the replicate pans was treated with enzyme, and the soil in one pan was used as a control.

Each day, the pans were mechanically aerated and the moisture content was checked and adjusted as necessary. Moisture content was maintained at about 10 percent throughout the study by daily adjustments and by enclosing the pans in a high-humidity chamber. The test was truncated after 11 days.

Sampling and Analysis. Subsamples from each pan were removed and analyzed for PAHs at time 0 to demonstrate homogeneity. Thereafter, subsamples were removed six more times in 11 days to measure the rate of degradation. Following mechanical aeration (mixing), 10-g subsamples were removed from each pan for PAH analysis. For moisture content analysis, 20-g subsamples were removed from each pan.

RESULTS AND DISCUSSION

Analysis. Table 1 shows the PAH results, corrected to a dry-weight basis. The bench test results show degradation of the target PAHs by an order of magnitude in 11 days. Benzo(a)anthracene (see Figure 1) was degraded from 4,7000 µg/L to 560 µg/L. Chrysene and fluoranthene (see Figures 2 and 3, respectively) were also degraded quickly. PAH concentrations were still decreasing when the test was truncated.

TABLE 1. PAH results, corrected to dry-weight basis for chrysene, fluoranthene, and benzo(a)anthracene

	Chrysene (mg/kg - dry weight)								
	Day 0	Day 2	Day 4	Day 7	Day 8		Day 9	Day 10	Day 11
Treated	4.5	5.3	5.1	5.0	4.9	more	2.5	1.3	0.45
Treated Rep	4.7	4.9	4.3	4.3	3.5	enzyme	3.2	1.5	0.59
Control	4.7	4.9	5.0	4.7	5.1	added	NA	3.9	4.6
	Fluoranthene (mg/kg - dry weight)								
	Day 0	Day 2	Day 4	Day 7	Day 8		Day 9	Day 10	Day 11
Treated	9.2	11.	8.9	8.1	7.8	more	5.2	<2.2	0.65
Treated Rep	9.5	10.	7.7	6.9	5.6	enzyme	5.8	<2.2	0.94
Control	9.6	10.	8.9	8.6	8.5	added	NA	8.6	8.2
	Benzo(a)anthracene (mg/kg - dry weight)								
	Day 0	Day 2	Day 4	Day 7	Day 8		Day 9	Day 10	Day 11
Treated	4.5	5.0	4.6	4.7	4.5	more	2.9	0.92	0.48
Treated Rep	4.7	4.7	4.0	4.1	3.4	enzyme	3.7	1.1	0.63
Control	4.7	4.7	4.9	4.6	4.7	added	NA	4.3	4.3

NA = not analyzed

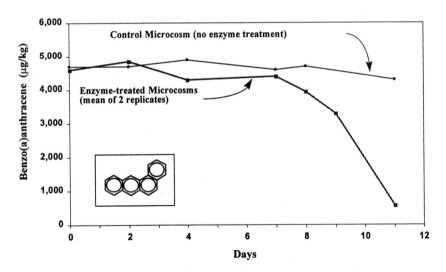

FIGURE 1. Bench-scale test results for benzo(a)anthracene in MGP-contaminated soil

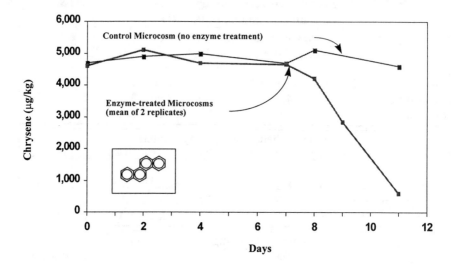

**FIGURE 2. Bench-scale test results for chrysene
in MGP-contaminated soil**

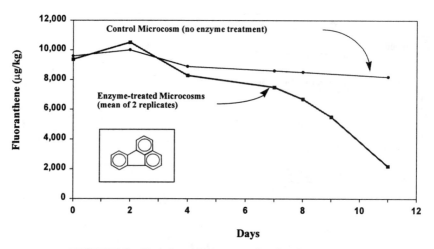

**FIGURE 3. Bench-scale test results for fluoranthene
in MGP-contaminated soil**

CONCLUSIONS

The pan-scale bench test was designed to demonstrate the efficacy and rate of remediation of PAHs in soil by direct application of enzymes. The effects of volatilization and *in situ* biodegradation were accounted for by a control microcosm. Degradation of the target 4-ringed PAH compounds by approximately one order of magnitude occurred within 11 days, and the PAH concentrations were still decreasing when the test was truncated.

REACTIVE WALL COLUMN TESTING TO SUPPORT REMEDIATION
DECISION MAKING

Alex Rafalovich / Metcalf & Eddy, Chico, CA; Stephanie O'Hannesin /
EnviroMetal Technologies Inc., Guelph, Ontario, Canada; Greg Haling / Metcalf
& Eddy, Beale AFB, CA; Carol Gaudette / Beale AFB, CA; Wendy Clark /
Laguna Construction Company, Yuba City, CA;

ABSTRACT: Site 17, located on Beale AFB, CA, is a site with complex hydrogeology, characterized by strong stream-aquifer interaction, dense soils, and shallow bedrock. The site's groundwater contains high levels of trichloroethylene (TCE) and 1,1,2,2 tetrachloroethane (1,1,2,2-TeCA). Concentrations have exceeded 200 ppm for TCE and 480 ppm for 1,1,2,2-TeCA. The complexity of the site has led the Air Force to evaluate a range of different potential approaches to cleanup the site as rapidly and cost-effectively as possible. To determine the design and potential cost-effectiveness of passive reactive wall technology at this site, bench-scale treatability testing was conducted using groundwater from the site and 100% granular iron.

A flow rate of 55 cm/day was used to determine degradation half-lives for the chlorinated solvents and their daughter products, as well as the potential for inorganic mineral precipitation. The results of the column testing were taken in combination with available groundwater flow data at the site to determine the optimal configuration and thickness of a field-scale iron-reactive wall system.

SITE DESCRIPTION

Site 17 is an area adjacent to Best Slough, Beale AFB, CA, that has been used for the disposal of drums containing chlorinated solvents. This practice has lead to the contamination of the shallow groundwater. Numerous investigations have shown that the local geology consists of alluvial interbedded silts and sands within the top 10 feet (3 m) of soil (Law Environmental, 1995; URS Greiner Inc, 1996). The site is underlain by two relatively thin, shallow aquifers separated by a low permeability zone, which locally acts as an aquitard. Saturated soils are first encountered at approximately 12 feet (3.7 m) below ground surface (bgs). Soils at the site are underlain by indurated gray siltstone and sandstone located approximately 30 feet (9.1 m) bgs. A plan view of the site is depicted as Figure 1. Local geologic cross-sections showing the subsurface lithology are shown in Figures 2 and 3

Local groundwater gradient direction ranges from S30W to S13E. This fluctuation appears to be influenced to a large degree by adjacent surface water in Best Slough. Hydraulic conductivity estimates for the site range from $7x10^{-4}$cm/s for the upper zone and $5x10^{-5}$cm/s for the lower zone.

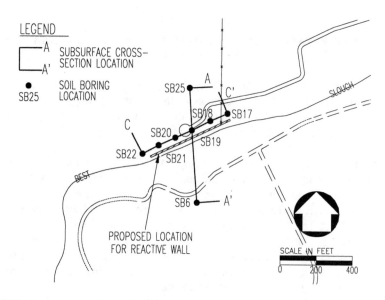

FIGURE 1. Site 17 Layout at Beale AFB, CA (After Laguna Construction Company, 1997)

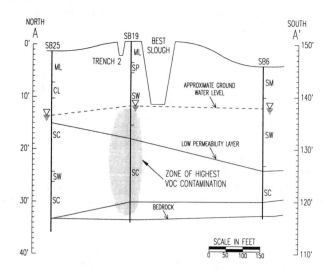

FIGURE 2. Cross-Section A-A' (After Laguna Construction Company, 1997)

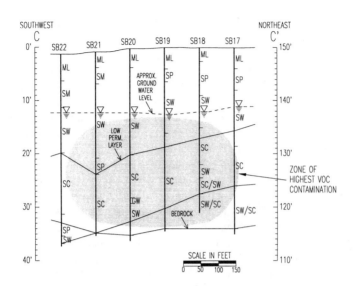

FIGURE 3. Cross-Section C-C' (After Laguna Construction Company, 1997)

RESULTS OF BENCH SCALE TESTING

In order to obtain the necessary design parameters for full scale remediation, bench scale testing was undertaken. The test provided degradation rates of each VOC present in the site water, the possible generation of chlorinated breakdown products and the possible inorganic geochemical changes that may occur as the groundwater passes through the granular iron material.

Methodology. Ground water was collected from the site and shipped to the University of Waterloo for the column treatability study. The main organic compounds detected in the site water were trichloroethylene (TCE; 252 mg/L) and 1,1,2,2-tetrachloroethane (1122TeCA; 488 mg/L). Lower levels of perchloroethylene (PCE), 1,2-dichloroethane (12DCA), 1,1,2-trichloroethane (112TCA), cis-1,2-dichloroethylene (cDCE), trans-1,2-dichloroethylene (tDCE), 1,1-dichloroethylene (11DCE), vinyl chloride (VC) and traces of other VOCs were also detected.

At a flow velocity of 55 cm/day (residence time of 44.2 hr) and a temperature of 22 C, the site water was pumped into a Plexiglas^{TM} column with a length of 100 cm and an internal diameter of 3.8 cm which was packed with 100% granular iron (Peerless Metal & Abrasives). Seven sampling ports were positioned along the length of the column. These ports, as well as the influent and effluent, were sampled over time until steady state profiles were achieved. Inorganic parameters such as redox potential (Eh), pH, cations and anions were also collected. Analyses for the VOC compounds were performed using gas chromatograph with either an electron capture or photoionization detectors. Inorganic analyses were completed using inductively coupled plasma and/or ion chromatography.

Observed VOC Degradation. Using the flow velocity, the distance along the column was converted to time and the degradation rate constants were calculated for the organic compounds, using a first-order kinetic model. Half lives (the time required to remove one-half of the contaminant mass) were determined from the degradation rate.

A total of 46 pore volumes of water had passed through the column, steady state concentration profiles for the two highest concentration compounds (TCE, 1122-TeCA) are shown for the 100% granular iron column in Figure 4. With initial concentrations of 252 and 488 mg/L for TCE and 1122-TeCA, respectively, an initial increase in concentration was observed at the first sampling port. This increase was attributed to sampling artifacts resulting from the extremely high organic loading (755 mg/L) at the influent end of the column. However, a steady decline in concentration was observed from these peak values with effluent concentrations of 10,690 and 177 µg/L for TCE and 1122-TeCA, respectively. The observed decline in concentration with distance reflects the rate at which the organic compounds are degrading within the iron. Due to the dechlorination of TCE, and 1122-TeCA, increases in concentrations were observed for the DCE isomers and VC. With a lower flow velocity, peak concentrations and subsequent declines in concentrations occurred for tDCE, cDCE and VC. No detectable amounts of 11DCA or 111TCA as breakdown compounds were observed over the entire sampling period.

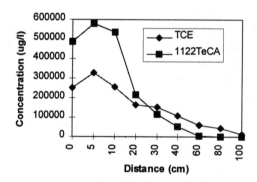

FIGURE 4. Degradation Curves for Contaminants of Concern

Half lives for TCE, 1122TeCA, PCE, cDCE, tDCE, and VC were calculated by setting the initial concentration (C_o) equal to the peak concentration, giving values of 9.5 , 3.4, 7.6, 18.0, 22.0, and 17.6 hr, respectively. Half lives for TCE, PCE, DCE isomers and VC were much larger than in waters tested previously with lower influent concentrations (Gilham and O'Hannesin, 1994) These longer half-lives may result from increased competition for reactive sites on the iron surface. The rates of conversion of TCE and PCE to cDCE are typically 20%, thus conversions of 1122TeCA have been attributed to 30% of tDCE and 30% of cDCE.

In summary, the column test data indicates that most of the VOCs present can be degraded within a few days residence times in granular iron. This reduction is quantified and related to possible field application in the Design Implications section of this paper.

Inorganic Parameters. The redox potential indicated slightly reducing conditions, while pH declined from values of 7.0 to 5.3. Most likely the pH did not increase, typically observed with most site waters (Gilham, 1996), due to the high concentration of dissolved organics in the groundwater.

Comparison of column influent and effluent results showed that significant additional chloride was generated as a result of degradation of the high influent VOC concentrations. Both manganese and sulphate concentrations increased slightly in the column effluent. The total iron concentrations increased substantially to 137 mg/L as the groundwater passed through the granular metallic iron, attributed to the low pH conditions. Calcium concentrations decreased by about 3 mg/L in the effluent, while the alkalinity declines by about 50 mg/L. The moderate declines in calcium and alkalinity concentrations are likely due to the formation of carbonate minerals.

DESIGN IMPLICATIONS OF BENCH SCALE TESTING

Using the results of the bench scale testing, in-situ flow velocities, in-situ contaminant distributions, and treatment goals, a reactive wall geometry can be selected that will achieve the needed remediation. In order to determine the required wall thickness, a simple contaminant transport/transformation model was run for four different initial conditions: 1) Centerline conditions in the Upper Aquifer Zone; 2) Centerline conditions in the Lower Aquifer Zone; 3) Conditions 30 m laterally away from centerline in Upper Aquifer Zone; 4) Conditions 30 m laterally away from centerline in Lower Aquifer Zone. Figure 5 is an example analysis for the Centerline of the upper zone. Treatment zone thickness was estimated using a target effluent concentration of 0.5 ug/l for all contaminants.

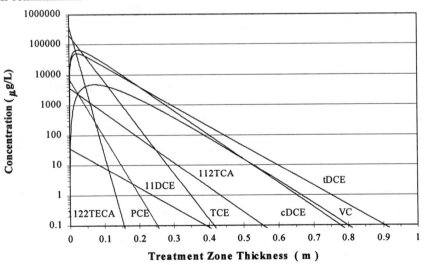

FIGURE 5. Reactive Wall Thickness as a Function of Target Effluent Concentration

By defining the magnitude of influent concentrations laterally along the wall, the wall's configuration can be optimized so that large wall thickness are only used where influent concentrations are highest.

Using similar analyses for the other zones, the following wall thickness requirements were determined:

Zone	Centerline Thickness	Thickness +/- 30 Meters from Centerline
Upper	1 m	0.7 m
Lower	0.12 m	0.09 m

SUMMARY

The following important conclusions were developed from this study:

- Most of the VOCs present at site 17 can be degraded within a few days residence times in granular iron;

- Half lives for TCE, 1122TeCA, PCE, cDCE, tDCE, VC were found to be 9.5, 3.4, 7.6, 18.0, 22.0, and 17.6 hr, respectively. These half-lives were much larger than in waters tested previously with lower influent concentrations;

- During the bench scale testing, water pH declined from values of 7.0 to 5.3. The decrease in pH suggests a lower potential for wall clogging due to precipitate formation, but also created dissolved iron concentrations of 137 mg/l;

- The reactive wall configuration was varied to use differing thickness for different locations, depending on influent contaminant levels.

REFERENCES

Gilham, R.W., 1996 In-Situ Treatment of Groundwater: Metal-Enhanced Degradation of Chlorinated Organic Contaminants, In: M.M. Aral (ed.), Advances in Groundwater Pollution Control and Remediation, Kluwer Academic Publishers, pp. 249-274.

Gilham, R.W., and S. O'Hannesin, 1994. "Enhanced Degradation of Halogenated Aliphatics by Zero-Valent Iron", Ground Water, 32: 958-967.

Laguna Construction Company, 1997. Site Characterization Summary Informal Technical Information Report for Site OT-17, Beale AFB, CA.

Law Environmental, 1995. Site Characterization Summary Informal Technical Information Report for Site 17, Beale AFB, CA.

URS Greiner Inc. 1996. Draft Engineering Evaluation / Cost Analysis, Site 17, Beale AFB, CA.

TCE DIFFUSION THROUGH CLAY BARRIERS CONTAINING GRANULAR IRON

L. Major[1], R.W. Gillham and C.J. Warren
(University of Waterloo, Waterloo, ON, Canada)
([1]Currently with Pacific Groundwater Group, Seattle, WA)

ABSTRACT: Laboratory tests were conducted in the Fall of 1996 and 1997 to determine if granular iron could be combined with bentonite clay to create a reactive hydraulic and diffusion barrier. One dimensional diffusive flux of trichloroethene (TCE) across simulated liners containing different amounts and types of iron was studied using stainless steel diffusion cells. The results indicated that the iron was effective at limiting the flux of TCE to values that were between 2% and 46% of the flux across a liner that contained no iron. Results from the long term diffusion experiment suggested that interactions between the iron and clay caused changes to both materials within the liner and could affect the long term performance of the barriers.

INTRODUCTION

Hydraulic barriers are often used to prevent groundwater contamination by limiting the advective transport out of source zones that contain high concentrations of chlorinated contaminants. In the absence of advective transport, diffusion becomes the dominant transport mechanism. The diffusive flux of contaminants out of these source zones is controlled by the concentration gradient across the barrier, and the effective diffusion coefficient and porosity of the barrier material. In the past, natural, thick clay deposits were used as hydraulic barriers but the use of much thinner engineered barriers, such as geosynthetic clay liners or geomembranes, has been increasing (Daniel, 1993). Although this reduction in thickness makes the engineered barriers easier to handle during construction, it also increases the concentration gradient across the barrier and therefore increases the diffusive flux out of the source zone.

Incorporating a reactive material that degrades the contaminant as it diffuses through the barrier could eliminate the mass transport out of the source zone and prevent groundwater contamination. Gillham and O'Hannessin (1994) have shown that zero-valent iron degrades chlorinated organics through the process of reductive dechlorination and more recent work (Gillham et al., 1997) has shown that iron plated with a second metal, such as nickel, degrades these compounds at faster rates.

Objective. The objective of this study was to determine whether granular iron, incorporated into a geosynthetic clay liner, would be an effective means of reducing the diffusive flux of chlorinated organic contaminants across thin hydraulic barriers.

MATERIALS AND METHODS

The clay material used in these experiments was a sodium-saturated bentonite clay commonly used in the manufacture of geosynthetic clay liners. The granular iron, with a grain size of 25 to 60 mesh, was obtained from Masterbuilders in Cleveland, Ohio. Tests were performed using the iron as received and also using the iron plated with about 0.1% nickel (enhanced iron) (Gawrilov, 1979). The two organic compounds used were toluene and trichloroethene (TCE). Toluene was used as a non-reactive organic tracer and TCE was the chlorinated organic expected to degrade in the presence of both the regular and enhanced iron. All water used in the experiment was distilled and passed through an Easypure UV filter to remove any trace amounts of residual organic compounds.

Four stainless steel diffusion cells, similar in design to those used by Luber (1992) and Warren et al (1996), were used to test four combinations of iron and bentonite. Each of the four cells contained approximately the same mass of bentonite, 5.3 g, but different amounts and types of iron. The control cell (Cell C) contained no iron, Cell R contained 1.5 g of regular iron, Cell E1 contained 1.5 g of enhanced iron, and Cell E2 contained 0.16g of enhanced iron. The iron and bentonite were combined dry and then placed in the solids compartment of each cell. In each cell, the liner material was held between two porous stainless steel disks to ensure contact with the solution reservoirs located on either side. Once the solids were in place, the cells were sealed and flushed with CO_2 and the liner materials were saturated using organic free water over a period of 2 to 4 days.

Diffusion Testing. At the start of the diffusion experiment, the source reservoir of each cell was flushed with source water containing toluene and TCE. The source reservoirs were then connected to a 9L glass carboy containing the same solution that was used for flushing. The source solution was circulated from the glass carboy through the reservoirs and back to the carboy at a flow rate of 0.08 mL/min.

Toluene and TCE diffused from the source reservoirs across the liners and into the sample reservoirs causing the concentrations in the sample reservoirs to increase. The sample reservoir of each cell was drained under gravity every 24 hours for 80 days and then flushed with organic-free water to set the concentration back to zero. For direct comparison, all four cells were sampled and flushed within a 15-minute period. The source carboy was sampled at the same time as the cells and the concentrations of toluene and TCE were analyzed in all samples. At the conclusion of the diffusion experiment, the solution reservoirs of each cell were flushed with organic-free water and were stored in that state for one year. A second diffusion experiment, identical to the first, was then conducted over a period of 27 days.

RESULTS AND DISCUSSION

First Diffusion Experiment. The concentration of toluene measured in the large source carboy is plotted as a function of time in Figure 1. Over the final 40 days of the experiment, the concentration of toluene in the source carboy was 49 (+/- 7) mg/L. The concentrations of toluene measured in the water drained from the sample reservoirs of each cell are also plotted with time in Figure 1. From Day 40 to Day 80, experimental and analytical variability were minimized giving relatively steady concentration data. The steady state toluene flux for each cell, listed in Table 1, was calculated using the average of all the concentrations of toluene that were measured for that cell between Day 40 and Day 80. Cells C, R, and E2 had similar concentrations at steady state of 5.3 mg/L, 5.1 mg/L, and 5.6 mg/L respectively, while the concentration in Cell E1 was lower at 2.9 mg/L.

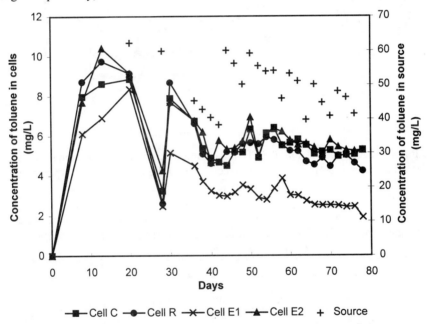

FIGURE 1. Concentration of toluene in the source carboy and in the water drained from the sample reservoirs of the four cells during the first diffusion experiment, plotted with respect to time.

The concentration of TCE in the source carboy is plotted with time in Figure 2. The average source concentration between Day 40 and Day 80 was 45 (+/- 6) mg/L. The concentration of TCE measured in the water drained from the sample reservoir of each cell is also plotted in Figure 2 and the steady state flux values and corresponding concentrations are listed in Table 1. The highest concentration of TCE (average value of 6.1 mg/L) was measured in the sample reservoir of the cell that contained no iron, Cell C. The lowest concentrations of

TCE occurred in the sample reservoir of the cell containing the higher amount of enhanced iron, E2, with an average value of 0.094 mg/L at steady state. The average values for the other two cells were intermediate, at 1.37 mg/L in Cell R and 2.82 mg/L in Cell E2.

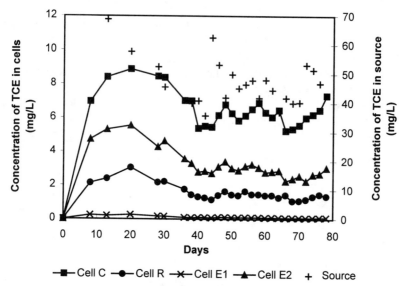

FIGURE 2. Concentration of TCE in the source carboy and in the water drained from the sample reservoirs of the four cells during the first diffusion experiment, plotted with respect to time.

TABLE 1. Average values of concentrations (Day 40 to Day 80) measured in the sample reservoirs during the first diffusion experiment, with the corresponding calculated and relative flux values.

	Toluene			TCE		
	Conc.	Flux	Relative Flux	Conc.	Flux	Relative Flux
	mg/L	mg m^{-2} s^{-1}	%	mg/L	mg m^{-2} s^{-1}	%
Cell C	5.34	3.96 x 10^{-4}	100	6.10	4.52 x 10^{-4}	100
Cell R	5.10	3.78 x 10^{-4}	96	1.37	1.02 x 10^{-4}	22
Cell E1	2.90	2.15 x 10^{-4}	54	0.094	7.02 x 10^{-6}	2
Cell E2	5.61	4.16 x 10^{-4}	105	2.82	2.09 x 10^{-4}	46

The relative flux across each liner, also listed in Table 1, was calculated by dividing the average of the concentrations in the sample reservoir of each cell by the average of the concentrations in the sample reservoir of the control cell at steady state. Experimental variability among the four cells was assessed by comparing the relative flux values for the non-reacting compound, toluene. The toluene results show good agreement between Cells C, R, and E2, with relative flux values between 96% and 105%. The concentration of toluene at steady state in the sample reservoir of Cell E1 was only 54% of the concentration measured in the control cell.

Table 1 clearly shows that the presence of the iron resulted in a lower flux of TCE across the liner. The most effective liner was in Cell E1, which contained 30% by weight enhanced iron. This cell had a concentration at steady state that was only 2% of the concentration measured in the control cell. Cell E1 was also the cell with the lower toluene flux. However, if the toluene results are used to adjust the TCE results by dividing the concentration of TCE at steady state by 0.54, the performance of this liner is not significantly compromised (increases to 3%). The liner mixture containing 30% by weight regular iron reduced the flux of TCE to 22% of the control, while the flux across the liner that contained 3% by weight reactive material was 54% of the control value.

Second Diffusion Experiment. The results from the second diffusion experiment, conducted one year after the initial test, were used to determine if the reactivity of the iron had changed over that time. The average values of the concentrations of toluene and TCE that were measured in the sample reservoirs once steady state transport was achieved, along with the relative flux values for each cell, are listed in Table 2. The results for toluene showed a decrease in the relative flux of this compound across the liners in Cells R and E1, which both contained 30% by weight reactive material while the relative flux of toluene across the liner that was only 3% by weight iron did not appear to change between the two experiments. The TCE results for the three cells containing iron indicate that the reactivity of the iron changed in all three cells between the first and second tests. The relative flux into the sample reservoirs of both the enhanced iron cells (E1 and E2) increased, indicating a loss of reactivity. The reactivity of the regular iron improved between the two experiments, indicated by a decrease in the relative flux of TCE across the liner in Cell R.

TABLE 2. Relative steady state flux across each liner in the second diffusion experiment

	Toluene	TCE
Cell C	100 %	100%
Cell R	76%	8%
Cell E1	37%	2%
Cell E2	100%	87%

CONCLUSIONS

The first diffusion experiment clearly demonstrated that the combination of bentonite and granular iron created a reactive hydraulic barrier capable of reducing the diffusive flux of TCE. The second diffusion experiment demonstrated that interactions between the iron and clay caused changes to occur within the liners. These changes resulted in different fluxes of both the non-reactive and reactive compounds, suggesting that both the iron and clay were affected. These interactions should be studied in more detail to gain a better understanding of the long-term potential of reactive hydraulic barriers.

ACKNOWLEDGEMENTS

This work was funded by an NSERC Motorola ETI Industrial Chair awarded to Dr. R.W. Gillham at the University of Waterloo, the Solvents-in-Groundwater Consortium, a grant from the Waterloo Centre for Groundwater Research and an Ontario Graduate Scholarship. All work was performed at the University of Waterloo under the supervision of Drs. Gillham and Warren.

REFERENCES

Daniel, D.E. 1993. "Landfills and Impoundments" In D.E. Daniel (Ed.), *Geotechnical Practice for Waste Disposal*, pp. 97-112. Chapman and Hall, London.

Gawrilov, G.G. 1979. Chemical Electrolytes in Nickel-Plating. H.E. Warne and Co Ltd., Cornwall, Great Britain.

Gillham, R.W., S.F. O'Hannessin, M.S. Odziemkowski, R.A. Garcia-Delgado, R.M. Focht, W.H. Matulewicz, and J.E. Rhodes. 1997. "Enhanced Degradationof VOC's: Laboratory and Pilot-Scale Field Demonstration" In *1997 International Containment Technology Conference Proceedings, St. Petersburgh, Florida, Feb. 9-12,* pp. 858-864.

Luber, M. 1992. "Diffusion of Chlorinated Organic Compounds through Synthetic Landfill Liners" University of Waterloo, Waterloo, ON.

Warren, C.J., S.F. O'Hannessin, and R.W. Gillham. 1996. "Degradation of Trichloroethene (TCE) by Zero-Valent Iron Incorporated Into Geosynthetic Liner Clay. Final Report for Albarrie Canada Limited" University of Waterloo, Waterloo, ON.

Gillham, R.W. and S.F. O'Hannessin. 1994. "Enhanced degradation of halogenated aliphatics by zero-valent iron" *Ground Water*. 32:958-967.

ENHANCED ZERO-VALENT IRON DEGRADATION OF CHLORINATED SOLVENTS USING ULTRASONIC ENERGY

Nancy E. Ruiz (University of Central Florida, Orlando, FL, USA)
Debra R. Reinhart (University of Central Florida, Orlando, FL, USA)
Christian A. Clausen (University of Central Florida, Orlando, FL, USA)
Cherie L. Geiger (University of Central Florida, Orlando, FL, USA)
Nancy Lau (University of Central Florida, Orlando, FL, USA)

Abstract: Permeable iron barriers have gained popularity in the past decade as a near-passive in situ remediation technology for halogenated organic solvents. A continuing problem is the loss of iron reactivity over time, most probably due to a build up of corrosion products or other precipitates on the iron surface. If these materials can be removed, the lifetime of an iron barrier may be significantly extended. Batch studies exposed coarse iron filings in deoxygenated natural groundwater to 330 W-Hr of ultrasonic energy prior to contact with trichloroethylene (TCE). Initial surface conditions of the iron influenced the improvement observed in the first-order rate constant. Specifically, the more obstructed the surface, the greater the impact of ultrasonic energy. While acid-washed iron exhibited a rate constant within one percent of its unsonicated counterpart, unwashed/as received iron demonstrated a 24% increase. Reactive material from a related column study yielded a near-initial rate constant after sonication. Sonication increased the fraction of nonchlorinated species among the daughter products, even if the TCE degradation rate constant changed very little. Increases in iron surface areas after sonication also suggest that ultrasound may be a candidate technique for restoring iron activity.

INTRODUCTION

In light of the pervasive nature of chlorinated solvent groundwater contamination and the potential for a passive remediation methodology afforded by permeable iron barriers, further study is imperative to explore their merits and minimize their deficiencies. Degradation rates increase with the degree of chlorination, increasing iron mass to solution volume ratio, and increasing iron surface area to mass ratio (Gillham and O'Hannesin, 1994, Vogan et al., 1994). Beside iron concentration, factors influencing rates include the treatment the iron receives prior to use, initial contaminant concentration, and the oxygen level and buffering capacity of the sample solution (Sivavec and Horney, 1995).

Reduction in dechlorination rates and flow problems have been linked to deposition of material, mainly carbonates in highly alkaline water, on the iron surface (Mackenzie et al., 1995). Reactivity also decreases as corrosion occurs on the surface of the iron (Johnson and Tratnyek, 1994). Two walls placed in series, one with large iron chips to act as an oxygen sink while maintaining good flow conditions and a smaller wall to provide dechlorination, may aid in preventing these problems (Mackenzie et al., 1995). A more direct approach which cleans the iron surface and restores reactivity would preclude the need for a second wall. The application of ultrasonic energy may be such a tool.

Ultrasonics refers to periodic stress waves that occur at frequencies in excess of 20 kHz. At their upper extreme, ultrasonic frequencies are so high that their extremely short wavelengths are comparable to the agitation of molecules caused by heat (Heuter and Bolt, 1955). Effects produced by ultrasonic waves arise from cavitation, which occurs in those regions of a liquid which are subjected to rapidly alternating pressures of high amplitude. These high pressures are relieved by the radiation of intense shock waves (Brown and Goodman, 1965). Ultrasound has degraded organic compounds (Bhatnagar and Cheung, 1994), suggesting a potent remedial approach for contaminated groundwater. However, its more common application as a cleaning process may provide an opportunity to enhance a zero-valent iron wall through surface maintenance. Suslick and Casadonle (1987) observed that acid-washed nickel particles were several hundred-fold more active in hydrogenation reactions after being sonicated.

Objective. This work employed ultrasonic energy to restore the reactivity of a degraded iron surface. Through batch studies, the capacity of iron in various conditions to degrade dissolved trichloroethylene (TCE) in natural groundwater with and without the intermittent application of ultrasonic energy was examined.

MATERIALS AND METHODS

Reagents were obtained from Fisher Scientific, all at least 99 percent purity and used as received. Coarse uncrushed iron filings were obtained from Peerless Corporation (Cincinnati, OH). Natural groundwater was obtained from the East coast of Central Florida.

Description of Experiments. An experiment was a set of two duplicate 1.0-L Tedlar® bags (a total of four bags) of identically processed iron which received different amounts of ultrasound. A bag contained 25g of iron, measured prior to any treatment, emplaced with 500 mL of natural groundwater, and purged with nitrogen gas for 30 minutes. The bag was then either sonicated for two hours (330 W-Hr) or immediately dosed with a stock aqueous solution to yield a TCE concentration of approximately 14 mg/L. A 600-W 20-kHz Branson Ultrasonic water bath for sonication events. After dosing, the bags were placed on a shaker table set at a rate of 160 shakes per minute to ensure adequate mixing and remained at ambient conditions, about 22°C, throughout a 28-day experiment.

Iron Treatment. Iron filings were used untreated or washed with a five percent w/w H_2SO_4 solution and rinsed with deionized water. To further study the effect of sonication on deteriorated iron, a column from an associated study was sacrificed after treating 300+ pore volumes of TCE in deionized water. The reactive material was composed of 20 percent acid-washed iron and 80 percent coarse sand, by weight. Material retrieved from the entrance end of the column, was emplaced 125g per bag, to yield an effective 25g iron per bag.

Groundwater Chemistry. Major physical and chemical properties for the natural groundwater used in this study are summarized in Table 1. "Field" indicates parameters which were measured with a Corning Checkmate M90 field instrument with interchangeable probes.

TABLE 1. Major physical and chemical parameters of groundwater from East coast of Central Florida.

Parameter	Value	Units	Method
pH	7.32	---	Field
Dissolved Oxygen	0.1	mg/L	Field
Conductivity	1923	μS	Field
Ca	81.4	mg/L	EPA 6010
Fe	0.059	mg/L	EPA 6010
Mg	57.7	mg/L	EPA 6010
SO_4^{2-}	71	mg/L	EPA 300
Total Alkalinity	289	mg/L as $CaCO_3$	EPA 310.1
Total Dissolved Solids	1100	mg/L	EPA 160.1

Sample Collection. Single samples were removed via a side valve on the bag septum using a gas-tight syringe and placed with zero headspace into 12-mL glass vials with Teflon-lined septa screw caps. The samples were then parafilmed and immediately refrigerated at 4°C until analyzed.

Organic Analysis. Liquid samples were analyzed for TCE and daughter products (dichloroethylene isomers, vinyl chloride, and ethylene) following EPA Method 624. A five-mL portion of a sample was injected with 1.0 μL of internal standard (5000 mg/L bromochloromethane in methanol) then transferred to a purge vial. Helium was bubbled through the vial at 22°C for 11 minutes to transfer the TCE onto a Vocarb 3000 trap. The trap desorb time was four minutes at 250°C and the bake time was seven minutes at 260°C. A Hewlett-Packard gas chromatograph Model 5890 equipped with a flame-ionization detector (FID) and a 0.25-mm i.d., 60-m long Vocol capillary column was programmed for a three-minute hold at 60°C, a 15°C/min rise to 180°C, and a three-minute hold at that final temperature.

Specific Surface Area. Filings were analyzed for specific surface area before and after sonication. Samples were rinsed with acetone and dried with nitrogen. To prevent oxidation, sonicated and acid-washed samples were stored under nitrogen in parafilmed screw-cap glass vials. Measurements were taken using a Porous Materials Inc. (PMI) Brunauer-Emmett Teller (BET) Sorptometer Model 201 (Ithaca, NY) using nitrogen as the adsorbate.

Scanning Electron Micrographs (SEM). The morphology of the iron surface and deposits before and after sonication were examined via SEM. Samples were prepared with a gold-palladium mixture applied using a Hummer VI-A Sputtering

System (EM Corporation, Chestnut Hill, MA) with an argon plasma then observed with an Amray 1810 Scanning Electron Microscope (Bedford, MA).

RESULTS AND DISCUSSION

First-Order Rate Constant Impact. Table 2 summarizes average first-order rate constants and the ratio of sonicated to unsonicated rate constants. The relative standard deviation for the rate constants ranged from one to 12 percent. Rate constants were calculated after 24 hours of exposure to TCE to allow equilibration with the iron surface. All correlation coefficients exceeded 0.99. Blank bags (no iron) exhibited a two to three percent decrease in TCE concentration over 14 days.

TABLE 2. First-order rate constants for TCE degradation in natural groundwater by coarse iron filings.

Iron Condition	1st Order Rate Constant, hr^{-1} (x 10^{-3})		Ratio, US/No US
	No US	US	
Unwashed	2.70	3.36	1.24
Acid Washed	4.74	4.68	0.99
Column Material	2.20	4.78	2.17

Exposure to ultrasonic energy generated improvement in first-order degradation rate constants, with one exception. Sonicated acid-washed iron yielded a rate constant within one percent of its unsonicated baseline rate constant. Unwashed iron rate constants, while not as great as acid-washed rate constants, increased approximately 24 percent after sonication. The greatest improvement was observed for the column material, where sonication restored reaction rates to near-initial (acid-washed) conditions. These post-sonication increases suggest that iron with an occluded surface can benefit from exposure to ultrasonic energy.

Nonchlorinated Fraction of Products. Because of ultrasound's capacity to activate catalyst surfaces and the observed increases in rate constants, the question arose whether sonication changes the degradation process or simply speeds it up. To learn more, the nonchlorinated fraction of C2 degradation products (expressed as equivalent TCE concentration) were plotted against half-life reductions in TCE. These plots disregarded time and focused on TCE transformation. For acid-washed iron, sonication increased the nonchlorinated fraction about five percent relative to unsonicated iron, substantially greater than the slight difference in post-sonication rate constants.

A more striking contrast was observed for unwashed filings, where the nonchlorinated fraction increased by about 10 percent, as illustrated in Figure 1. The lines in the figure represent approximate best fits. The divergence in C2 product composition due to sonication was noted as early as one half-life of TCE loss. Thus, exposure to ultrasound seems to make the degradation process more efficient as well as more rapid.

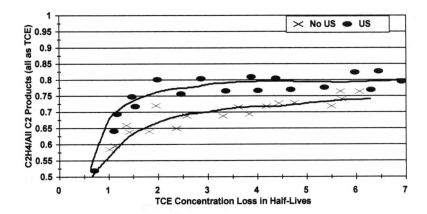

FIGURE 1. Ethylene fraction of C2 degradation products as related to TCE loss in half-lives for coarse iron filings.

Specific Surface Area. Table 3 summarizes specific surface area for different iron samples pre- and post-sonication. All correlation coefficients exceeded 0.99.

TABLE 3. Specific surface area for coarse iron filings resulting from pretreatment and sonication, m^2/g.

Iron Condition	Unsonicated	Sonicated	Ratio US/No US
Unwashed/As Received	0.77	2.07	2.68
Acid-Washed	1.92	2.91	1.52
Soaked Several Months in Natural Groundwater	11.31	7.49	0.66

Specific surface area seems related to first-order rate constants. For unsonicated samples, acid-washed iron exhibited specific surface area 2.49 times and a rate constant 1.76 times that of unwashed iron. For sonicated samples, acid-washed iron exhibited specific surface area 1.40 times and a rate constant 1.39 times that of unwashed iron. Acid washing increased specific surface area 149 percent for unsonicated iron. Filings soaked long-term in groundwater displayed surface area almost 14 times that of unwashed iron. Sonication increased iron specific surface area at least 50 percent, with the exception of the groundwater-soaked iron, which decreased 34 percent.

Surface Morphology (SEM Analyses). At 400X magnification, unwashed iron featured a dense concentration of large masses. Sonication reduced both the size and prevalence of surface debris. Acid-washed iron was substantially etched and pitted, and more so post-sonication. Groundwater-soaked filings featured structures including filaments and rosettes which obscured the underlying bulk material. Sonication yielded a surface similar to acid-washed iron.

CONCLUSION

Sonication positively impacts iron's degradation of chlorinated solvents, probably most directly linked to an increase in specific surface area, which is achieved by removing deposited materials and/or etching the surface. For some degraded irons, first-order rate constants can be restored to near initial values. Even where rate constant differences are minimal, the nonchlorinated fraction of degradation products is significantly increased. These effects suggest that iron can benefit from exposure to ultrasonic energy.

REFERENCES

Bhatnagar, A. and H. M. Cheung. 1994. "Sonochemical Destruction of Chlorinated C1 and C2 Volatile Organic Compounds in Dilute Aqueous Solution," *Environmental Science and Technology*, 28(8):1481-1486.

Brown, B. and J. E. Goodman. 1965. *High Intensity Ultrasonics*. London:Iliffe.

Gillham, R. W. and S. F. O'Hannesin. 1994. "Enhanced Degradation of Halogenated Aliphatics by Zero-Valent Iron," *Ground Water*, 32:958-967.

Heuter, T. F. and R. H. Bolt. 1955. *Sonics.* New York:Wiley.

Johnson, T. L. and P. G. Tratnyek. 1995. "Dechlorination of Carbon Tetrachloride by Iron Metal: the Role of Competing Corrosion Reactions," preprinted extended abstract, presented before the Division of Environmental Chemistry, American Chemical Society, Anaheim, CA, April 2-7.

Mackenzie, P. D., S. S. Baghel, G. R. Eykholt, D. P. Horney, J. J. Salvo, and T. M. Sivavec. 1995. "Pilot-Scale Demonstration of Chlorinated Ethene Reduction by Iron Metal: Factors Affecting Iron Lifetime," preprinted extended abstract, presented before the Division of Environmental Chemistry, American Chemical Society, Anaheim, CA, April 2-7.

Suslick, K.S. and D. J. Casadonle. 1987. Heterogeneous Sonocatalysis with Nickel Powder, *Journal of the American Chemical Society*, 109: 3459.

Vogan, J. L., J. K. Seaberg, B. Gnabasik, and S. O'Hannesin. 1994. "Evaluation of In situ Groundwater Remediation by Metal Enhanced Reductive -Dehalogenation - Laboratory Column Studies and Groundwater Flow Modeling," presented at the 87th Annual Meeting and Exhibition of the Air and Waste Management Association, Cincinnati OH, June 19-24.

WALL-AND-CURTAIN FOR PASSIVE COLLECTION/TREATMENT OF CONTAMINANT PLUMES

David R. Lee (AECL, Chalk River ON), David J.A Smyth (Dept. of Earth Sciences, University of Waterloo, Waterloo ON), Steve G. Shikaze (Environment Canada, National Water Research Institute, Burlington ON), Robin Jowett (Waterloo Barrier Inc., Rockwood ON), Dale S. Hartwig (AECL, Chalk River, ON) and Claire Milloy (Deep River Science Academy, Deep River ON)

ABSTRACT

A novel passive system has been designed to intercept and treat contaminated groundwater in a 12-m thick sandy aquifer. The contaminant plume, which contains strontium-90 (^{90}Sr), occupies the lower portion of the aquifer. The system is a combination of a cut-off wall, a reactive curtain and a passive drain. It will reduce cost, improve hydraulic control and provide more accurate compliance monitoring over existing subsurface treatment methods. Field application, planned for 1998, is designed to mitigate the anticipated discharge of ^{90}Sr to receiving surface waters. The reactive curtain, in this case granular clinoptilolite, is predicted to retain ^{90}Sr for over 20 years, or longer if additional material is used. The curtain is backed by a sheet-piling cut-off wall (Waterloo Barrier®) installed perpendicular to the plume axis and sealed with grout to the underlying bedrock. Perforated pipe within the curtain maintains a uniform head across the reactive ma terial. A weir controls the water level in the pipe, thus maintaining a subtle zone of depression and regulating the width of capture. Contaminant-free water in the upper portion of the aquifer is diverted from the treatment system by an upgradient horizontal drain in which hydraulic head is controlled by a separate weir. This controls the depth of capture and conserves treatment media.

Numerical simulations of various configurations of the wall-and-curtain illustrated its flexibility and permitted several options of the wall and the clinoptilolite curtain to be compared. Lab-scale modelling in a large sandbox demonstrated the integrated system and showed that a range of capture zones were achievable with an emplaced wall-and-curtain.

INTRODUCTION

This paper outlines a treatment system to be implemented in 1998 to mitigate the impact of a groundwater plume containing ^{90}Sr. Treatment of contaminated groundwater to a condition that satisfies drinking water quality guidelines will be achieved prior to its discharge to surface water. The leading edge of the contaminated groundwater plume passes through a subsurface zone of granular clinoptilolite where ^{90}Sr is removed from solution by sorption and retained for an extended period. Clinoptilolite is a zeolite, which is a hydrated aluminosilicate mineral with large surface area and sorptive capacity. The system design has generic applicability and combines three existing technologies: cut-off wall, reactive treatment and passive tile drain. The curtain of clinoptilolite is backed by a steel sheet pile cut-off wall and is connected to a drain. Monitoring and control is performed at two weirs with adjustable heights, allowing the groundwater capture zone to be manipulated. The weirs also provide single-point monitoring of effluent concentration and flow. Alternative designs considered did not offer the degree of control and monitoring that has become the standard in the nuclear industry.

SITE DESCRIPTION / BACKGROUND

The Lake 233 aquifer near Chalk River, ON has been relatively well characterized by previous studies (Parsons 1963, Killey and Munch, 1987). This aquifer has 10-18 m of saturated sands comprised of very fine sand, fine sand, fine-medium sand, minor inter-stratified silty sand and, in some areas, a basal till (Killey and Munch, 1987). Hydraulic conductivities range 2 orders of magnitude: 1.2×10^{-2} to 4×10^{-4} cm/s. Within the sand units only, hydraulic conductivities fall in a narrower range: 1.2×10^{-2} to 2×10^{-3} cm/s. Average linear groundwater velocities near the discharge zone have been estimated to be of the order of 0.2 m/day and the hydraulic gradient is about 0.02. The sands and silty sands lie in horizontal beds nearly parallel to the direction of the hydraulic gradient with the result that groundwater moves without complicated diversions due to variations in permeability. Ground penetrating radar and confirmatory drilling indicated that crystalline bedrock, at the proposed treatment site adjacent to the swamp, is relatively flat and lies at 10 to 12 m depth..

The ^{90}Sr plume is 20 m wide and occupies the lower half of the aquifer. Unchecked, the plume could result in undesirable concentrations of ^{90}Sr in Duke Swamp. The leading edge of the ^{90}Sr plume has been moving at a rate of about 10 m/a, which is 7% of the groundwater speed (Killey and Munch, 1987). The Sr^{+2} cation is sorbed on mineral surfaces and amorphous iron coatings as it exchanges with other cations.

APPROACH TO MITIGATION

The objective of this program is to ensure that groundwater discharge does not jeopardize the quality of receiving surface waters. Strontium-90 levels should be maintained below the drinking water standard of 5 Bq/L (Health Canada, 1996).

Capture and control of the ^{90}Sr plume could be achieved using various technologies and approaches. A key factor in the selection of a remedial system, whether *in situ* or *ex situ*, is the general reliance on sorptive removal of ^{90}Sr from water using clinoptilolite. Fuhrmann et al. (1995) and Cantrell (1996) have documented the efficiency of sorption by clinoptilolite in treating ^{90}Sr-contaminated waters to a quality acceptable for drinking. Sorption is, however, not chemical specific, and competition for sorption sites on clinoptilolite with cations such as Ca^{+2} may be significant. The shallow groundwater in the vicinity of the plume has low total dissolved solids, and Ca concentrations are generally about 10 mg/L. Thus, conditions are quite favourable for ^{90}Sr sorption by clinoptilolite.

Plume control and capture could be established using the pump-and-treat method. This would be an active solution, and would require on-going care and maintenance. Furthermore, to ensure capture of the plume, pumping of a considerable volume of uncontaminated groundwater would occur, which would result in the accelerated use of the clinoptilolite treatment media.

Plume interception and treatment could also be achieved using *in situ* permeable reactive walls or funnel-and-gate® systems. Although these systems could function passively for a period, the treatment media would need to be removed and replaced periodically before ^{90}Sr breakthrough occurred.

An alternate approach, which draws on some of the passive advantages of permeable reactive walls and funnel-and-gate® systems, combines a permeable curtain placed upgradient of an impermeable wall or barrier (Figure 1). In a manner similar to a system at Oak Ridge, TN (Barton et al., 1997) a lower ground surface elevation downgradient of the system enables the hydraulic head within the permeable curtain to be maintained passively using an adjustable weir, or control point, and a subsurface drain. The hydraulic head within the drain is at an elevation lower than the original water table at the wall, and is used to create a depression of head within the permeable curtain. The size of the capture zone of groundwater encountering the wall-and-curtain is controlled by regulating the difference in hydraulic head between the permeable curtain and the ambient groundwater system. Thus, capture is marginally larger than a contaminant plume. Because all groundwater captured by the system passes through the weir, flow rate and water quality can be monitored accurately.

The wall-and-curtain offers two options for treatment of contaminated groundwater. Coarse treatment media (clinoptilolite) can be used for backfill in the permeable curtain. In this way, all groundwater captured by the system is treated before it enters the drain. Alternately, treatment could be achieved within the drain system by passing all collected water through canisters containing treatment media before discharge. The use of clinoptilolite in the subsurface in the permeable curtain allows sorptive removal of ^{90}Sr to occur before aeration of the captured water occurs in the drain and weir system. Aeration may promote chemical reactions such as the precipitation of iron oxyhydroxides, which could clog the pore spaces or coat the clinoptilolite in the canisters. In other applications, however, replacement of canisters may be operationally easier to manage than replacement of the permeable curtain.

LABORATORY AND NUMERICAL MODELLING

A physical model of the wall-and-curtain was constructed in a sandbox aquifer. A drain, with a control weir, was incorporated downgradient of the permeable curtain. A series of tracer tests demonstrated that by varying hydraulic head in the permeable curtain, a range of plume widths and depths could be captured without changing the physical structure. Then, a horizontal drain was inserted in the flow system some distance upgradient of the wall-and-curtain. Trials showed that water from shallow depths could be captured before encountering the permeable curtain. Such a drain could be used to divert uncontaminated groundwater from shallow depths around a field-scale treatment system.

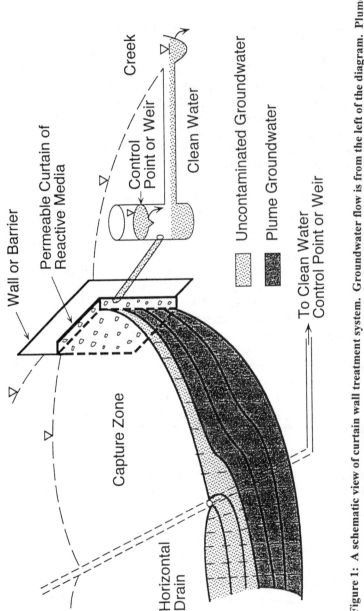

Figure 1: A schematic view of curtain wall treatment system. Groundwater flow is from the left of the diagram. Plume capture is achieved passively and treatment is achieved within the permeable curtain. Shallow clean groundwater is intercepted by an upgradient drain and diverted around the treatment system. Hydraulic head levels within the permeable curtain and drain are regulated by adjustable weirs situated in a manhole and connected by subsurface piping to an existing surface water body.

Numerical simulations were conducted using a three-dimensional, variably saturated flow and transport model (Therrien and Sudicky, 1996). The computational domain was established to represent a portion of the aquifer several times larger than the areal extent of the leading portion of the plume. The domain covered an area of dimensions 250 x 250 m, and the aquifer was assigned a thickness of 13 m. Constant hydraulic head values within the aquifer of 12 and 9 m were specified for the upgradient (x=250m) and downgradient (x=0) faces of the domain, respectively. The top surface was assigned a specific fluid flux of 0.3 m/year, and the lateral and bottom boundaries were assumed to be impermeable. The hydraulic conductivity and porosity of the aquifer were 1.5 x 10^{-2} cm/s and 0.33, respectively, with the exception of a near surface fine sand layer which was assigned values of 1.5 x 10^{-3} cm/s for hydraulic conductivity and 0.40 for porosity. The low-permeability barrier wall and associated permeable curtain were located 50 m from the downgradient face and equidistant from the lateral boundaries of the domain. The low permeability cutoff wall was assumed to have a hydraulic conductivity of 1.5 x 10^{-10} cm/s and the thickness of the permeable curtain was 2.5 m. Identical pressure head-saturation and relative permeability-saturation characteristics were used for the aquifer, the layer of fine sand, and the permeable curtain.

A series of simulations were conducted to evaluate the influence on capture-zone characteristics of variations in hydraulic conductivity and hydraulic head within the permeable curtain, and the influence of the length dimensions of the wall-and-curtain perpendicular to groundwater flow. For a wall 30 m in length and a permeable curtain 14 m in length (hydraulic conductivity 1.5 x 10^{-1} cm/sec), the width of the capture zone could be maintained from more than 50 m (hydraulic head at 1.0 m below the pre-existing water table), to approximately 35 m (hydraulic head at 0.6 m). Additional simulations showed that performance was similar even when the hydraulic conductivity of the curtain was equal to that of the aquifer.

As noted in the description of clinoptilolite, treatment capacity is consumed by the competitive sorption of common cations such as Ca^{+2}. To optimize treatment efficiency, it is advantageous to capture minimal amounts of groundwater containing no ^{90}Sr. One way to achieve this is through control of the characteristics and dimensions of the wall-and-curtain. A second option is based on the fact that the ^{90}Sr plume exists in the deeper portions of the aquifer and is overlain by a zone of ^{90}Sr-free groundwater. Preliminary modelling indicated that it may be possible to intercept some of the shallow uncontaminated groundwater and divert it around the treatment curtain through the use of a horizontal drain upgradient of the wall-and-curtain. The hydraulic head in the horizontal drain would be maintained below the pre-existing water level in the aquifer through the use of an adjustable weir. The modelling suggest that the performance of the drain is very sensitive to variations in hydraulic head.

DESIGN AND INSTALLATION

   Installation of the wall-and-curtain is scheduled for 1998. It will have the following features:

a) The low-permeability barrier or cut-off wall will be constructed of sealable-joint steel sheet piling (Waterloo Barrier®). The wall will be emplaced from surface through the aquifer to bedrock. The barrier will be sealed to bedrock using pressure-grouting from the base of the cavities associated with each joint. The joints are spaced approximately at 0.6 m intervals. The sealant will be a cementitious grout with a silica-fume modifier. Sheet-piling barrier installation and sealing techniques have been demonstrated and performance proven in previous experimental and commercial applications (Starr et al, 1992; Smyth et al., 1997). It is anticipated that bulk hydraulic conductivity of the barrier will be less than 10^{-8} cm/s. The barrier will be 30 m in length, and will straddle the extended centre-line of the contaminant plume.

b) The permeable curtain will consist of coarse granular clinoptilolite, with a hydraulic conductivity of 10^{-2} cm/s or greater. The curtain will have a length of 14 m, and be 2.5 m thick. The curtain will have a concrete footing approximately 0.3 m thick to improve the seal to bedrock and to facilitate dry emplacement of clinoptilolite. The curtain will extend from ground surface to the concrete footing. Both the footing and the clinoptilolite will be installed in an excavated trench within a temporary sheet-pile enclosure upgradient of the low-permeability cut-off wall. The upgradient face of the enclosure will be removed from the ground once the clinoptilolite has been emplaced.

c) Ten vertical well screens will be installed at 0.6 m intervals on the upgradient face of the sheet piling barrier from surface to 10 m depth. These will be connected in manifold to a horizontal drain installed at a depth of approximately 1.5 m below the pre-existing water table level also along the upgradient face of the wall. The drain will be connected with piping to an adjustable weir, or control point, downgradient of the wall-and-curtain. All collected water will exit through the pipe and weir to facilitate capture zone control and accurate monitoring of water quality and flow.

d) A second horizontal drain will be installed 1 m below the watertable upgradient of the wall-and-curtain. This drain will also be connected by piping to an adjustable weir, and will be used to divert shallow, uncontaminated groundwater around the treatment system. The diverted groundwater will be monitored closely, and the weir will be adjusted to ensure that contaminated groundwater does not bypass the treatment system.

   It is anticipated that *in situ* removal of ^{90}Sr from the groundwater will occur within the permeable clinoptilolite curtain for more than two decades. This prediction relies on calculations using data from plume measurements, laboratory experiments and comparisons with other investigations (Sterling, 1995; Fuhrmann et al., 1995; Cantrell, 1996). The retardation factor, R, for a contaminant moving through a porous medium with groundwater flow is defined by the equation:

$$R = V_L/V_C = 1 + \rho_b/n * Kd \qquad (1)$$

where: V_L is the average linear velocity of groundwater, V_C is the velocity of the contaminant, ρ_b/n is the ratio of the bulk density of the porous medium to its bulk porosity, and Kd is the distribution coefficient. Sterling (1995) measured R values in laboratory column tests using granular, calcium-saturated clinoptilolite of 5,000 and 2,200. Corresponding Kd^{Sr} values are 1,700 and 745 mL/g, respectively. These compare to Kd^{Sr} values of 630 (Fuhrmann et al., 1995) and 2,600 mL/g (Cantrell, 1996) determined from laboratory tests using groundwaters from two Department of Energy facilities in the United States. The differences are a function of the concentrations of competing ions in the different waters, laboratory testing by batch or column methods, and differing characteristics of sources of clinoptilolite. Accounting for the width (35 m) and depth (10 m) of the capture zone, the volumetric porosity of the clinoptilolite, and the natural groundwater velocity (0.20 m/day), it is estimated that the permeable curtain will be flushed by groundwater approximately 85 times per year. Assuming vertically averaged plume concentrations, and incorporating the range of retardation factors determined by Sterling (1995), the time for arrival of unacceptable concentrations of ^{90}Sr to reach the drain is estimated to be several decades.

Clearly, using the permeable curtain for treatment in perpetuity will require that the clinoptilolite be replaced or that additional volumes of clinoptilolite be placed so that flow from the permeable curtain is treated before water is released downgradient.

SUMMARY

A versatile concept is presented for passive plume mitigation. It has been designed and tested with lab and numerical modelling. The initial application in 1998 is intended to prevent the discharge of a ^{90}Sr plume on the property of AECL's Chalk River Laboratories. Key components are an impermeable cut-off wall, a reactive and permeable curtain of treatment media and a drain leading to a topographic low. The advantages are lower cost and greater hydraulic control compared with other methods. This method may not be applicable where conditions are adverse to installation of cut-off walls or where effluent cannot be channeled to a downgradient surface water body or to a subsurface distribution gallery.

REFERENCES

Barton, W. D., P.M. Craig and W. C. Stone. 1997. Two passive groundwater treatment installations at DOE facilities. In Proceedings 1997 International Containment Technology Conference and Exhibition, Feb. 9-12, St. Petersburg. Sponsored by U. S. Department of Energy, DuPont Company and U. S. Environmental Protection Agency. pp 827-834.

Cantrell, K.J. 1996. A permeable reactive wall composed of clinoptilolite for containment of SR-90 in Hanford groundwater. In Proceedings of the

International Topical Meeting on Nuclear and Hazardous Waste Management Spectrum '96. Seattle, Washington, pp. 1358-1365.

Fuhrmann, M., D. Aloysius and I. Zhou. 1995. Permeable, subsurface sorbent barrier for ^{90}Sr: laboratory studies of natural and synthetic materials. Waste Management, Vol. 15, No. 7, pp. 485-493.

Health Canada. 1996. Guidelines for Canadian drinking water quality. 6th ed. Supply and Services Canada, Hull, Quebec.

Killey, R.W.D. and J.H. Munch. 1987. Radiostrontium migration from a 1953-54 liquid release to a sand aquifer. Water Poll. Res. J. Canada, Vol. 22, No. 1, pp. 107-128.

Parsons, P.J. 1963. Migration from a disposal of radioactive liquid in sands. Health Phys., Vol. 9, pp. 333-342.

Smyth, D., M. Gamble and R. Jowett. 1997. Sealable joint steel sheet piling for groundwater control and remediation. Case histories. In Proceedings 1997 International Containment Technology Conference and Exhibition, Feb. 9-12, St. Petersburg. Sponsored by U. S. Department of Energy, DuPont Company and U.S. Environmental Protection Agency. pp 206-214.

Starr, R.C., J.A. Cherry and S. E. Vales. 1992. A new type of steel sheet piling with sealed joints for groundwater pollution control. 45th Canadian Geotechnical Conference, October 26-28.

Sterling, S. 1995. Strontium removal from groundwater using ion exchange with clinoptilolite. B.Sc. thesis. Civil Engineering, Queen's University. 64pp.

Therrien R. and Sudicky, E.A., 1996. Three-dimensional analysis of variably-saturated flow and solute transport in discretely-fractured porous media. Journal of Contaminant Hydrology, v. 23, pp. 1-44.

REACTANT SAND-FRACKING PILOT TEST RESULTS

Donald L. Marcus (EMCON, Burbank, California)
James Farrell, (University of Arizona)

EMCON, in conjunction with the property owner and under license from EnviroMetals Technology, Inc. (ETI), has completed a Reactant Sand-Fracking (RSF) pilot study at a former aerospace facility in Newbury Park, California. RSF uses hydraulic fracturing equipment to emplace zero valent iron reactive media into a chlorinated solvent and metal impacted, fractured, bedrock aquifer. EMCON has applied for a patent for the technology (Marcus, 1994).

The purpose of the RSF pilot study was to install reactive media in to an impacted bedrock aquifer and to measure the effectiveness of in-situ treatment of VOCs and metals in groundwater. Specific goals of the program were to demonstrate the effective placement of reactive media into fractured bedrock; to measure the chemical changes in the groundwater after the media was installed, and to measure the changes in well capacity.

The RSF pilot study demonstrated that

- Zero valent iron foam has the physical and chemical properties necessary to effectively treat VOCs and metals in groundwater when inserted under high pressures into fractured bedrock
- Iron foam reactive media can be placed in bedrock using high pressure hydraulic fracturing equipment
- Well capacity is increased by improving hydraulic conductivity through hydraulic fracturing and proppant injection

The scope of work consisted of the following activities.

- Reactive proppant development, evaluating both zinc and zero-valent iron proppants
- Reactive proppant bench scale testing
- Borehole preparation, including drilling a new borehole
- Pre-RSF testing
  - Surface geophysics and seismic refraction and resistivity surveys
  - Borehole geophysics: induction, caliper, and acoustic televiewer logging
  - Borehole video camera log
  - Chemical analysis of discrete groundwater samples from the identified fracture zones
  - Packer permeability testing
- Data evaluation and selection of three fracture zones for RSF injection
- Hydraulic fracturing of the selected fracture zones before RSF injection
- RSF injection
- Postinjection testing
  - Surface geophysics and resistivity survey
  - Borehole geophysics: induction, caliper, and acoustic televiewer
  - Chemical analysis of discrete groundwater samples from the treated fracture zones

  – Packer permeability testing
* Well installation, well development, and pump testing
* Continuing monitoring of the treatment's effectiveness

The RSF treatment well was installed near a former leachfield that had been used for wastewater disposal in the late 1950s and early 1960s. Wastewaters were discharged from a nearby plating shop into the leachfield, which was installed above weathered volcanic bedrock. The waste stream contained concentrations of VOCs (trichloroethene [TCE], tetrachloroethene [PCE], and 1,1,1-trichloroethane [1,1,1-TCA]) and metals (hexavalent chromium [Cr^{+6}] containing chromate compounds, cadmium, and copper). These chemicals percolated down into the bedrock aquifer and have been detected in concentrations up to 180 mg/L of TCE and 35 mg/L of Cr^{+6} in groundwater samples obtained from monitoring well E-1 located 42 feet from the RSF test well.

A half-mile-long groundwater plume extends away from the site into a groundwater basin used for a water supply. Although wells installed in the weathered zone of bedrock can produce significant groundwater, wells installed in the deeper unweathered bedrock are generally poor producers of groundwater (Marcus, 1993). The plume is currently contained by a 2.5-million-gallon-per-month groundwater pump-and-treat system consisting of eight extraction wells. The low productivity of the leachfield extraction well has made it a small contributor to the system. Because of the high chemical concentrations in the source area, the October 1993 U.S. Environmental Protection Agency Statement of Basis for the site) references the technical impracticability of groundwater restoration at DNAPL sites (USEPA, October 1993), and requires pumping and treatment in perpetuity, or until technology can remediate the source areas. In it's ultimate implementation the RSF methodology can potentially be used to create a series of overlapping a passive treatment zones (Starr and Cherry, 1994) for *in situ* treatment of impacted groundwater. The passive *in situ* treatment zone will potentially offer a low maintenance alternative to the scenario of containment by pumping and treating forever.

To obtain reactive media for the study, EMCON tested media manufactured by Teleflex's Sermatec division and by Cercona of America, Inc., Dayton, Ohio. Sermatec manufactures reactive proppants of zinc-alloy-coated aluminosilicate. Cercona's proppants are composed of a proprietary iron foam material. Bench test results indicated that the iron foam proppants are more reactive and have greater longevity than the zinc-coated proppants

The Cercofoam® proppant design enables emplacement by hydraulic injection. Testing requires the proppants to withstand closure pressures up to 3,000 psi and pumping rates up to 1,000 gallons per minute without catastrophic failure or significant loss of reactivity.

Bench tests were conducted at the University of Waterloo Center for Groundwater Research (ETI, May 1996), University of Arizona, and USEPA Athens laboratory. University of Arizona testing (Farrell and Kason, January 1997) demonstrated the effectiveness of the iron foam proppants at dechlorinating VOCs such as TCE and PCE, achieving up to 99 percent reductions. Influent concentrations of up to 700 mg/L TCE were treated in the column test. Reactivity was sustained for over 20 months in the test, which identified reductive dehalogenation from both biotic and abiotic processes. The existence of the abiotic process is documented by changes to pH, redox potential, and first order breakdown of PCE with a half-life of 400 to 2,000 minutes. The existence of biotic processes is supported by the non first order, rapid breakdown of TCE, with a half-life between 200 and 400 minutes.

For the RSF test, fractures were identified in the borehole by a downhole geophysical survey. Selected fractures were discretely sampled using the SimulProbe Aquifiler zero purge sampler and chemically tested to verify that they were likely to be pathways for migrating chemicals of concern. The groundwater samples were brought to the surface in a pressurized canister, eliminating the possibility of cross-contamination or loss of volatile. The Aquifiler, equipped with YSI 600XL sensor array, also took real-time *in situ* measurements of conductivity, pH, oxidation-reduction potential, dissolved oxygen, and depth. The pretest results were consistent with those of a shallow aquifer with oxidizing groundwater conditions.

Pre-RSF chemical testing detected TCE and other VOCs in the samples at concentrations up to 7,210 ug/L. Concentrations of dissolved chromium were between 0.017 and 2.7 mg/L—up to 1 order of magnitude less than expected (based on historical data from adjacent wells). This suggested some dilution of borehole groundwater from the tap water added to the borehole to enable acoustic logging. Additional dilution occurred from hydraulic conductivity testing and hydrofracturing activities but was considered to have been offset by advection and diffusion of *in situ* groundwater into the borehole.

Three discrete fractures (42, 57, and 87 feet below ground surface) were subjected to hydraulic conductivity testing using packer injection techniques. Pretesting data showed that the fractures had very low hydraulic conductivities, ranging from 10^{-5} to 10^{-6} cm/sec. Northeast Water Production, Inc., (Sterling, Massachusetts), treated three fractured zones using its custom hydrofracturing equipment. The treated zones were isolated with high-pressure liquid inflatable packers. A polysaccharide viscosifier suspended the proppants in the fracking fluid. The fluid was injected under high pressures (up to 2,500 psi) into the fractures until the desired quantity of proppants was emplaced, or refusal occurred. Between 300 and 1,700 pounds of 1 millimeter diameter reactive proppants, were injected in each zone to treat the groundwater. A datalogger and pressure transducers monitored water levels during injection and recovery. Ion-selective electrodes monitored bromide tracer concentrations *in situ*. The monitoring data obtained from adjacent observation wells verified that fracking fluids reached at least 42 feet from the treatment well during hydrofracturing.

Reductive dehalogenation using zero-valent metals and *in situ* reduction and precipitation of metals are University of Waterloo ETI proprietary processes. After the reductive dehalogenation reaction proceeds to completion, most of the chlorinated VOCs are converted to innocuous products such as nonchlorinated hydrocarbons and chloride ions (Gillham, 1993). As shown on Figure 1 below, the concentrations of many of the chlorinated VOCs decreased up to 98 percent based on the results of pre- and post-RSF treatment analyses. Chlorinated VOCs that were noted to decrease included TCE, PCE, 1,1,1-TCA, 1,1-DCA and 1,1-DCE. Increases in concentrations of cis-1,2-DCE and chloroform suggest that the rate of transformation of the parent compounds to these daughter products is higher than the rate of destruction of the daughter products.

Figure 1 Reactant Sand-Fracking Pilot Test
CVOC Concentrations Detected in Groundwater
Well E-39B Before and After Treatment

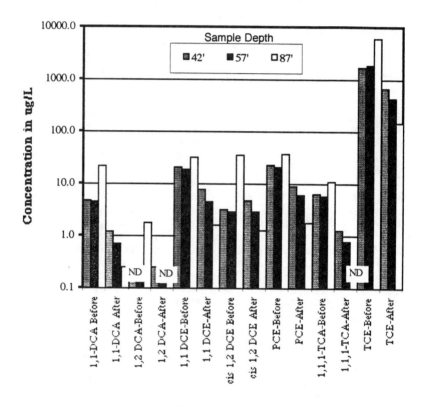

Reductive dehalogenation is indicated by the highly reducing groundwater conditions with a redox potential as low as -500 millivolts. In addition, the presence of reactant groundwater conditions associated with the agent has caused the reduction and precipitation of oxidized species such as Cr^{+6}, Cu^{+2}, nitrate, and sulfate (Blowes and Ptacek, 1992; Vogan, 1993). Although increases in iron and zinc concentrations have been noted, the levels detected are all below the maximum contaminant levels for the metals.

Follow-up packer tests, borehole and surface geophysics, and discrete chemical sampling for VOCs and metals were conducted to verify that the fractures had been effectively treated. The packer test results indicate that the hydraulic conductivity of the fractures increased up to two orders of magnitude after RSF treatment. After RSF injection and post tests were completed, the extraction well was reinstalled and the guar breakdown products and fine sediment were pumped from the well. The developed fluids were monitored for bromide tracer using ion-selective electrodes. This monitoring verified that well development had recovered the center of mass of the injected fluids. Pump tests performed during the well development, indicated a significant improvement in well productivity, from a negligible capacity prior to treatment to about 1.5 gallons per minute afterward. Nearly all of the preliminary results appear positive, suggesting that RSF could provide a long-term alternative and/or augmentation to conventional pumping and treatment at the site.

As a next step in the evaluation of RSF, it could be feasible to install overlapping zones of fractures containing reactive media to create permeable reactive barriers for the containment and treatment of impacted groundwater at deep and fractured bedrock sites.

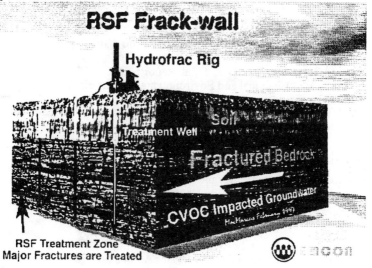

REFERENCES

Blowes, D.W., and C.J. Ptacek, 1992. Geochemical Remediation of Groundwater by Permeable Reactive Walls: Removal of Chromate by Reaction with Iron-Bearing Solids." Proceedings of the Subsurface Restoration Conference, 3rd International Conference on Groundwater Quality Research, June 21-24, Dallas, Texas, pp 214-216.

EMCON, March 1997. Reactant Sand-Fracking Pilot Study Workplan. Talley Facility, Newbury Park, California.

EnviroMetal Technologies Inc., May 1996, Report on the Laboratory Treatability Tests Conducted on the Ground Water from the Talley Facility.

Farrell, J., and Mark Kason, 1997. Final Report: EMCON Iron Column Test. Department of Chemical and Environmental Engineering, U. of Arizona.

Gillham, R., 1993. US Patent, Reductive Dehalogenation of Chlorinated Volatile Organic Compounds, November.

Marcus, D., 1993. Physical and Chemical Factors Affecting the Migration of TCE and Chromium in a Fractured Bedrock Aquifer. Proceedings of Hazmacon, San Jose, California.

Marcus, D., 1994. In Situ Treatment of Groundwater Contaminants, Patent Application Serial No. 08/310,223

Smith, Stuart A., 1989. Manual of Hydraulic Fracturing for Well Stimulation and Geologic Studies. National Groundwater Association Special Publication.

Starr, R. and Cherry,J. 1994. In Situ Remediation of Contaminated Groundwater: The Funnel and Gate System. *Groundwater*, 32:3, pp 465.

USEPA, October 1993. OSWER Directive Technical Impracticality of the Restoration of Groundwater at DNAPL Sites..

Vogan, J., 1993. Treatment of nitrate in groundwater using zero valent iron. MS Thesis, University of Waterloo, Ontario.

ENHANCING THE REACTIVITY OF PERMEABLE BARRIER MEDIA

Arun R. Gavaskar, Bruce M. Sass, Eric Drescher, Lydia Cumming,
Daniel Giammar, and Neeraj Gupta (Battelle, Columbus, Ohio, USA)

Abstract: Due to a galvanic corrosion effect, trichloroethylene (TCE) degraded
significantly faster in a bench-scale column containing an iron-copper granular
mixture as with iron alone. When a long-term test was conducted, however, the
reaction rate of iron-copper declined, indicating that this medium was used up
faster. A weak acid was passed through the two used media (following the long-
term test) to dissolve precipitates and regenerate the media.

INTRODUCTION

The objectives of this study were to investigate ways of enhancing the
reactivity of fresh and used reactive metal media. Reactive metals, particularly
granular zero-valent iron, are finding application in permeable barriers for
groundwater remediation. The reactivity of an in-situ permeable barrier medium
is measured in terms of its reaction rate with target contaminants. In this study,
trichloroethylene (TCE) was used as the target contaminant.

In an effort to enhance the reactivity of fresh iron medium, a second metal
was added. Previous research (Muftikian et al., 1995; Orth and McKenzie, 1995;
Appleton, 1996) has looked at the reactivity of bimetallics. The second metal
(e.g., palladium or nickel) usually has a catalytic effect on the reactivity of the
first metal, iron. Generally, the second metal is electroplated on the first to obtain
the reactivity enhancement.

In this study, a combination of metals that form a galvanic couple was
used to enhance reactivity. Individual granules of the two metals were mixed
together in a simple mixture without resorting to electroplating. Electroplating
adds significantly to the cost of the reactive medium, several tons of which may
be required in a field barrier. The two metals in the mixture were selected on the
basis of the galvanic series in seawater shown below:

> *More anodic (corroded end)* ← *magnesium, zinc, aluminum, cadmium, cast iron,
> lead, tin, copper, nickel, silver, gold, platinum* → *More cathodic (protected end)*.

Galvanic couples work on the principle that when two dissimilar metals are
electrically connected in an electrolyte, current flows from the more noble metal
(cathode) to the less noble metal (anode). The potential difference between the
two metals causes the anodic metal to release electrons to the electrolyte. In this
process, the anodic metal is preferentially sacrificed or corroded. This increased
electron activity held the prospect of a higher degradation (reduction) rate of TCE
dissolved in groundwater in the vicinity of the metal couple.

Iron was one of the metals used in this study because it is highly reducing
and relatively inexpensive. The second metal had to have the following qualities:

- ❏ It had to be more cathodic than iron so that the less expensive metal (iron) would be sacrificed in the barrier.
- ❏ To generate sufficient potential difference, it had to be relatively far from iron in the galvanic series.
- ❏ It had to be relatively benign in the environment.
- ❏ It had to be relatively inexpensive.

Copper appeared to meet all these qualities to some degree and was the second metal selected. A lower proportion of copper (25% by mass) than iron was used in the mixture because copper, as the cathode, is conserved, while iron is sacrificed upon exposure to groundwater.

$$Anode: \quad Fe \rightarrow Fe^{2+} + 2e^- \quad E^0 = 0.447 \ V$$
$$Cathode: \quad Cu^{2+} + 2e^- \rightarrow Cu \quad E^0 = 0.342 \ V$$

To enhance the reactivity of used iron and iron-copper media (after long-term exposure to groundwater), an acid solution was passed through the two media. An acidic environment held the possibility of dissolving several types of precipitates that might have formed on the metal surfaces. Use of an inorganic acid, such as hydrochloric or nitric, was avoided because a strong acid would probably attack the metal medium itself. A weak organic acid, acetic, was used because of its potential to dissolve several types of precipitates without significantly attacking the metal medium.

MATERIALS AND METHODS

Two 5-cm diameter glass columns were set up side by side as shown in Figure 1. The "Fe" column contained granular iron whereas the "Fe-Cu" column contained a mixture of iron and copper granules. Copper constituted 25% by mass of the mixture. Zero-valent iron (–8+40 mesh) was obtained from Peerless Metal Powders, Inc. and copper (–30+50 mesh) from SCM Metals, Inc. The metals were packed to a height of 45 cm in each column.

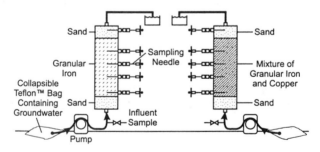

Figure 1. Columns for comparing the performance of two reactive media.

Water spiked with 2 to 3 mg/L TCE was passed upward through both columns at approximately the same flow rate of 3.5 mL/min each (equivalent to an average linear velocity of 4 m/day). Water withdrawn from sample ports along the columns was analyzed for TCE and other organics using a gas chromatograph (GC) with a flame-ionization detector (FID). Oxidation-reduction potential (ORP) was measured using a platinum electrode and a Ag/AgCl reference. Core samples of the used media were collected and analyzed by scanning electron microscopy (SEM) and x-ray diffraction (XRD) to determine the types of precipitates formed after prolonged exposure to the groundwater

RESULTS AND CONCLUSIONS

Table 1 compares the reactivities of iron and iron-copper media at steady state. Steady state is indicated by steady TCE concentration profiles in the columns and was reached after approximately 40 pore volumes of TCE-spiked groundwater had passed through. The first order reaction rate constant, k, was approximately twice as high in iron-copper as in iron. Thus, the half-life of TCE in iron-copper was significantly shorter than its half-life in iron alone. This indicated that individual iron and copper granules could be mixed together to obtain faster TCE degradation without resorting to electroplating.

Table 1. Degradation rates of TCE in iron and iron-copper media

Pore Volumes	Medium	Rate Constant, k* (hr^{-1})	First–Order Goodness of Fit, r^2	TCE Half-Life** (hrs)
37	Iron	0.34	0.981	2.0
41	Iron-Copper	0.62	0.928	1.1
44	Iron	0.34	0.969	2.0
49	Iron-Copper	0.59	0.977	1.2
52	Iron	0.19	0.928	3.7
58	Iron-Copper	0.52	0.976	1.3
60	Iron	0.22	0.994	3.2
65	Iron-Copper	0.48	0.995	1.4
72	Iron	0.27	0.998	2.6
80	Iron-Copper	0.52	0.998	1.3

* k is obtained from the first order rate equation, $C = C_0 \cdot (e^{-kt})$, where C and C_0 are the TCE concentrations at times t and zero, respectively.
** 2 to 3 mg/L TCE spiked into deionized water. Groundwater was used in all subsequent tests.

In additional tests with local groundwater, the partially dechlorinated byproduct cis-1,2 dichloroethene (DCE) was generated in both columns, indicating similar TCE reduction pathways in both media. However, DCE degraded much faster in iron-copper than in iron (Figure 2a).

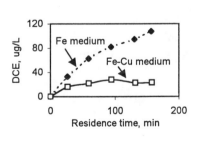
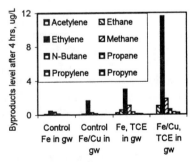

Figure 2. (a) Chlorinated and (b) non-chlorinated byproducts

Batch tests were conducted to identify other reaction byproducts. Both iron and iron-copper media generated similar byproducts as shown in Figure 2b.

The presence of ethene and ethane, as well as acetylene and propyne, in both media indicates that hydrogenolysis as well as reductive beta-elimination mechanisms (Roberts et al., 1996) are operative in both media. All byproducts were generated much earlier in iron-copper than in iron, as illustrated in the 4-hour snapshot in Figure 2b. Indications are that the iron-copper galvanic couple makes TCE degrade faster through pathways similar to those with iron alone.

Table 2. Behavior of Inorganic Constituents in Groundwater

Pore Volumes (PV)	Iron (mg/L)		Copper (mg/L)		Calcium (mg/L)		Alkalinity (mg/L)	
	Fe Column	Fe/Cu Column	Fe Column	Fe/Cu Column	Fe Column	Fe/Cu Column	Fe Column	Fe/Cu Column
200 PV:								
Influent	0.2	0.2	0.022	0.014	113	118	347	353
Effluent	4.0	8.4	0.026	0.025	37	51	149	186
1,000 PV:								
Influent	0.4	0.4	0.017	0.017	119	119	348	348
Effluent	13.3	15.0	0.024	0.023	67	78	261	272

Another indication that the galvanic corrosion mechanism is the cause of the increased reactivity is obtained from the fact that the level of dissolved iron introduced into the groundwater effluent from the iron-copper column was greater than that from the iron column (see Table 2). On the other hand, copper levels in the groundwater effluent from either column were low and remained unchanged. Iron, being more anodic than copper, dissolves preferentially in the Fe-Cu column. The elevated levels of iron in the effluent may be a minor environmental concern, because iron is subject to a secondary drinking water standard of 0.3 mg/L. However, these iron levels were still within the upper part of their range in natural groundwaters. Interestingly, dissolved iron levels in the effluent increased over time (1,000 pore volumes) in both columns. There was virtually no difference in the pH and ORP profiles in the two media (Figure 4).

Table 3. Effect of relative particle size on degradation rate

Medium	TCE Half-Life, min
Iron (-8+40 mesh) & Copper (-30+50 mesh)	27
Iron (-8+40 mesh) & Copper (-8+40 mesh)	55
Iron, -8+40 mesh	49

When iron-copper mixtures in the same proportion by mass, but with different sizes of copper were compared in two columns, the relative particle size of the two metals was found to affect reactivity (Table 3). When both iron and copper were coarse-grained, the reaction rate was not as high as when the copper was finer than the iron. This indicates that contact between the two metals, a factor essential for galvanic corrosion, is improved when copper particles are smaller than the iron particles. A wider range of particle sizes may affect the hydraulic flow characteristics within the reactive medium. Tracer tests (see Figure 4) conducted in the two columns using bromide as a conservative tracer, show a tendency for greater dispersion (and residence time distribution) when iron and copper were dissimilar rather than similar in size. This is probably due to a greater number of preferential pathways in the medium with the wider particle size distribution. The tailing of the tracer may also indicate greater adsorption of bromide on the fine copper particles.

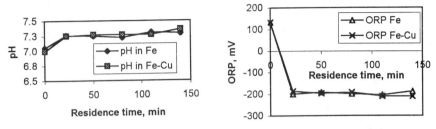

Figure 3. The (a) pH and (b) ORP profiles for the two media

To investigate the ability of the iron-copper medium to sustain its reactivity over the long term, 1,200 pore volumes of groundwater were run through each column. A higher flow rate of 7 mL/min was used for this long-term test. For each sampling event, however, the flow rate was temporarily brought back down and steadied at 3.5 mL/min, as in the short-term test. Figure 5 shows the result of this long-term test. The higher reactivity of the iron-copper column could not be maintained over the long term. This indicates that, faster degradation of TCE was accompanied by faster reactions with other groundwater constituents. Eventually, at around 600 pore volumes, the reactivity of the metals mixture fell below that of iron alone.

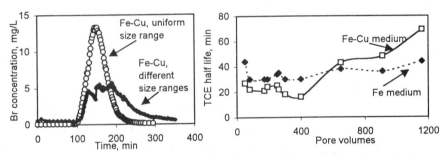

Figure 4. Tracer residence time distribution.   **Figure 5. Long term test.**

When samples of the used iron (after exposure to 1,200 pore volumes of groundwater) were collected, SEM and XRD indicated the presence of iron oxy-hydroxide and iron carbonate hydroxide precipitates on the iron near the influent end. Carbonate precipitates (calcite and aragonite) were indicated in the bulk iron throughout the column. In an effort to improve the reactivity of the two used media, a 0.01 M acetic acid solution (with a pH of 3.8 when prepared in groundwater) was passed through the two columns following the long-term test. At this concentration, the acid was expected to target mainly the carbonate and, to some degree, the hydroxide precipitates. About 4 pore volumes of acid solution was used in the first injection and 5 pore volumes in the second. The effluent pH in the two columns was around 5.7 after the first injection and 5.3 after the second. Following each acid injection, sampling was conducted after sufficient TCE-spiked groundwater was again passed through the two columns to achieve steady state.

As seen in Figure 6, the reactivities of both media improved after the acid treatment. Interestingly, the improvement was significant in the first few inches of the two columns, where one might expect a higher proportion of iron hydroxides. This was indicated by the significant fall in concentrations measured in the first port of each column. Increased levels of dissolved iron were noticed in the acid treatment effluent from both columns. Figure 6 shows that more acid was required to restore the reactivity of the iron-copper medium than to restore the reactivity of the iron.

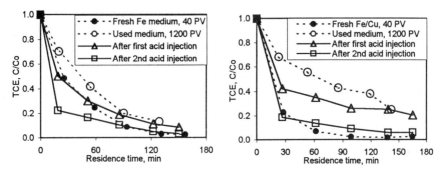

Figure 6. Regeneration of (a) iron and (b) iron-copper media with acetic acid.

REFERENCES

Appleton, E.L. 1996. "A nickel-iron wall against contaminated groundwater." *Environmental Science and Technology*, 30(12): 536A-539A.

Roberts, A.L., L.A. Totten, W.A. Arnold, D.R. Burris, and T.J. Campbell. 1996. "Reductive elimination of chlorinated ethylenes by zero-valent metals." *Environmental Science and Technology*, 30(8): 2654-2659.

Muftikian, R., Q. Fernando, and N. Korte. 1995. "A method for the rapid dechlorination of low molecular weight chlorinated hydrocarbons in water." *Water Research*, 29(10): 2434-2439.

Orth, R.G. and D.E. McKenzie. 1995. "Reductive dechlorination of chlorinated alkanes and alkenes by iron metal and metal mixtures". Extended Abstracts from the Special Symposium: *Emerging Technologies in Hazardous Waste Management VII*, p. 50. American Chemical Society, Atlanta, GA.

BENCH-SCALE TRACER TESTS FOR EVALUATING HYDRAULIC PERFORMANCE OF PERMEABLE BARRIER MEDIA

Lydia Cumming (Battelle, Columbus, Ohio)
Bruce Sass, Arun Gavaskar, Eric Drescher, Travis Williamson, Melody Drescher
(Battelle, Columbus, Ohio)

ABSTRACT: During a bench-scale column study, tracer tests were used to evaluate the hydraulic performance of two reactive metals media. Potassium bromide was used as the relatively conservative tracer. A bromide-selective electrode and a reference electrode were used to continuously monitor the effluent during the tests. Trichloroethylene (TCE)-contaminated groundwater was run for an extended period of time through the columns. Tracer tests were conducted in the initial and final stages of the run (approximately 1100 pore volumes) to evaluate changes in porosity and residence time caused by corrosion and precipitate build-up in the columns.

INTRODUCTION

An important step in designing a permeable barrier to treat chlorinated-solvent contaminated groundwater is to identify and screen candidate barrier media. The permeable barrier media selected must have a particle size large enough to ensure required hydraulic capture by the permeable barrier but have a small enough particle size ensure a sufficient surface area to react with the contaminant. It is expected that the peak performance within the permeable barrier will occur when the reactive metals particles are relatively new and few mineral precipitates have deposited on them. As the reactive medium corrodes and the pore spaces become clogged with mineral precipitates, its performance may decline. Tracer tests with a conservative tracer (one that does not interact with the solid surface of the porous matrix) can be used to determine residence time distribution and average residence time of the groundwater flowing through a bench-scale column filled with permeable barrier media.

Objective. The purpose of this paper is to present the results of bench-scale tracer tests for evaluating the hydraulic performance of permeable barrier media. The objective of the tracer tests were (1) to compare the differences in the hydraulic performance between two columns containing two different reactive metals media and (2) monitor changes in hydraulic performance after long-term exposure to groundwater.

MATERIALS AND METHODS

Two columns 56-cm in length with associated sampling ports were used. Column 1 was packed with –8 to +40 mesh granular iron from Peerless Metal Powders, Inc. A 5.08-cm section of Ottawa sand was placed above and below the reactive medium to maximize the flow distribution throughout the column.

Column 2 was packed with –8 to +40 mesh granular iron (75% by mass) and -30 to +50 mesh copper (25% by mass) from SCM Metals, Inc. A 5.08-cm section of sand was placed above and below the reactive medium as in the first column. Teflon™ and Viton™ tubing was used to supply groundwater collected from an uncontaminated aquifer to both columns. A peristaltic pump was used to pump the groundwater to the inlet of the columns. The groundwater was contained in collapsible Teflon™ bags to minimize headspace and volatilization of the organic contaminants spiked into the groundwater. The groundwater was circulated from bottom to top to simulate flowrates likely to be encountered in subsurface aquifers. A tee valve located between the peristaltic pump and the column inlet was used to inject the tracer.

A bromide-selective electrode and a reference electrode, manufactured by Microelectrodes, Inc., were used to continuously monitor the bromide concentrations in the effluent. A flow-through cell, connected in-line to the effluent tubing, had two ports designed to match the 2.5-mm diameter of the electrodes. A Teflon™-coated stir bar was placed inside the sample cell ensure sufficient mixing. The sample cell was held in place over a magnetic stirrer with a piece of corrugated cardboard situated in between to act as insulation. The electrodes were connected to a Corning® 350 pH meter.

A 1000mgBr/L standard solution was prepared by adding 1.49g potassium bromide to 1000-ml of deionized water in a volumetric flask. 5-ml of the standard solution was added to 50-ml of groundwater to create a tracer concentration of 100mgBr/L. Groundwater was used to create the calibration solutions used with the meter. The calibration solutions were created within the expected concentration range.

Procedure. TCE-contaminated groundwater in concentrations of 2 to 3 mg/L was pumped to the columns at approximately 3.5ml/min flowrate. To conduct the tracer test, the pump was shut down and the inlet valve was turned to the closed position. A 30-ml glass syringe, filled with 100mgBr/L tracer concentration, was injected into the influent for 9 minutes to maintain the 3.5ml/min flowrate. The pump was turned on and the inlet valve was turned to the open position to restart the contaminated groundwater flow through the column. The effluent was continuously monitored with the bromide-selective and reference electrodes placed in the in-line 5-ml flow through cell. The test continued until the concentration reading returned to baseline. A calibration curve constructed from the measurements of known solutions before and after the test was used to convert the millivolt (mV) readings into actual bromide concentrations. Bromide concentration versus time was plotted on a graph. Effective porosity of the media and average residence times were then determined based on the tracer test results.

Initially, large shifts in calibration curves and random fluctuations were observed in the mV potential as the tests progressed. This was attributed to a build-up of deposits on the bromide sensor under the strong reducing conditions created by the reactive metals media. Frequent electrode maintenance was required to increase the stability of the electrodes and obtain reliable readings.

RESULTS AND CONCLUSIONS

Comparison of the hydraulic performance of the two relatively new media.

Figure 1 depicts the tracer profiles of the two columns after being exposed to more than 300 pore volumes (PV) of TCE-contaminated groundwater. Column 1, which contained the –8 to +40 mesh granular iron, shows a higher peak bromide concentration and less tracer dispersion than Column 2, which contained the -8 to +40 mesh iron/-30 to +50 mesh copper mixture. At relatively low concentrations, the bromide should be conservative with respect to the liquid and solid matrix. However, mechanical dispersion occurs as a consequence of local variations in velocity around some mean velocity of flow (Bear, 1979). Because Column 2 contained a wider range of grain sizes than Column 1, more mixing probably resulted in Column 2, causing greater dispersion.

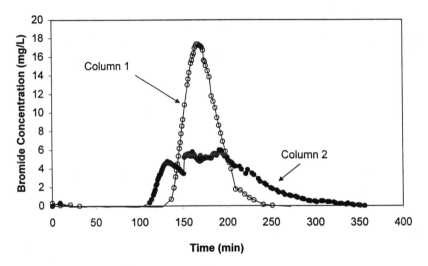

Figure 1. Tracer profiles of Column 1 and Column 2 after being exposed to 300 PV of TCE-contaminated groundwater.

In addition, a shorter average residence time was observed in Column 1. The shorter residence time was probably the result of the smaller porosity of Column 1. The total porosity of the sand and each media was calculated based upon the weight and density of the material added to the column. The total porosity of the reactive media placed in Column 1 was 0.593; the total porosity of the Column 2 reactive media was 0.624. Given the wider particle size distribution in Column 2, a lower porosity would be expected in this column. However, packing and particle shape also influence porosity (Bear, 1979). Table 1 details the some of the parameters measured during the tracer tests.

TABLE 1. Parameters measured during the tracer tests

Column	PV	Average Residence Time[1]	Maximum Conc.[2] (mg/L)	Effective Porosity
1	300	170	17.4	0.592
2	300	179	6.0	0.627
1	1100	171	12.4	0.599

ND = Not Determined.
[1] Average residence times were normalized to 3.6ml/min flowrate for comparison.
[2] Baseline corrected.

The uneven appearance of the tracer profile for Column 2 may be partly due to the low concentrations of tracer being close to the detection limit of the bromide-selective electrode. In addition, the highly reducing conditions of the effluent from Column 2 had a greater effect on electrode performance. Precipitate build-up on the surface of the electrode caused large shifts in calibration curves and erratic readings. Therefore, an oxidizer was introduced into the flow-through cell to address the reducing condition effect and improve the electrode performance.

Changes in hydraulic performance over time. Figure 2 shows the tracer profiles of Column 1 after 300 and 1100 PV.

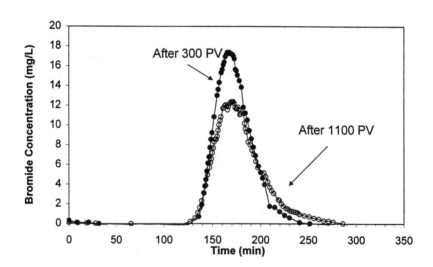

Figure 2. Tracer profiles in Column 1 after 300 and 1100 pore volumes

No significant change occurred in the average residence time and effective porosity of Column 1 between 300 and 1100 PV. However, slightly greater dispersion was observed. This was probably a result of the precipitate created by the contaminated groundwater in the interstitial pore spaces. Both media turned slightly gray in appearance over time. SEM photos of the samples taken from the media of both columns identified these precipitates as iron oxy-hydroxide, iron carbonate hydroxide, and carbonate precipitates. 1100 PV may be too short a time period for significant precipitation and flow changes to develop, and perhaps at over 2500 to 5000 PV one may observe noticeable differences. The slightly longer tailing after 1100 PV may indicate higher absorption of bromide on hydroxide and carbonate precipitates.

Field Application. Based on the laboratory experience using continuous ion-selective electrodes, a successful field application was conducted at the former Naval Airstation Moffett Field, California. The field tracer test was used to evaluate the flow direction and velocity through the permeable iron cell (gate) of the funnel-and-gate type permeable barrier (Gavaskar, et al., 1997). Figure 3 shows the tracer profile of one of the monitoring points in the permeable barrier during the tracer test.

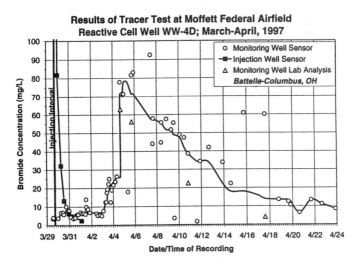

Figure 3. Results of the tracer monitoring in a Moffett Federal Airfield reactive cell well.

Discussion. The tracer profiles showed the differences in grain size distribution and effective porosity between the two reactive metals used in the column studies. Although precipitate build-up was observed and identified in the columns after long-term exposure to TCE-contaminated groundwater and its constituents, the tracer tests revealed that the hydraulic parameters such as porosity and residence time were not significantly affected in the time period (1100 PV). During longer time periods, the effects of precipitation may be more noticeable.

Difficulties encountered during these tracer tests include maintaining electrode performance in strongly reducing conditions of effluent. This resulted in large shifts in calibration curves and lowered sensitivity. An oxidizer was introduced into the flow-through cell to decrease the reducing conditions. Although this succeeded in minimizing the shift in the calibration curve, the electrode still lost some sensitivity at low-end concentrations. Because of the significant cost in replacing electrodes, special handling and maintenance is required; however, the ability to conduct continuous monitoring with electrodes improves tracer profiles as opposed to discrete interval sampling. In a field tracer test, monitoring of a bromide tracer with bromide-selective electrodes enabled a successful hydraulic evaluation of a permeable barrier.

REFERENCE

Bear, Jacob. 1979. *Hydraulics of Groundwater*. McGraw-Hill Inc. New York.

Gavaskar, Arun, Bruce Sass, Neeraj Gupta, and Jim Hicks. 1997. *Field Tracer Application to Evaluate the Hydraulic Performance of the Pilot-Scale Permeable Barrier at Moffett Federal Airfield*. Prepared for the Naval Facilities Engineering Service Center, Port Hueneme, California by Battelle.

DESIGN AND CONSTRUCTION OF VERTICAL HYDRAULIC FRACTURE PLACED IRON REACTIVE WALLS

Grant Hocking, Samuel L. Wells and Rafael I. Ospina

Golder Sierra LLC, Atlanta, GA, USA.

ABSTRACT: The conventional means of installing iron reactive walls for remediation of ground water contaminated with chlorinated solvents is by shoring and excavation or continuous trenching. Recently orientated vertical hydraulic fracturing has been used as an alternate mode of placing permeable iron reactive walls in situ, at modest depth and with minimal site impact. The design activities and construction of permeable iron reactive walls installed in permeable sands and gravel are described. The design methodology involves collecting site characterization data and performing specific field and laboratory tests to quantify iron reactivity and iron/gel required properties. The fracture geometry is computed in real time by active resistivity and the wall's hydraulic effectiveness is quantified by high resolution hydraulic pulse interference tests.

INTRODUCTION

Zero valent metals have been known to abiotic degrade certain compounds; such as pesticides as described by Sweeny and Fisher (1972), and halogenated compounds as detailed in Gillham and O'Hannesin (1994). In the case of zero valent iron, the abiotic degradation of halogenated aliphatics can be approximated by a first order reduction process. The compounds are progressively degraded to daughter products and eventually broken down into ethanes and ethenes.

In situ permeable iron reactive walls have been placed at a number of sites to abiotic degrade chlorinated solvents in groundwater. Iron reactive walls have significant advantages over conventional technologies for remediating chlorinated solvent contaminated groundwater; with the prime advantages being that the system is passive, significantly cheaper than alternatives and has minimal operation and maintenance. It is a simple process, that has been proven both in the laboratory and the field. Site characterization and laboratory bench scale studies are sufficient to design and construct a permeable iron reactive wall.

ORIENTATED VERTICAL HYDRAULIC FRACTURING

Hydraulic fracturing field experiments in unconsolidated sediments have demonstrated, (Hocking, 1996), that a) vertical fractures can be placed at any required azimuth or bearing, and b) by injection in multiple well heads, continuous coalesced fractures are formed. The technology involves initiating the fracture at the correct orientation at depth and by controlled injection a continuous wall is created, see Figure 1. The hydraulic fracture iron reactive wall is constructed by injecting through multiple well heads spaced along the wall alignment. The iron reactive wall installation is monitored in real time during injection to determine it's geometrical extent and to ensure fracture coalescence or overlap occurs. The quantities of iron reactive mixture injected are continuously monitored to ensure sufficient reactive iron is injected through the individual well heads.

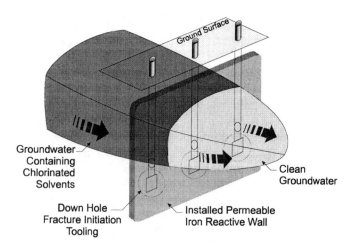

Groundwater Containing Chlorinated Solvents

Clean Groundwater

Down Hole Fracture Initiation Tooling

Installed Permeable Iron Reactive Wall

FIGURE 1. Hydraulic Fracture Iron Reactive Wall.

The gel is injected into the formation and carries the iron filings to the extremes of the fracture. The gel is a water based cross link gel, hydroxypropylguar (HPG), which is a natural polymer used in the food industry as a thickener. HPG is used in the process because it has minimal impact on the iron's reactivity and upon degradation leaves an extremely low residue. The gel is water soluble in the uncross linked state, and water insoluble in the cross linked state. Cross linked, the gel can be extremely viscous, ensuring the iron filings remain suspended during the installation of the wall. An enzyme breaker is added during the initial mixing to controllably degrade the viscous cross linked gel down to water and sugars. Once the gel is degraded, a permeable iron reactive wall remains with a wall thickness typically of 3" to 4" in sand and gravel formations.

DESIGN METHODOLOGY

An overview of the design methodology for a hydraulic fracture placed permeable iron reactive wall is given in Figure 2. First the site is characterized to determine the detailed geology, hydrogeology, contaminant and groundwater chemistry data. Next a risk assessment of the site is conducted to assess the need for remedial action at the site. If remedial action is required a focused feasibility study is generally conducted to determine what is the most appropriate form of the remedial activity. In this case a permeable iron reactive wall has been selected due to it's cost benefit and effective remediation of the site. The risk assessment along with regulatory requirements provides the basis for selecting either target effluent contaminant levels or risk reduction factors as deemed most appropriate for the performance of the reactive wall.

The design criteria for the permeable iron reactive wall are quantified to ensure the wall is designed and constructed to meet the risk reduction or target effluent levels as set in the risk assessment. These design criteria also address issues regarding impact on groundwater flow regimes, variability of input parameters on system performance, construction quality assurance, long term monitoring and health and safety.

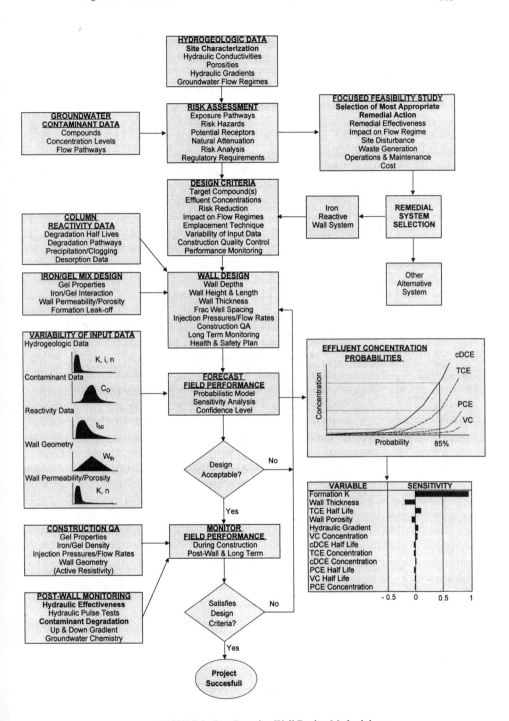

FIGURE 2. Iron Reactive Wall Design Methodology

The wall design activity requires additional data over conventional site characterization data; namely, column reactivity data and iron/gel mix design data. These data are generated from laboratory tests conducted on site groundwater and soils. The abiotic reduction of trichloroethene (TCE), tetrachloroethene (PCE), vinyl chloride (VC) and isomers of dichlorethene (DCE) by zero valent iron metal has been described by a number of workers, (Gillham and O'Hannesin, 1994 and Matheson and Tratnyek, 1994). Laboratory column tests utilizing site groundwater quantify the degradation reaction rates and degradation pathways (daughter products) of the particular contaminant species in the presence of iron filings. From detailed chemical analysis of column effluent water, additional issues are addressed such as potential precipitation and clogging of the wall.

A probabilistic model is utilized in the design process to enable confidence limits to be placed on wall performance and to quantify the impact of parameter variability on overall system performance. Probabilistic distributions are assigned to all of the system's parameters based on their expected variability. Not only are site charterization data; such as, hydraulic conductivity, porosity and hydraulic gradient, system parameters, but so are wall thickness, wall porosity and contaminant degradation data. The probabilistic analysis determines confident levels of system performance; such as contaminant degradation or overall risk reduction, and also ranks each parameter's impact on system performance from the most to the least sensitive, as illustrated in Figure 2.

Computed probabilities of contaminant degradation are illustrated in Figure 3 for the contaminants TCE and PCE for a particular reactive wall system. The contaminant degradation is plotted as a percentage based on the influent contaminant concentration. From the probability distributions, the confidence levels can be computed for the system performance on a specie by specie basis or as an overall system risk reduction factor. In Figure 3, the ratio of contaminant effluent to influent concentration are shown for a 85% confidence level for a particular reactive wall system.

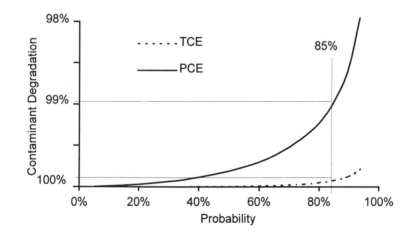

FIGURE 3. Probabilistic Forecast of Reactive Wall Performance.

CONSTRUCTION

The construction activities for a permeable iron reactive wall installed orthogonal to the natural groundwater flow direction involve installing groundwater monitoring wells, both up and down gradient and down hole resistivity receivers for tracking wall geometry during injection. After completion of these activities a pre-wall groundwater sampling round is completed along with pulse interference tests between monitoring wells.

The hydraulic fracture place reactive wall is constructed from a series of 6.25" polyvinylchloride (PVC) casings installed along the wall alignment and grouted in place. Each casing is cut and initiated with a controlled vertical fracture at the required azimuth orientation by a special down hole tool, see Figure 4. Upon initiation of the controlled fracture, the tool is withdrawn and a packer set. Injection well heads are connected to the packer assembly and injection hoses connected to the pumping unit.

FIGURE 4. Mixing and Pumping Unit and Down Hole FracTool.

The iron filings are loaded in hoppers and feed to the mixing and blending system via 12" inclined screw conveyors, see Figure 4. The gel is mixed and blended with the iron filings in the uncross linked state with sufficient mechanical agitation to ensure the iron filings remain evenly distributed. Precise in line density measurements are collected to ensure the mix is of the correct consistency and to provide a detailed record of iron injected. Following mixing and blending, the iron gel mix is pumped to a plunger hydraulic fracturing pump for injection into the prepared well heads. The gel and iron filings are feed to the pumping unit and cross linked in line, to form a highly viscous cross linked gel of a specific gravity typically around 2.

The iron reactive wall installation is monitored in real time to determine it's geometrical extent and ensure it is constructed as planned. Monitoring of the wall installation involves recording and control equipment to maintain pressures, flow rates, densities and volumes injected and to determine detailed fracture geometry. Down hole resistivity receivers are monitored to detect changes in the induced voltages by the propagation of the fracturing fluid. From these induced voltages and utilizing an incremental inverse integral model, the fracture geometry can be quantified during the installation process.

The hydraulic continuity of the wall is quantified by pulse interference tests, (Johnson et. al., 1966), with pulse source wells on one side of the wall and high precision receiver transducers on the opposite side. The test involves a cyclic injection of fluid into the source well and high precision measurement of the pressure pulse in a neighboring well. The time delay and attenuation of the hydraulic pulse enables the hydraulic effectiveness of the wall to be assessed.

CONCLUSIONS

Permeable iron reactive walls have been constructed in highly permeable sand and gravel formations down to depths greater than 50'. The geometry of the wall can be delineated by electrical active resistivity and the wall's hydraulic effectiveness determined from hydraulic pulse interference tests. The gel used to transport the iron filings and propagate the hydraulic fracture has minimal impact on the iron's reactivity and the wall's permeability. The prime benefits of the fracturing installation process are cost savings over alternate installation techniques, minimal impact on groundwater flow regimes, minimal waste volumes generated, limited site disturbance and deep application of the technology.

REFERENCES

Gillham, R. W., and S. F. O'Hannesin. 1994. "Enhanced Degradation of Halogenated Aliphatics by Zero-Valent Iron", *Ground Water*, Vol. 32, No. 6, pp958-967.

Hocking, G. 1996. "Azimuth Control of Hydraulic Fractures in Weakly Cemented Sediments". 2nd North American Rock Mechanics Symp, Montreal, A.A. Balkema, Rotterdam, pp1043-1048.

Johnson, C. R., R. A. Greenhorn and E. G. Woods. 1966. "Pulse-Testing: A New Method for Describing Reservoir Flow Properties Between Wells". *JPT*, Vol. 237, pp1599-1604, Trans., AIME.

Matheson, L. J., and P. G. Tratnyek. 1994. "Reductive Dehalogenation of Chlorinated Methanes by Iron Metal". *Environ. Sci. Technol.*, Vol. 28, pp2045-2053.

Sweeny, K. H., and J. R. Fisher. 1972. "Reductive Degradation of Halogenated Pesticides". U.S. Patent No. 3,640,821.

METAL ENHANCED REDUCTIVE DEHALOGENATION BARRIER WALL DESIGN

Darrin J. Wray, US Air Force, Hill AFB, Ogden, Utah, USA
Michael J. McFarland, Ph.D., Utah State University, Logan, Utah, USA

ABSTRACT: Metal Enhanced Reductive Dehalogenation (MERD) is an innovative chemical process used to remediate ground water contaminated with halogenated organic compounds. MERD employs zero valent iron to catalyze the removal of the halogenated substituents from organic compounds under oxygen limited conditions. The resulting compounds are less toxic and may be rapidly mineralized by subsurface microorganisms.

In the present effort, the MERD technology was evaluated for field scale application at Hill Air Force Base, Utah (Hill AFB). After consideration of site conditions, a MERD system consisting of 400 feet (121.9 m) of sheet pile funnel wall and a 20-foot (6.1 m) gate was designed and evaluated. The gate consists of three sections; a precipitation zone, the reactive iron cell, and a dispersion zone. In addition to the funnel wall and gate design, a ground water monitoring system of two-inch multi-screened wells was also incorporated into the system to evaluate the effectiveness of the MERD technology. The cost of constructing the MERD system was estimated at $922,500 dollars. The system has a design life of 30 years and an associated operations and maintenance cost of approximately $1,278,000 dollars.

INTRODUCTION

The chemical trichloroethene (TCE) has been used extensively as an industrial-grade solvent at Department of Defense (DOD) facilities. In the past, disposal practices were not regulated to prevent ground water from becoming contaminated with TCE. Therefore, there are currently many TCE-contaminated ground water aquifers at DOD sites. Traditional approaches to remediate these contaminated aquifers have generally been based on conventional pump-and-treat methods. Since these traditional methods are both expensive and are often labor intensive, more cost-effective and efficient remediation technologies are constantly being sought. One technology which has proven to be more cost-effective and efficient in remediating TCE-contaminated aquifers is Metal Enhanced Reductive Dehalogenation (MERD).

Hill Air Force Base (Hill AFB), Utah is a US Air Force installation that is currently attempting to remediate TCE-contaminated ground water at its 1100 Zone site. The ability to implement a low cost, low-maintenance, *in-situ* remediation system, such as MERD, would be a benefit to the Hill AFB ground water restoration program. The MERD system would effectively treat the contaminated ground water and control the plume indefinitely without any energy input, minimal maintenance and relatively low initial cost when compared to traditional pump-and-treat systems. This report contains the design and cost information for a MERD system that could potentially be installed at the 1100 Zone site.

Objective. The objective of this report is to determine if a MERD system represents a viable alternative for remediating the TCE contaminated ground water at Hill AFB's 1100 Zone site. The two primary goals of this effort were to: 1) Design a full-scale MERD treatment system for the 1100 Zone site TCE plume and 2) prepare a cost estimate of the designed system to evaluate the economics of implementation and long-term operation and maintenance.

Site Description. The 1100 Zone site at Hill AFB, Utah contains a maximum ground water TCE concentration of 33 parts per billion. The contaminated ground water begins approximately 10 feet (3 m) below ground surface and is confined by a clay layer approximately 40 feet (12.2 m) below ground surface. The contaminated plume is approximately 300 feet (91.4 m) wide and 1400 feet (426.7 m) long while the ground water velocity is in the range of 0.03 to 0.3 feet (0.91 cm to 9.1 cm) per day.

MATERIAL AND METHODS

The effectiveness of the MERD process is dependent on the control of ground water flow through a funnel-and-gate system. The funnel-and-gate system consists of low hydraulic conductivity cutoff walls (funnels) with gaps that contain in-situ reactors or gates (Figure 1).

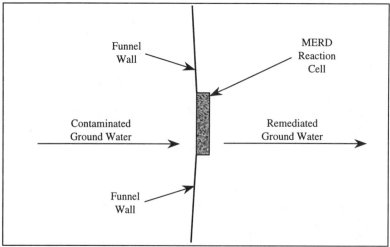

**Figure 1. Basic funnel and gate system for ground water remediation.
(Adapted from Starr and Cherry, 1993)**

The first step to be considered when designing a funnel-and-gate system is its location. The funnel system should be located so that all of the contaminated ground water flows through the gate. This is a significant engineering challenge since ground water flow directions vary over time (Starr and Cherry, 1993). Secondly, the residence time of contaminated ground water in the gate must be sufficient so that the desired reduction in concentration is achieved. Residence time requirements depend on the influent concentration, the required effluent concentration and the reaction rate.

Finally, if possible, the contamination zone or source area should be isolated from uncontaminated groundwater through the use of cutoff walls. These walls not only reduce the amount of uncontaminated water that comes into contact with the source area but also increases the retention time of contaminated ground water in the MERD system by reducing the volumetric flow through the gate. This design yields the greatest retention time for a given length of cutoff wall, and generates a capture zone that is insensitive to ground water flow variations (Starr and Cherry, 1993).

Reactor or Gate Design. The reactor for the MERD process consists of the gate portion of the funnel-and gate system. The reactor is basically an iron filled trench through which the TCE contaminated ground water must pass. If the MERD system is properly designed and constructed, the TCE in the ground water reacts with the iron and is dechlorinated allowing contaminant-free water to leave the reaction cell (Figure 2).

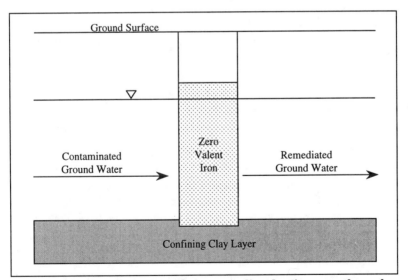

Figure 2. Contaminant plume being remediated as it moves through a permeable reaction cell. (Adapted from O'Hannesin and Gillham, 1992)

One of the most important parameters to consider in the reaction cell design is the ground water retention time. The retention time, which is the average time that a unit of contaminated ground water spends within the gate, governs the extent to which ground water treatment occurs. A method involving chemical half-lives is used to estimate the required retention time. For degradation processes that proceed by first order reactions, such as TCE degradation by the MERD process, the number of half-lives necessary for reducing the concentration to a given effluent amount can be estimated by Equation 1 (Starr and Cherry, 1993):

$$N_{1/2} = \frac{\left[\ln\left(\frac{C_{effluent}}{C_{influent}}\right)\right]}{0.69} \tag{1}$$

where: $N_{1/2}$ = number of half lives required
        $C_{effluent}$ = effluent contaminant concentration
        $C_{influent}$ = influent contaminant concentration
        0.69   = $\ln(1/2)$

Therefore, knowledge of the influent contaminant and regulatory clean up concentration (*i.e.,* effluent concentration) allows prediction of the minimum number of half-lives (or time) that the ground water must remain within the gate. Furthermore, knowledge of the average ground water flow velocity then can be used to physically size the gate.

Precipitation Plugging. Another critical design concern for the MERD process is the potential of chemical precipitation (and its associated plugging) within the reactive wall of iron. Results from a full-scale MERD wall in Colorado reported a porosity reduction of 13 percent over the first year of operation (Gallant, 1997). However, another study simulating reactive wall performance by Sivavec (1995) predicted that the porosity would only decrease by approximately 10 percent over a 10 year period (Wilson, 1995).

It is suspected that the range of porosity loss reported in the literature is due to differences in the geochemistry of the ground waters being treated. Therefore, unless site specific geochemistry information is known for the ground water of interest, it is impossible to predict the magnitude of precipitate plugging that will occur. This plugging problem affects the long-term effectiveness of a MERD system and should be considered in the design.

RESULTS AND DISCUSSION

Many sites across the country have shown that the MERD technology is capable of remediating TCE-contaminated ground water to below regulatory mandated concentration levels. Due to the fact that the MERD technology is an *in-situ*, low maintenance treatment alternative, it is being considered at the 1100 Zone site at Hill AFB. This document presents data that can be used to design a

MERD remediation system for the TCE-contaminated ground water at the 1100 Zone site. Based on available data, a MERD system having a 3-foot by 20-foot (0.9 m to 6.1 m) reactive cell with 200 feet (61 m) of funnel wall on each side is sufficient to capture and treat the contaminated ground water. The cost of constructing the 1100 Zone MERD treatment system is approximately $922,500 dollars. Based on a design life of 30-years, the operation and maintenance costs are estimated at $1,278,000 dollars.

REFERENCES

Gallant, William A. and B. Myller, 1997. *The Results of a Zero-valence Metal Reactive Wall Demonstration at Lowry AFB, Colorado* Air and Waste Management Association's Annual Meeting and Exhibition, Toronto, Ontario, Canada. June 8 - 13, 1997. 21p.

Gillham, Robert W. and S. F. O'Hannesin, 1994. "Enhanced Degradation of Halogenated Aliphatics by Zero-Valence Iron" *Ground water*, May 20, 1994.

Johnson, Timothy L., Michelle M. Scherer, and Paul G. Tratnyek, 1996. "Kinetics of Halogenated Organic Compound Degradation by Iron Metal" *Environmental Science and Technology,* Vol. 30, No. 8, 1996. pp. 2634-2640.

Matheson, Leah J. and P. G. Tratnyek, 1994. "Reductive Dehalogenation of Chlorinated Methanes by Iron Metal" *Environmental Science and Technology*, April 1994.

O'Hannesin, Stephanie F. and R. W. Gillham, 1992. *A Permeable Reaction Wall for In Situ Degradation of Halogenated Organic Compounds.* 45th Canadian Geotechnical Society Conference, Toronto, Ontario. October 25 - 28. 5p.

Sivavec, Timothy M. and David P. Horney, 1995. *Reductive Dechlorination of Chlorinated Ethenes by Iron Metal.* Extended Abstracts for the American Chemical Society Special Symposium, Anaheim, CA. April 2 - 7, 1995. 4 p.

Starr, Robert C. and J. A. Cherry, 1993. "In Situ Remediation of Contaminated Ground
water: The Funnel-and-Gate System" *Ground Water*, May 1993.

United States Air Force (USAF), 1996. *Draft Technical Protocol for Evaluating Natural Attenuation of Chlorinated Solvents in Ground water.*

Wilson, Elizabeth K., 1995. "Zero-Valent Metals Provide Possible Solution to Ground Water Problems" *Chemical and Engineering News* July 3, 1995. pp. 19-22.

PHYSICAL CONTAINMENT OF A DNAPL SOURCE

Steve G. Brown (CH2M HILL, Salt Lake City, Utah)
Paul C. Betts and Steve T. Hicken (Hill Air Force Base, Utah)
James R. Schneider (CH2M HILL, Denver, Colorado)

Abstract: The Selected Remedy for the Source Area of Operable Unit 2 at Hill Air Force Base, Utah, stipulated containment of the pure-phase trichloroethylene contamination. The physical setting was a primary factor in the selection of the remedial design approach and technology. In-place-mixed methods were selected for the wall construction because of the irregular topography, small area of the site, and the need for the installation to reach depths of 90 feet (27.4 m) below ground surface. Bench-scale compatibility studies were performed on three candidate commercial bentonite clays for use in the containment wall construction including compatibility screening tests and long-term tests for evaluation of changed soil properties when exposed to the contaminated ground water. Construction of the approximately 1,500 foot (457 m) long wall was started in June 1996 and completed in December 1996. As a result of intensive treatability study investigations following construction, a pool of DNAPL was discovered outside the containment wall. This result points out the difficulty of locating all the DNAPL at a site. After 8 years of intensive investigation and study, a significant pool of DNAPL was not identified. The presence of this pool renders the containment wall largely ineffective in preventing the downgradient transport of DNAPL and dissolved phase contaminants.

INTRODUCTION

Hill Air Force Base (HAFB) is located in northern Utah, approximately 25 miles (40 Km) north of Salt Lake City, Utah. As a result of industrial processes for aircraft and missile maintenance at HAFB, up to 110,000 gallons (419,100 L) of waste chlorinated solvents (primarily trichloroethylene (TCE)) were disposed between 1967 and 1975 at Operable Unit 2 (OU2). Interim pump and treat remedial actions have recovered approximately 33,000 gallons (125,730 L) of DNAPL from the source area (Oolman et al, 1995). Contaminant recovery is primarily from the large pools of DNAPL retained at the source area. A large dissolved phase contaminant plume has developed down the steep terraced hillside to the alluvial aquifer in the Weber River Valley floor. The dissolved plume contaminants were found in the shallow, unconfined ground water system. Restoration of the dissolved plume is expected to require 30 years or more.

Objective. The objective was to design and install a source encircling containment wall in accordance with the Selected Remedy for the site (USEPA, 1996). The application of containment wall technologies to the OU2 DNAPL site involved particular consideration of the location and extent of the DNAPL, the physical constraints of the site, the location on an ancient landslide, and the chemical compatibility of the wall materials with the contamination.

Physical Setting. The source area and dissolved plume at OU2 are located on a hillside that is the site of a former landslide of several million cubic meters in size. In the source area, the DNAPL migrated rapidly downward through the Provo Formation, an alluvial terrace deposit consisting primarily of sand and gravel, until its progress was impeded by the Alpine Formation, a low permeability deltaic deposit composed primarily of silty clay with numerous thin silt lenses and occasional thin silty sand lenses. Subsurface investigations in the Source Area indicate the thickness of the Provo Formation ranges from 20 to 50 feet (6.1 m to 15 m). The thickness of the Alpine Formation deposits is estimated to range from 100 to 200 feet (30.5 m to 61 m). Numerous seeps and springs occur along the hillside. Traversing the mid-height of this hillside escarpment is the Davis-Weber Canal, a privately owned concrete-lined canal. Localized areas of the hillside show evidence of recent slope movement.

Remedial Strategy. The objective of the containment is to reduce the short-term risk associated with the source. The long-term plan for the site is to monitor the development of promising contaminant mass reduction technologies with the purpose of implementing an appropriately capable technology when it is practical.

Several technologies are available for mass reduction at DNAPL sites. However, it is generally accepted that these technologies are in the treatability study phase and have not yet been shown to be commercially viable technologies for restoring DNAPL zones in aquifers to drinking water standards.

Freeze and McWhorter, 1997, present a qualitative framework for understanding the level of mass-removal efficiency required to achieve long-term risk reduction for DNAPL in low-permeability soils. The distribution of the contaminant mass between the soil media (fractures, matrix) and phase (aqueous, sorbed, or NAPL) greatly influences the efficiency of mass removal. In general, it is expected the distribution of TCE mass in low-permeability soils will be primarily in the form of sorbed contamination on the soil phase and as NAPL in the soil fractures. This mass provides the source for continued long-term dissolved contamination in the water phase.

Given the very large quantities (assume initial quantity of 30,000 gallons (114,300 L)) of DNAPL at OU2, even a mass reduction technology with overall removal efficiencies as high as 95 percent would leave a significant long-term source quantity (1,500 gallons (5,715 L)) of DNAPL remaining in the subsurface. The corresponding reduced aqueous concentration would likely be above 5,000 µg/l and would not result in significant changes in the expected long-term risk and site operational cost requirements.

CONTAINMENT SYSTEM DESIGN
Objectives. The purpose of the containment wall is to hinder horizontal migration of highly contaminated ground water and the source area DNAPL pools. The containment wall provides two protective functions: a low permeability physical

barrier to lateral contaminant migration and a hydraulic barrier when ground water extraction within the containment wall produces an inward gradient. The containment wall will also reduce ground water flow into the source area thereby reducing the amount of ground water to be extracted and treated.

Containment Wall Alignment and Depth. The wall alignment encircles approximately 2 acres of the Source Area and includes the area where TCE concentrations were greater than 100 ppb. The target elevation for the bottom of the wall was established at elevation 4620 feet (1408.2 m) MSL placing the bottom of the wall at approximately 15 feet (4.6 m) below the elevation of the deepest known TCE contamination at the site to provide containment. At this depth, the containment wall will be keyed a minimum of 35 feet (10.6 m) into the low permeability Alpine clay soils. Also, the wall will be completed below the bottom of the landslide which is estimated to be approximately 55 feet (16.7 m) below the existing ground surface in the Source Area. The resulting wall depths would range between 67 and 92 feet (20.4 m to 28 m).

Design Alternatives Evaluation. Alternatives considering during development of the containment wall design included the following structures:

- sheet pile walls with grouted joints (Waterloo Barrier™ (cold rolled 7.5 mm thickness with internal sealable cavity), or hot rolled piles with external sealable cavity)
- conventional slurry wall
- in-place mixed (IMP) slurry wall

Based on results of a field sheet pile drivability study (Montgomery Watson, 1993), the probability that sheet piles could not be driven without damage to the target elevation of 4620 feet (1408.2 m) at all locations was high. In addition, sealable sheet piles were not commercially available in the wall thickness required to drive to the required depth at this site. An uncertainty with sheet piles is the integrity of the inter-sheet seal which can be compromised during installation, and the damage may not be correctable in all cases.

The IPM method allows soil-bentonite walls to be constructed in-place, leading to significant reductions in the requirements for excavation, out-of-trench mixing, and contaminated materials handling. Deep applications of IPM generally involve a large crane with a bank of turning augers and mixing paddles that drill the soil as well as mix the injected reagents with the cuttings. Continuity is produced by overlapping adjacent columns of mixed soil to form a continuous wall. The IPM method is suited for use in a confined area where the wall alignment has tight curves, and existing structures and facilities are in close proximity. The target depths required at the site are within the capability of commercial IPM equipment.

The bentonite content of the injected slurry is generally limited to six percent bentonite by weight, because of the increased difficulty with pumping thicker slurry's. A six percent bentonite slurry mixture translates to a maximum two percent bentonite by dry weight of soil-in-place. The permeability of an IPM wall is generally slightly higher than, but competitive with, the conventional slurry wall in most soil conditions. This is because the consistency of the barrier mixture is controlled by the potentially variable native soil incorporated into the mixture. The native soil needs to be suitable for both auger installation and for use as the barrier backfill.

The IPM method was selected for the design because of the ability to produce a wall having low-permeability properties comparable to a conventional slurry wall, the ability to achieve depths of 90 feet (27.4 m) below ground surface, and the ability to negotiate the corners and constraints of a small site with existing structures.

Compatibility of Backfill with Site Contaminated Ground Water. Chemical compatibility bench-scale tests were performed to evaluate the effects of site contaminated ground water on the geotechnical properties of various barrier wall mix design recipes. Four commercially available high swelling sodium bentonite clays were used. The bentonite was mixed with native clay samples and a composited native sand sample. The bentonite-sand sample was included to represent worst-case conditions for the upper portion of the wall that is constructed through the sand deposits of the Provo Formation.

Compatibility testing included initial compatibility screening, blow-out testing, and long-term permeability testing. Long-term flexible wall permeability tests were performed using an initial permeant of site tap water followed by contaminated ground water saturated with dissolved TCE at a concentration over 100 ppm. The samples were prepared with a 2 percent concentration of bentonite by dry weight of soil basis. The testing results indicated the following:

- Blow-out or piping of the containment wall slurry material is not likely even under the highest potential hydraulic gradient expected during operation of the facility.
- The permeability of sand matrix soil mixed with 2 percent bentonite ranged from 6.5×10^{-7} cm/sec to 1.4×10^{-6} cm/sec when permeated with water.
- Sand matrix mixes showed an increase in permeability when permeated with three pore volumes of contaminated site ground water. Permeability measurements, after passing three pore volumes, ranged from 6×10^{-7} cm/sec to 3×10^{-6} cm/sec. The increase in permeability ranged from 30 percent to 360 percent for the three bentonite soils tested.
- The permeability of clay matrix soil, mixed with 2 percent bentonite by dry weight, varied between 6×10^{-9} cm/sec and 8×10^{-9} cm/sec for the three mixes when permeated with water.
- The clay matrix mixes show no increase in permeability when permeated with three pore volumes of contaminated ground water

flow. The final permeability ranged from 2.5×10^{-9} cm/sec to 6×10^{-9} cm/sec.

- The long-term permeability tests indicate the clay-bentonite samples performed well when subject to highly concentrated site ground water. The sand-bentonite samples were significantly affected by the contaminated ground water. The sand-bentonite tests represented a worst-case condition that is not expected to be encountered in the field. However, the sand-bentonite tests do point to the need to maintain an inward gradient through the more vulnerable upper portion of the containment wall. This will reduce the potential for contact with highly concentrated site ground water.

IN-PLACE MIXED WALL CONSTRUCTION

The approximately 1500 foot (457 m) long IPM wall construction was completed in 39 working days, for an average of approximately 3,000 square feet (278 sq. m) per day. The IPM rig consisted of four overlapping mixing shafts, with bottom bentonite slurry discharge, capable of creating a wall with a minimum thickness of 24 inches (0.61 m). The mixing assembly was mounted on a Manitowoc 4100 Series II crane. The mixing augers and blades blend the injected bentonite-water slurry with the soil. The bentonite slurry was mixed in a high shear colloidal mixing plant and delivered to a holding tank by positive displacement pumps.

In order to achieve an in-place 2 percent bentonite by dry weight of native soil, the injected bentonite slurry would need to have a minimum 6 percent bentonite by weight. The slurry injected during construction ranged from 7 percent to 9 percent bentonite. The corresponding estimated volume of bentonite slurry to achieve the minimum specified 2 percent bentonite ranged from approximately 92 to 73 gallons (350 L to 278 L) per foot of augured area for each primary stroke.

The IPM wall was constructed using a sequence of overlapping primary and secondary auger strokes, such that no soil within the specified alignment and depth remained unmixed. The location of each stroke was surveyed and the coordinates recorded. If an auger stroke experienced lateral drift during penetration, and field measurements indicated that overlap was not achieved at full depth, a tertiary stroke was added between the primary and secondary strokes to complete the mixing.

CONTAINMENT WALL PERFORMANCE

Long-term operation of the containment wall involves controlled ground water removal from the interior of the wall to maintain an inward gradient to the source area. Twenty-one piezometers, installed along the containment wall, will be used to verify that an inward gradient is maintained into the source area. While there is evidence that the containment wall is performing adequately as a low permeability barrier, the effectiveness of the wall is not an issue at the present time. As a result of intensive treatability study investigations following construction, a pool of DNAPL was discovered outside the containment wall. This result points out the difficulty of locating all the DNAPL at a site, a recurring problem at many DNAPL sites. After 8 years of intensive investigation and study, a significant pool of DNAPL had not been identified. The presence of this pool renders the containment wall largely ineffective in preventing the downgradient transport of DNAPL and dissolved phase contaminants.

SUMMARY AND CONCLUSIONS

The process for selection and design of a containment system for a DNAPL source area has been presented. The IPM method was selected for construction of the wall, leading to significant reductions in the requirements for excavation, out-of-trench mixing, and contaminated materials handling over conventional slurry wall construction methods. The IPM method was found to be suited for use in a confined area where the wall alignment has tight curves, existing structures are in close proximity, and wall depths in excess of 70 feet (21.3 m) are required. The installation of the IPM wall at OU2 was completed in December 1996. The containment wall has performed adequately as low permeability barrier. However, the wall has been rendered largely ineffective for preventing the downgradient transport of DNAPL and dissolved phase contaminants due to the discovery of a DNAPL pool outside the containment wall. This result reaffirms the difficulty of locating all the DNAPL at a site. After 8 years of intensive remedial investigation and study, a significant pool of DNAPL had not been identified.

REFERENCES

CH2M HILL, *Final Remedial Action Work Plan for Operable Unit 2, Schedule A and B Construction.* Hill Air Force Base, Ogden Air Logistics Center, December 1997.

Freeze, R.A., and McWhorter, D.B., "A Framework for Assessing Risk Reduction Due to DNAPL Mass Removal from Low Permeability Soils." *Journal of Ground Water*, Vol. 35, No. 1, January-February 1997.

Montgomery Watson, *Sheet Pile Constructability Evaluation Report Operable Units 1, 2, and 4,* Hill Air Force Base, Ogden Air Logistics Center, October 1993.

Oolman, T., Godard, S.T., Pope, G.A., Jin, M., and Kirchner, K., *DNAPL Flow Behavior in a Contaminated Aquifer: Evaluation of Field Data,* Journal of Ground Water Monitoring and Remediation, Vol., 15, No. 4, 1995.

Radian Corporation, *Final Addendum to the Remedial Investigation Report for Operable Unit 2, Site WP07, SS21.* Hill Air Force Base, Ogden Air Logistics Center, August 1993.

USEPA, 1996, *Record of Decision for Operable Unit 2, Hill Air Force Base, Ogden Air Logistics Center*, U.S. Environmental Protection Agency, Region VIII, September 1996.

DEMONSTRATION OF A PERMEABLE BARRIER TECHNOLOGY
FOR PENTACHLOROPHENOL-CONTAMINATED GROUNDWATER

Jason D. Cole, Sandra Woods, Kenneth Williamson and David Roberts
(Dept. of Civil, Construction, & Environmental Engineering,
Oregon State University, Corvallis, Oregon)

ABSTRACT: The development and implementation of a pilot-scale *in-situ* permeable barrier for the bioremediation of pentachlorophenol-contaminated groundwater at a wood preserving facility is being investigated. A permeable reactor was constructed to fit within a large diameter well. Arranged in series, a cylindrical reactor 24" x 36" (0.61 x 0.91m) (diameter x height) was partitioned to provide one anaerobic and two aerobic treatment zones. Mixing zones precede each biologically active zone to provide the opportunity for nutrient injection and gas lift mixing. A mixed microbial consortia supported on ceramic saddles was used to inoculate both the anaerobic and aerobic treatment zones. Environmental conditions in the anaerobic and aerobic treatment zones are monitored with two continuous flow cells capable of pH and oxidation/reduction potential measurements. Aqueous samples are collected from twenty eight sampling points within the reactor and allow for the spacial and temporal characterization of biological removal processes.

INTRODUCTION

The down-borehole reactor is a passive, in-situ, permeable biological reactor that allows the introduction of nutrients and other chemicals to a subsurface biological population. Equipped with sensors, nutrient delivery, and mixing systems, the reactor is installed in a large diameter well screened over an interval of the contaminated aquifer. Biodegradation of the aqueous phase organic compounds occurs over the length of the reactor in a combination of anaerobic and aerobic biological zones. The goal of this research is to demonstrate the applicability of a permeable barrier for the bioremediation of groundwaters contaminated with pentachlorophenol (PCP). The ultimate goal is to develop this technology for the treatment of groundwaters contaminated with complex waste mixtures.

Objective. Pentachlorophenol (PCP) is a fully chlorinated aromatic compound that is susceptible to reductive dechlorination, a common anaerobic bio-transformation process that results in production of lesser chlorinated phenols. Reductive dechlorination of PCP can be rapid. However, as the number of chlorines decrease, rates of reductive dechlorination also decrease (Nicholson et al., 1992; Mohn and Teidje, 1992). Conversely, aerobic biotransformation rates tend to increase with decreasing chlorination and may result in complete mineralization (Roberts, 1997; Saber and Crawford, 1985; Valo et al., 1989;

Cassidy et al., 1997). The process configuration used in this study capitalizes on these process characteristics and is shown in Figure 1. The development and demonstration of a down-borehole reactor employing sequential anaerobic/aerobic biological processes for PCP-contaminated groundwater is the focus of this study.

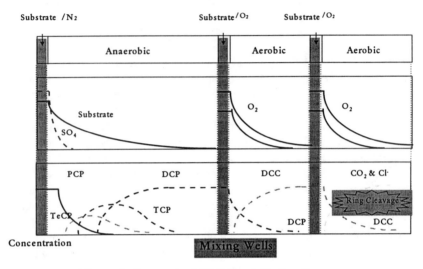

Figure 1. Conceptual PCP Remediation Process

Site Description. The permeable barrier reactor was selected for demonstration at an active wood preserving facility in Eugene, Oregon. The facility began operation in the mid 1950s and applied PCP in a medium aromatic treating oil to telephone poles. Several process variations over the years occurred but without change in treatment chemicals. Operational practices and several accidental spills resulted in contamination of the underlying aquifer with PCP and its carrier oil. Subsurface remedial action measures center around four groundwater recovery wells. Groundwater contamination off site has been mitigated by the reversal of local groundwater flow. Free oil is removed from the water surface by skim pumps when required. Water removed from the aquifer is treated by granular activated carbon while, recovered oil is returned to the process.

The reactor test site is located between two of the site's groundwater recovery wells. The aquifer on site is a shallow semi-confined structure comprised of two major geologic units. The upper unit averages 10 feet in thickness and is characterized as a dense yet permeable clay formation. Underlying the clay and ranging in thickness, are well sorted sands and gravel. Groundwater elevations vary seasonally and range from 5 to 15 feet (1.5 to 4.6 m) below ground surface (bgs). Well logs of the adjacent recovery and monitoring wells show a lower aquitard present 25 feet (7.6 m) bgs. In March of 1996, a 24" (0.61 m) diameter well was installed in a protected concrete vault on site. The well was drilled to a depth of 25 feet (7.6 m) and was screened over a three foot (0.91 m) section within the lower gravel section of the aquifer.

MATERIALS AND METHODS

A custom fabricated permeable barrier reactor was designed and constructed. The cylindrical unit is constructed of modular partitions and treatment cells. In it's current configuration, the unit is assembled to operate with three biologically active zones. Growth within these zones is supported on ceramic saddles that possess both high surface area and hydraulic conductivity. Each zone is separated by one inch wide vertical partitions that serve as nutrient supply and mixing areas. Nutrient addition consists of continuous low flow injection of a highly concentrated aqueous feed solution. Periodic agitation of the treatment zone influent is conducted by a gas lift mixing scheme. Inert gas or oxygen is used in the mixing regime depending on the desired environmental condition of the biological zone. All nutrient supply systems are isolated to allow for independent operation regardless of location within the reactor. Figure 2 shows the reactor in plan view while, Figure 3 presents a cross section of the unit.

The reactor is equipped with a pneumatic sampling system and 28 discrete sample points. Collection locations allow for the spacial and temporal characterization of biological removal processes. Sensing the need to continuously monitor environmental conditions within the reactor's biological treatment zones, two recirculating sample loops were installed. Water is continuously removed from the first and second treatment zones of the reactor and pumped to the surface with a peristaltic pump. Tubing resistant to oxygen diffusion was selected for the sample collection lines to characterize the groundwater within the treatment zones. Water from each zone is pumped through a flow cell and re-injected into the reactor. Low flow rates and equivalent mass removal and injection with the continuous loop design minimizes preferential flow through the reactor. The flow cells were designed to allow the use of three ion selective electrodes. Signal manipulation and data storage for the electrodes in each cell are handled by a 16 channel data logger. The system is configured for real time measurement of temperature, pH, and oxidation/reduction potentials (E_H) in two treatment zones.

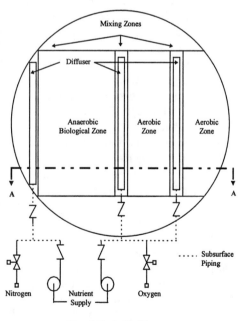

Figure 2. Reactor Plan View

Sampling and Analysis.
Aqueous samples collected from the well and reactor were analyzed for chlorophenol concentration by capillary gas chromatography. Samples were acetylated and extracted into hexane using a modification of the method developed by Voss et al. (1981) and the National Council of the Paper Industry for Air and Stream Improvement (1981). Chlorophenols were quantified on a Hewlett Packard 6890 gas chromatograph with a ^{63}Ni Electron Capture Detector (ECD).

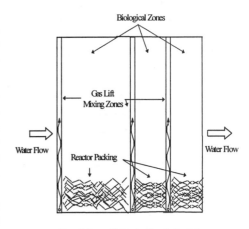

Figure 3. Permeable Reactor Cross Section A-A

Separation of chlorophenol congeners was accomplished on a DB-5 fused silica capillary column (30m x 320µm I.D. x 0.25µm film; J & W Scientific, Folsom, CA). Anion measurement of selected samples was accomplished with a Dionex 2000I ion chromatograph.

RESULTS AND DISCUSSION

Chlorophenols present in the reactor well have been measured on a weekly basis beginning in mid February of 1997. The aquifer system is semi-confined and highly susceptible to precipitation events. Water table fluctuation within the reactor well are large and over the course of the year have varied as much as 10 feet. Water table variation correlates well with observed PCP concentrations as values ranged between 0.2 and 3.0 mg/L over a 9 month period. These numbers correspond with lows observed in the dry summer months and highs in the typically wet season. Similar trends were also observed with nitrate and sulfate concentrations.

The reactor system was installed in the well during the first week of August, 1997. Initially, the reactor was installed without the packing and inoculum. In the absence of packing, the hydraulic retention time in the well was approximately 24 hours. During the first week of December, 1997, the reactor was removed from the borehole and was packed and inoculated with 50/50 (vol.) mixture of anaerobic digester and activated sludges. The reactor was installed and chlorophenols, major anions, pH, and E_H values were monitored. The initial reactor installation occurred in the absence of electron donor and substrate addition. Approximately 5 weeks of data collection showed little change in chlorophenol concentrations. Data collected from the flow cells also indicated no major changes in the overall environmental conditions within the well. The pH of the system in both the anaerobic and aerobic zones held stable at 7.3 while the redox potential showed anoxic conditions with an average measurement of -140 mV (Ag/AgCl reference).

Based on the successful transformation of ^{14}C-PCP to ^{14}CO$_2$ (Roberts, 1997) in the laboratory, imitation vanilla flavoring was selected as the primary carbon source and electron donor. Imitation vanilla flavoring contains several benzene derivatives including guaiacol (*o*-methoxyphenol), ethyl vanillin (3-ethoxy-4-hydroxybenzaldehyde), and sodium benzoate in propylene glycol. It was supplied to the reactor at a carbonaceous oxygen demand level of approximately 100 mg/L beginning during the second week in January of 1998.

Control of the oxidation reduction potential has been evaluated under three conditions to date. Without mixing gas addition, redox potential within the well and presumably the aquifer measure -140 mV (Ag/AgCl reference). When the mixing gas systems were activated to deliver gas for two seconds every 15 minutes, rapid changes to the environmental conditions were observed. Redox potentials in the aerobic treatment zone climbed steadily reaching a maximum value positive 340 mV within 96 hours of system activation. Despite the use of nitrogen for anaerobic treatment zone mixing, elevated redox measurements were also observed. Continued readings above the desired background instituted a reduction in gas supply duration. Each mixing systems was recalibrated to provide a momentary gas addition every 15 minutes. This option allowed for complete nutrient mixing at a minimum gas flow rate. Response to the supply system change was immediate and redox measurements fell sharply in both biological zones. Within 72 hours of the change, potentials in the anaerobic zone fell to -150 mV. The sudden drop was likely a factor of two forces: the reduced supply of oxidant to the aerobic zone and recharge from the aquifer structure. Currently, the effect of mixing and redox potential on PCP removal is being evaluated.

Engineering Significance. Bioremediation technology development has largely focused on strategies for remediating an individual compound or a closely related group of contaminants with a narrow range of physical, chemical, or biological characteristics. As a result, the successful remediation of sites contaminated with complex mixtures of metals and organic compounds requires the use of multiple technologies. While several groups of researchers are working toward the development of different types of permeable barrier technologies, the proposed technology differs in that a reactor containing many types of treatment processes in series is lowered into a prepared well. The modular design of the permeable reactor allows for the combination of physical/chemical and or biological treatment techniques that can be tailored for one specific contaminant or a waste mixture.

ACKNOWLEDGMENTS

Funding for this study was provided by the Office of Research and Development, U.S. Environmental Protection Agency, under agreement R-819751-01 through the Western Region Hazardous Substance Research Center. The content of this paper does not necessarily represent the views of the agency.

REFERENCES

Cassidy, M.B., K.W. Shaw, H. Lee, and J.T. Trevors. 1997. "Enhanced Mineralization of Pentachlorophenol by carrageenan-encapsulated *Pseudomonas sp.* UG30." *Applied Microbiol. Biotechnol. 47*:108-113.

Mohn, W.W, and J.M. Teidje. 1992. "Microbial Reductive Dehalogenation." *Microbial Reviews, 56*(3):482-507.

National Council of the Paper Industry for Air and Stream Improvement (NCASI). 1981. *Experience with the analysis of pulp mill effluents for chlorinated phenols using an acetic anhydride derivitization procedure.* Stream Improvement Technical Bulletin No. 347, June.

Nicholson, D.K., S.L. Woods, J.D. Istok, and D.C. Peek. 1992. "Reductive dechlorination of chlorophenols by a pentachlorophenol-acclimated methanogenic consortium." *Appl. Environ. Microbiol., 58*(7):2280-2286.

Roberts, D. 1997. "Down-Borehole Permeable Barrier Reactor: Verification of Complete Mineralization of Pentachlorophenol in a Sequential Anaerobic-Aerobic Process." M.S. Thesis, Oregon State University, Corvallis, OR.

Saber, D.L. and R.L. Crawford. 1985. "Isolation and Characterization of *Flavobacterium* Strains that Degrade Pentachlorophenol." *Appl. Environ. Microbiol. 50*:1512-1518.

Valo, R.J., M.M. Haggblom, and M.S. Salkinoja-Salonen. 1989. "Bioremediation of Chlorophenol Containing Simulated Groundwater by Immobilized Bacteria." *Water Res., 24*:253-258.

Voss, R. H., J. T. Wearing, and A. Wong. 1981. "A novel gas chromatographic method for the analysis of chlorinated phenolics in pulp mill effluents." p. 1059-1095. *In* L. H. Keith (ed.), *Advances in the identification and analysis of organic pollutants in water*, 2. Ann Arbor, Mich.

SITE CHARACTERIZATION TO AID IN THE DESIGN OF A PERMEABLE BARRIER AT DOVER AFB

Robert J. Janosy and James E. Hicks (Battelle, Columbus, Ohio)
Lt. Dennis O'Sullivan (Air Force Research Laboratory, Tyndall AFB, Florida)

ABSTRACT: Site characterization activities were performed to gather information for the design and subsequent installation of a permeable barrier. Through the use of cone penetrometer technology (CPT), accurate real-time hydrogeologic and geochemical information was gathered in an efficient manner. A total of 59 temporary sampling points were installed for collecting groundwater samples. Analyzed samples detected the presence of a shallow perchloroethylene (PCE) plume targeted for the permeable barrier demonstration with contamination levels above 5000 parts per billion (ppb).

INTRODUCTION

A cost-effective approach for remediating contaminated groundwater through the use of in situ permeable reactive barriers is generating increased interest. These permeable barriers essentially consist of a trench filled with a reactive medium, such as granular iron. As contaminated groundwater naturally flows through this treatment zone, the chlorinated solvents react with the reactive medium and passively degrade.

The U.S. Air Force Research Laboratory Air Base and Environmental Technology Division, Tyndall Air Force Base (AFB), Florida contracted Battelle, Columbus, Ohio in April 1997 to conduct a demonstration of a field pilot permeable barrier system. The goal of the study was to evaluate the performance using two different reactive media for treating a chlorinated solvent plume at Area 5, Dover AFB, Delaware. This project was funded by Strategic Environmental Research and Development Program (SERDP).

Objective. The main goal of the site characterization effort was to obtain hydrogeologic data and groundwater chemical data to assist in planning the design, construction, and performance evaluation of the proposed permeable barrier. Unlike conventional ex-situ technologies, such as pump-and-treat systems, in situ technologies are more dependent on site specific parameters. Therefore, it was critical that a full understanding of the site be gained before the barrier is installed in the ground.

SITE DESCRIPTION

Area 5 at Dover AFB was selected as the site for this pilot-scale field demonstration. The site was found to be a suitable location to test the permeable barrier technology because it contains groundwater contaminated with high concentration of chlorinated solvents, has a relatively deep and homogeneous aquifer, a competent aquitard, and a large accessible above-ground work space.

The near surface geology at Area 5, known as the Columbia Formation, consists primarily of 40 ft (12 m) of poorly sorted coarse to medium sand with minor amounts of gravel and occasional interbedded silt and clay lenses. An underlying firm, dense clay approximately 20 ft (6 m) thick forms a confining layer (aquitard) for the Columbia Aquifer. The saturated thickness of this surficial unconfined aquifer is about 20 to 23 ft (6 to 7 m). Groundwater levels typically range from 12 to 15 ft (3.5 to 4.5 m) below ground surface (bgs) and flow is to the southwest.

The groundwater contamination beneath Area 5 predominantly consists of volatile organic compounds (VOCs) and includes PCE, trichloroethylene (TCE), and *cis*-1,2 dichloroethylene (DCE). The source of these target contaminants is unknown, but is believed to be the result of potential releases from several sites. Before this characterization event, there was only one existing well that indicated the potential presence of a PCE plume in the area of interest.

METHODS

Site characterization was conducted in June 1997. Hydrogeologic data was collected from previous investigations (D&M, 1995), detailed subsurface tests using CPT, slug tests, and water level measurements. Geochemical data was acquired from analyzed groundwater samples collected at the site. Prior to any CPT work, proposed sampling locations were carefully staked for digging permit approval because many underground utilities existed at the site.

Cone Penetrometer Technology. Applied Research Associates (ARA) was hired by Battelle to provide a CPT rig for investigations in the parking lots around Building 639 where a shallow PCE plume was earlier suspected and crudely defined. Single or multiple CPT pushes were carried out at 23 locations on the southwest side and 5 locations on the northeast side of the building. At 20 locations, lithologic logs with soil classifications were generated from sleeve stress, tip stress, and pore pressure responses during CPT pushes. Pore pressure dissipation tests were performed at 20 locations for the purpose of determining aquifer hydraulic conductivity (K) data, a key parameter needed for the permeable barrier design. By measuring the drop in accumulated pore pressure during a CPT push at a certain depth over time, a K value can be calculated. Fourteen split spoon soil samples were collected from different depths at 5 locations using the CPT and described. Seven of these samples were analyzed for grain-size distribution, porosity, specific gravity, and cation exchange capacity for correlation with the generated lithologic logs.

Groundwater Sampling. After conducting hydrogeologic tests at several locations, the CPT then pushed temporary wells for the groundwater sampling effort. Each sampling point consisted of 1.0-inch ID CPT rods as well casing and a 2 or 3 ft (0.6 or 0.9 m) long, 0.5-inch (2.5-cm) diameter sacrificial PVC screen. The points were labeled T1, T2, etc. on the west side of Building 639 and A1, A2, etc., on the northeast side of the building. Sample depths were indicated as S, M, or D for shallow (15 to 20 ft (4.5 to 6 m)), medium (20 to 25 ft (6 to 7.5 m)), or deep (35 to 40 ft (9 to 10.5 m)), respectively. Water levels were recorded prior to and following sampling activities using a water level indicator tape. All wells including 6 existing wells, were purged at low flow rates until field parameters (pH, Eh, DO, and conductivity) stabilized. Purging and subsequent groundwater sampling for all the temporary sampling points were performed using a H_2OG Multiprobe Hydrolab fitted with a flow-through cell and a peristaltic pump. A submersible Grundfos pump was used to purge and sample existing 2-in (5-cm) wells. Groundwater collection techniques such as a bailer versus a submersible pump and before versus after purging were compared to determine the technique that provided the most accurate information. Purging was done at not more than 1 L/min and groundwater samples were collected at around 100 mL/min. The groundwater withdrawal rate was kept low to minimize dilution of target contaminant concentrations in an aquifer with slow moving groundwater and a vertically layered plume. Samples were collected for organic, inorganic, and field parameter chemical analyses. In addition, periodic duplicate and field blank samples were collected for quality assurance purposes. Samples collected for dissolved metals analysis were filtered in the field. Daily-collected samples were sent overnight to Battelle for real-time analysis and results were obtained the following morning. These results help delineate the plumes true shape and pinpoint new locations for additional CPT sampling. One third of all collected samples were sent to a certified laboratory for confirmation and additional testing.

Other Tests. Additional data on the aquifer hydraulic conductivity was determined from slug tests performed in five existing wells in Area 5. Three withdrawal-slug tests were conducted in each well to ensure test results were valid. A 1.5–inch (4-cm) diameter, 3 ft-long (0.9 m) PVC slug was used in the tests to displace an equal volume of water in the individual wells. An InSitu PTX-161D pressure transducer was used to monitor water-table responses and data was recorded using an InSitu SE 1000C datalogger.

The University of Delaware conducted a Ground Penetrating Radar (GPR) survey at the site for mapping the depth to the aquitard and delineating other shallow geologic features. Unfortunately, the results of the survey proved to be inconclusive, because the interference from Building 639 made the interpretation of the GPR signal difficult.

RESULTS AND DISCUSSION

Geology. CPT data were compiled and utilized in several ways. Lithologic log data in combination with results from grain size analysis of collected soil samples were used to generate seven geologic cross-sections. Figure 1 shows a NW-SE oriented cross-section through the proposed barrier location. The cross-sections helped to visually delineate lithologic variations in the Columbia Formation and locate the depth of the underlying clay aquitard. Interpretation of the data indicated the lithology in Area 5 consists of 40-50% medium sand, 20-30% fine sand, and about 10% each of coarse sand, gravel, and fines (silt and clay). Aside from a fairly continuous 2 to 3 ft (0.6 to 0.9 m) thick silt to clayey-silt zone present 13 to 18 ft (4 to 5.5 m) bgs near the water table interface, the lithology in this area is fairly homogeneous. The presence, absence, and continuity of this fine-grained layer may have an impact on the migration of contaminants into the saturated zone.

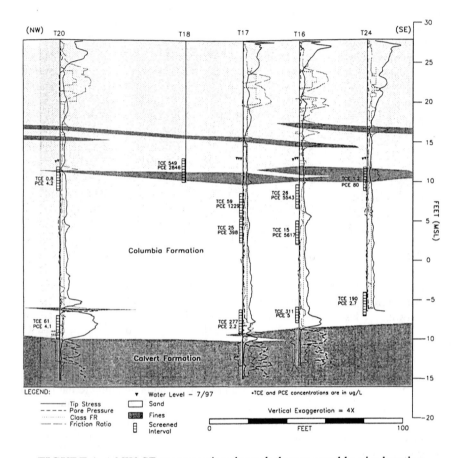

FIGURE 1. A NW-SE cross-section through the proposed barrier location.

The overall thickness of the Columbia Formation and depth to the underlying clay aquitard in Area 5, is 36 to 38 ft (11-11.5 m). This clay layer indicates a change in depositional settings and marks the transition of the next stratigraphic horizon, the Calvert Formation. The lithology of this competent and laterally continuous layer is clayey silt to predominantly silty clay.

Hydrogeology. Recorded water levels taken from the CPT sampling points were compared to those taken at nearby existing wells and historic measurements. The CPT based measurements were found to be a couple of inches lower than those taken from the existing wells suggesting that the CPT based data may not be reliable. This could be attributed to the fact that the CPT sampling points were temporary, uncompleted and undeveloped holes. Since the holes were only open for a short while, water levels may have not properly stabilized during the allotted time period. Also, slight mounding of the asphalt occurred during extraction of the sampling points (CPT push rods) and probably fouled the subsequent survey. With this in mind, only water level measurements taken from existing wells were used for modeling and design purposes. Comparison of plotted July 1997 water levels to posted historic water levels taken in May 1994 and December 1993 showed expected differences in water levels and flow directions attributed to seasonal fluctuations typical for Area 5. December 1993 water levels averaged 10.2 ft (3 m) above mean sea level (msl) and reflects low flow conditions. July 1997 data reflects intermediate conditions, and May 1994 reflects high flow conditions at 14.5 ft (4.4 m) above msl. Flow directions varied from west to southwest. From the available data showing variations in the saturated thickness of the Columbia aquifer, a hydraulic gradient of approximately 0.002 was calculated for the proposed permeable barrier site.

Porosity values for the aquifer media determined from 5 analyzed soil samples, showed a range of 23.7 to 34.1 percent with 32 percent as a representative value. Two analyzed soil samples taken from the aquitard showed porosity of 60.2 and 59.3 percent matching with their clayey content.

Hydraulic conductivity results from CPT dissipation tests conducted in the sandy aquifer were inconclusive since the pore pressure dissipated extremely fast as expected for a sandy medium. Results from dissipation tests conducted in the clay aquitard however showed that K values ranged from 0.003 ft/day (0.09 cm/day) to 0.045 ft/day (0.01 m/day). Slug test results from the five nearby existing wells showed a range of mean K values of 1.11 to 11.01 ft/day (0.3 to 3.3 m/day). Combined with data from previous investigations, the actual aquifer hydraulic conductivity of the permeable barrier site ranges from between 10 and 50 ft/day (3 and 15.2 m/day). With this K range, hydraulic gradients of 0.002 and porosity of 32%, the average linear flow velocity at the site is likely to range from 0.06 to 0.3 ft/day (0.02 to 0.09 m/day).

VOC Distribution. Maximum concentrations encountered during the sampling event were 5,617 ppb of PCE at T16M, 549 ppb of TCE at T18S, and 529 ppb of DCE at T12D. Chemical results obtained from all the analyzed groundwater samples were plotted and contoured to illustrate the true extent of contamination at the site. Figure 2 shows the identified shallow PCE plume targeted for the permeable barrier demonstration. The location of the proposed barrier is also depicted on the figure at the advancing front of the heavily concentrated part of the plume. PCE contamination is predominately confined to shallow portions of the aquifer with six locations reporting hits above 1000 ppb. TCE and cis-1,2-DCE contamination is higher at deeper depths. Relatively low levels of contamination were discovered from the five locations on the northeast side of Building 639. The highest detected VOC was 1206 ppb of PCE at A2S.

Field Parameters. Results showed a relatively high concentration of dissolved oxygen (DO) in shallow aquifer levels (approximately between 4.0 and 9.0 mg/L),

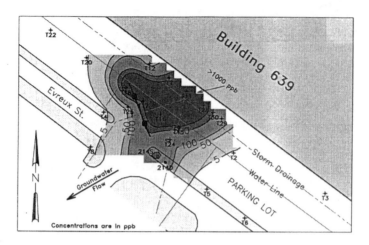

FIGURE 2. The extent of shallow PCE contamination at Building 639, DAFB. Also shown are the southwest sampling locations, underground utility lines, and the proposed barrier location at the leading edge of the higher concentrations.

but much lower at deeper levels (about 0.3 to 1.75 mg/L). These high levels may contribute to lower hydraulic conductivity of the reactive cell due to precipitated iron hydroxides as the groundwater reacts to the proposed iron media. Groundwater pH tends to be acidic (between 4.5 to 5.0) and in this range, the Eh appears somewhat insensitive to the observed DO levels.

Inorganics. Alkalinity , calcium, and magnesium occur in the groundwater at relatively low levels. Chloride concentrations tend to fall in the range of about 10 to 20 mg/L, high enough to make a chloride mass balance based on about 5 mg/L of chlorinated VOCs difficult. Redox-sensitive elements, such as iron (<5 parts per million - ppm) and manganese (<1 ppm) are relatively low.

DISCUSSION

Valuable information resulted from the site characterization effort and some challenges for the permeable barrier demonstration were discovered. The true extent of the PCE contamination was established and mapped. Also, the depth to the aquitard and the competency of this layer was confirmed. Another key finding is the variable groundwater flow direction due to seasonal fluctuations, which could prove challenging in capturing the PCE plume. The presence of underground utilities (water line, storm sewer line) near the preferred barrier location will have to be contended with and located accurately during installation of the system.

REFERENCES

Dames and Moore, 1995. *Draft West Management Unit Remedial Investigation: Dover Air Force Base, Dover, Delaware*. April.

IN SITU HORIZONTAL SUBSURFACE BARRIER TECHNOLOGY FOR SOIL REMEDIATIOM

David R. Muhlbaier (Soil Remediation Barriers Company, Aiken, SC)
Gregory M. Powers (RCS Corporation, Aiken, SC)

ABSTRACT: A method has been developed and tested in a pilot scale demonstration that will allow in situ placement of a horizontal subsurface barrier (HSSB) for remediation of contaminated soils. The process (Patent Pending) allows placement of various barrier materials under and around contaminated soil without significant disruption of the soil or movement of contaminants. The process can be used at various depths and in various soils to form a barrier for contaminants yet allow ground water to pass unimpeded. Used in conjunction with vertical barrier walls installed by current technology, total containment of contaminants can be achieved. Furthermore the HSSB process can utilize surface waters to flush contaminants through the remediation barrier for accelerated cleanup.

INTRODUCTION: Many locations in the United States are contaminated with hazardous and toxic substances. More than 200,000 sites are in need of remediation (U.S.EPA). These contaminants have permeated and adsorbed onto soils, diffused into interstitial spaces, and migrated to the underlying aquifers. In many cases, these contaminants have exhibited physical and chemical properties that make them difficult to remove from the environment. Immobile contaminants in subsurface sediments are difficult to access, certain contaminants may be resistant to normal subsurface chemical and biological degradation processes and others may strongly absorb on soil structures and have varying degrees of solubility in aqueous solutions. Work by other is in progress to develop permeable reactive barriers in soil (e.g. Sivanec, T.M. et al)

Hence there is an opportunity to provide a low cost system to perform in situ remediation. This document describes the process which has already demonstrated the ability to preferentially place in situ remediation materials in horizontal layers that can be used to stop, collect, or destroy environmental contaminants. Such barriers are illustrated in Figure 1. Multiple layers can be placed as shown and can be of the same or different materials to accelerate remediation.

The process can be used to place any pumpable material in subsurface layers. Hence, contaminants can be treated in situ by physical, chemical, or biological techniques. This is accomplished with the use of a porous barrier materials that allows groundwater movement through the barrier to accomplish treatment. For example, zero valance iron may be satisfactory in the treatment of certain contaminants. In addition, the method can be used to deliver nutrients and chemical reagents for biological and chemical methods of contaminant control. In certain cases where treatment material is not available, impermeable barriers could be used along with vertical barriers to prevent the further movement of the contaminant.

METHOD: The HSSB process employs preferential fracture of the soil by means of creating stress points at selected locations and then stressing the soil until failure (fracture) occurs. The stress is applied over a broad area causing the soil to fracture in the desired area (similar to the practice of scribing a line on glass and then stressing the glass along the scribe to cause it to fracture as desired). Pumpable material is then injected into the fracture area causing it to flow along the fracture plane to create a uniform and continuous barrier.

The HSSB process is truly an in situ treatment process. The soil phase remains in place at all times. Soil treatment occurs only at the barrier (unless the barrier is used to provides nutrients for bioremediation). The current installation process provides a compressive load on the soil above the barrier location during fracture which will prevent cracking of the overburden. Hence no enhanced movement of the contaminant is expected from this process. However, hydrofracing by conventional methods can be used to promote contaminant movement into the barrier. The barrier can be placed in saturated or unsaturated soil.   Limitations on depth of placement are expected to be from equipment limitations and not from process limitations.

DISCUSSION: The HSSB technology was recently employed in a pilot scale demonstration to install a barrier into undisturbed soil as illustrated in Figures 2, 3, and 4. The test site was a grassy field consisting of multiple layers of clayey-sand in the top 10 ft (3 m). First, bore holes were drilled into the soil in a hexagonal array, Figure 2. Next, undercuts were made into the soil with hydraulic jets to enhance stress concentrations in the soil (refer to Figure 3). Then, packers were installed into the bore holes at a consistent elevation, and inflated to seal off the bore holes from the atmosphere , Figure 3.

This allowed the application of air pressure past (through) the packer into the bottom of the bore hole. The soil was then preferentially stressed by the application of increasing air pressure at multiple points until the soil failed (fractured) at about 30 psi (200 kpa). Because of the application of the stress at the same time, at multiple points, and over a broad area, the soil fractured over a broad area. The fracture was quick and catastrophic because air was used as the fracture fluid. That is, once the soil began to fail, air quickly expanded into the soil crevices to broaden the area of the lift and increase the lifting force. The broader the fracture area, the less pressure is needed to lift the soil. Hence, catastrophic soil fracture occurred at the desired location. Vent holes were also used around the outside of the test area to limit the fracture area by relieving the air pressure, Figures 4. The vent holes performed well and allowed the air to escape to the atmosphere without noticeable lifting of the soil surface. After the fracture, air was escaping from each vent hole indicating a broad fracture area.

For this test, concrete grout was used to demonstrate the placement of a barrier. Any pumpable material can be placed in the fracture plane but the easiest to place is a lower viscosity Bingham fluid. Barriers can be created from solid particles such as iron filings by suspending them in a solution such as guar gum which will quickly decompose after placement in the soil and leave the solids as a porous barrier.

The recent pilot test successfully demonstrated the process for in situ placement of a HSSB. A barrier in the shape of a bowl, illustrated in Figure 4, was formed in situ in undisturbed soil with cement grout. The flat portion of the bowl was 8 ft. (2.4 m) in diameter and 8 ft. (2.4 m) below the soil surface. The sides of the bowl turned upward at a 20 to 40 degree angle to form a bowl with a total diameter of about 16 ft (4.8 m).

The barrier material at the bottom of the bowl was about 1/4-inch thick and the sides tapered to about 1/16-inch thick. Although 100% inspection of the bowl could not be achieved, excavation of the test site with multiple slices through the barrier showed a continuous barrier layer. No holes were found in the barrier except at the very edge where extreme thinning occurred. These results are very significant in that they demonstrate both the HSSB installation technique and the ability to produce a continuous barrier even with a very thin layer of grout. The HSSB can be made much thicker with the use of more barrier material. With thicker barriers, a continuous barrier can be ensured although not necessary for significant remediation benefits.

Current research is focused on the development of improved tools. The current method of installation is too expensive for large scale commercial use. It is believed that better tools can be developed to eliminate the bore holes and packers, and also allow easier pre-stressing of the soil. More theoretical work needs to be done to better understand the soil mechanics as related to the HSSB. Small scale test will also be conducted to develop more engineering data to allow efficient designs for specific field conditions. It is believed the HSSB can be made applicable to most contaminated sites.

CONCLUSIONS: The HSSB, as currently developed, can create an in situ horizontal subsurface layer of various materials in soil at various depths for the purpose of soil and groundwater remediation or contaminant containment. This HSSB technique is now in need of improved tools to reduce the cost of installation and increase commercial viability. The technique also requires correlation between the theory of implementation and new test data. This work will lead to expanding the data base for HSSB installation and provide engineering data to allow design of a full scale demonstration.

Monitoring of barrier location and performance is an issue common to all barriers. Sandia National Laboratory and Lawrence Livermore National Laboratory, (Daily,W and Ramirez, A) among others are performing research to improve the monitoring capability for subsurface barriers. The in situ HSSB will utilize such technologies when developed.

REFERENCES:

U.S. Environmental Protection Agency. 1996. *Cleaning Up The Nations Waste Sites: Market and Technology.* EPA-542-R-96-003

Sivavec, T.M.; Mackenzie, P.D; Horney, D.P.; and Baghal, S.S. " Redox-Active Media for Permeable Reactive Barriers" Published in International Containment Conference Proceedings, page 753-759. (February 1997, St. Petersburg, FL)

Daily,W and Ramirez, A. "A New Geophysical Method for Monitoring Emplacement of Subsurface Barriers." Published in International Containment Conference Proceedings, page 1053-1057. (February 1997, St. Petersburg, FL)

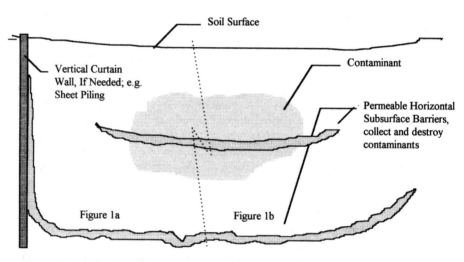

TYPICAL HSSB INSTALLATION FOR SOIL REMEDIATION

FIGURE 1

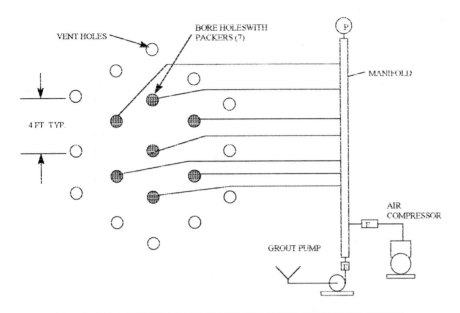

HORIZONTAL SUBSURFACE BARRIER TEST CONFIGURATION
FIGURE 2

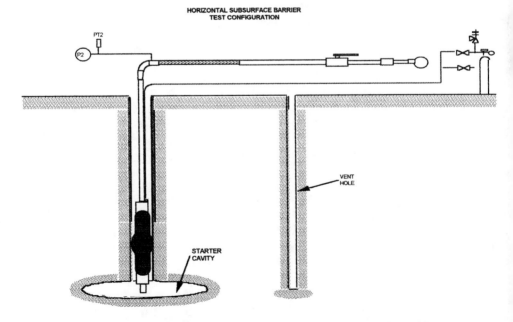

SECTION VIEW OF PACKERS, STARTER CAVITY AND VENT HOLE
FIGURE 3

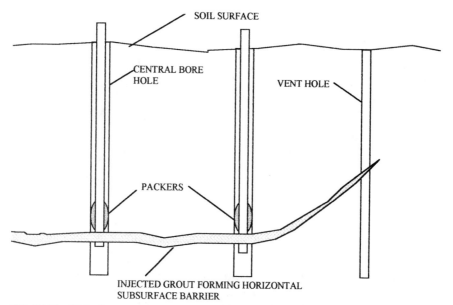

DIAGRAM OF IN SITU HAOIZONTAL SUBSURFACE BARRIER EMPLACEMENT
FIGURE 4

USE OF CONTINUOUS TRENCHING TECHNIQUE TO INSTALL IRON PERMEABLE BARRIERS

James R. Romer, EMCON (Medford, Oregon)
Stephanie O'Hannesin, EnviroMetal Technologies Inc. (Guelph, Ontario, Canada)

ABSTRACT: Describes construction of a funnel and gate system to prevent off-site migration of impacted groundwater at an Oregon aircraft maintenance facility. After computer modeling of groundwater hydraulics at the site, a design was selected calling for two 50-ft (15.2-m) gates and 650 ft (198 m) of funnel keyed into the underlying aquitard. The funnel consisted of a soil and bentonite (10 percent) slurry wall up to 34 ft (10.4 m) deep. Two gates were constructed, one consisting of two layers of 100 percent iron filings installed with a continuous trencher, the second consisting of a mixture of sand and iron filings installed with a trackhoe and drag box. The layered gate was offset from the funnel, then connected to it by driven sheet piles. Permeabilities of 10^{-7} cm/sec were achieved for the slurry wall. Direct placement of the iron filings met with mixed results.

INTRODUCTION

The first commercial permeable reactive barrier using granular iron to remediate volatile organic compounds (VOCs) in groundwater was installed in Sunnyvale, California, in January 1995. Since then, over 20 in situ treatment systems have been installed, including 11 full-scale installations in the United States and 1 in Northern Ireland. All have successfully met effluent groundwater criteria to date. Recent analyses of core samples obtained from two earlier iron installations have shown no significant effects of mineral precipitation or microbial activity on reactive zone performance (Vogan et al., 1998).

Construction of most full-scale systems to date has used excavation-based methods, to create either a continuous permeable barrier or a funnel and gate system, in which low-permeability funnel sections are used to direct groundwater toward a central reactive zone or gate (see the table). Funnel sections have been constructed using both slurry wall and sheet piling. Permeable treatment gate sections have been installed using removable sheet piling, prefabricated trench boxes, and driven caissons.

TABLE Installed full-scale in situ treatment systems

Continuous Permeable Barriers	Funnel and Gate
Private site, California (1995)	Private site, Kansas (1996)
Government site, North Carolina (1996)	Government site, Colorado (1996)
Private site, South Carolina (1997)	Private site, Oregon (1997)
Private site, New York, (1997)	Private site, Colorado (1997)

Continuous trenching techniques have been used to install the continuous permeable barriers listed above. The continuous trenching equipment uses a large cutting chain excavator system combined with a trench box and loading hopper. To install the permeable wall, the trencher moves along a trench line. As the cutting chain excavator removes the native soil, the granular iron is placed in the hopper and flows from the trench box into the excavated trench. The trench box walls extend to the width and depth of the trench.

A continuous trencher was recently used to install both the permeable and impermeable sections of a large funnel and gate system at a private facility in Oregon. This installation, described in detail below, illustrates the potential advantages and disadvantages of the permeable barrier construction method.

PROJECT DESCRIPTION

The project site is an aircraft maintenance facility in Oregon (see figure). Historical use of chlorinated solvents as degreasers resulted in a groundwater plume of tetrachloroethene, trichloroethene, and other degradation compounds extending off-site. An evaluation of on-site interim remedial measures (IRM) was conducted under Oregon's Voluntary Cleanup Program to prevent off-site migration of VOC-impacted groundwater.

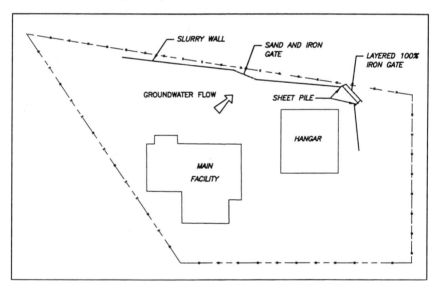

FIGURE Site layout showing location of funnel (slurry wall) and gate system

Approach. Results from 13 geotechnical borings on 50-ft (15.2-m) centers along the downgradient property boundary and additional water chemistry were combined with site data developed during the remedial investigation to evaluate three IRM scenarios: extraction wells, interception trench, and reactive wall/funnel and gate. VOC impacts up to 500 µg/L are limited to an upper water-bearing zone consisting of heterogeneous alluvial deposits, ranging from sandy silts to silty gravels, underlain by a fine-grain aquitard. The depth to the aquitard ranges from 24 to 34 ft (7.3 to 10.4 m) below grade. The depth to water in this zone ranges seasonally between 4 and 8 ft (1.2 to 2.4 m) below grade. The geometric mean of hydraulic conductivities from slug tests conducted across the site is 10^{-3} cm/sec, with variations of two orders of magnitude.

Because of the wide range of sediment sizes and resulting hydraulic conductivities across the site, using conventional extraction wells to effect hydraulic control of the site was not considered viable. The remaining options all required extensive excavation and significant capital costs. The additional water chemistry data indicated that either an extraction trench with aboveground treatment or a reactive wall or funnel and gate approach would work. The client decided to pursue a reactive media approach to minimize site disruption and long-term operation and maintenance costs.

Both a continuous reactive wall and a funnel and gate approach were evaluated. The funnel and gate approach was selected for two reasons: (1) because of heterogeneity of the site, water velocities at the wall could vary considerably, and if a conservative/worse case approach was taken, iron and associated costs would be too high; (2) gates were considered more feasible to maintain than a continuous wall. Several funnel and gate configurations were modeled using the two-dimensional model Flowpath™. The final design called for two 50-ft (15.2-m) gates, with a total of 650 ft (198 m) of funnel keyed into the underlying aquitard.

Installation of funnel. Construction of a soil-bentonite slurry wall using a continuous pass trencher to mix the dry bentonite into the native soils was selected over sheet piling or traditional slurry wall construction on the basis of cost, site disruption, and installation time. The slurry wall funnel was constructed by first digging a 2-ft (0.6-m)-wide and 3-ft (0.9-m)-deep trench along the proposed wall path. The trench was filled with bentonite, yielding a minimum of 10 percent by volume of bentonite in the slurry wall, which had a design hydraulic conductivity of 10^{-7} cm/sec. This mix design was based on laboratory-tested mixes using materials from the geotechnical boring program. Additional water was added as the machine mixed the soils in place. The process was quick and generated minimal soils that required disposal.

Depth to the aquitard was surveyed on the basis of the drilling program; however, the density of the confining unit was high enough to cause the machine to chatter when encountering the unit. Though this made it easy to keep the trenching arm at depth, it damaged the machinery, resulting in delays for repair. The soil-bentonite slurry wall was capped with reinforced concrete, allowing bearing capacity on top of the wall.

Installation of gate. The modeling indicated that a minimum design thickness of 14 in. (0.36 m) of 100 percent iron filings would be needed to achieve maximum contaminant levels. The trenching machine used for the funnel was modified with a box and hopper to install the 100 percent reactive iron filing gates. The original design called for connecting the gate to the slurry wall by cutting through the slurry wall at an angle 3 to 5 ft (0.9 to 1.5 m) from the end and allowing the slurry to "heal" into the gate, creating a seal. The viscosity of the slurry was low enough, however, that it followed the trencher into the gate area. Installation was terminated to avoid reducing permeability along the gate face.

Installation proceeded by offsetting the gate approximately 10 ft (3.1 m) downgradient of the slurry wall, for later connection to the funnel using driven sheet piles. Placement of iron using the trencher and drag box resulted in a gate thickness of only 9 in. (0.23 m). Formation pressure appears to have restricted the rate at which the filings emptied out of the box.

To obtain the required thickness of iron, a second parallel pass was made. The sheet piling connecting the first pass to the slurry wall was extended to connect the second pass as well, creating a layered gate with a combined iron thickness of 18 in. (0.46 m). Total installation time for this single 50-ft (15.2-m) gate exceeded the time required for 650 ft (198 m) of slurry wall.

The same installation approach was attempted for the second gate. However, the trenching machine met refusal in a cobble zone. Each end of the slurry wall was stabilized with driven sheet piling before soil was excavated from the gate area. A more traditional excavation method was employed than for the first gate, using a trackhoe with a 3-ft (0.9-m)-wide bucket and a 20-ft by 4-ft by 8-ft (9.2-m by 1.2-m by 2.4-m) drag box to stabilize the upper 8 ft (2.4 m) of soil. The gate was installed in three 20-ft (6.2-m) sections.

Once a section was excavated into the aquitard and the sloughed material removed, a mixture of sand and iron filings was placed. Mixing was accomplished in a concrete truck. The sand, of approved gradation, arrived on site in a standard concrete mixing truck. On excavation of the gate area, it was noted that the lower portion was more permeable than the upper. Depending on which portion of the gate was being filled, three to four bags of iron filings, 3,000 lb (1,361 kg) each, were emptied into the truck. After the sand and iron filings were mixed (at least 100 revolutions), the mixture was placed. In the lower portions of the trench, a tremie pipe was used to place the reactive mixture below the water table to keep it from segregating during settling.

Results. Grab samples of the slurry wall were collected during construction and Shelby tube samples collected at depth one month after installation. Results from laboratory permeability tests ranged from 2.3×10^{-8} to 8.6×10^{-7}, with a geometric mean of 1.4×10^{-7} cm/sec. Installation of four monitoring wells per gate is scheduled for late March 1998. One pair of wells will be located upgradient and one pair downgradient of each gate. They will be placed as close to the gates as possible and sampled for the full range of parameters recommended in Vogan (1997). The first round of sampling is scheduled for April 1998.

CONCLUSIONS

A continuous trencher can be a cost-effective and efficient method for installing funnel and gate structures in certain situations. Permeabilities of

10^{-7} cm/sec were achieved by in-place mixing of slurry walls up to 34 ft (10.4 m) deep. The trenching equipment is not designed to handle cobbles or partially cemented zones. Direct placement of iron filings by this method required two passes and connecting sheet piling, which increased both costs and installation time. Enhancements to this purely gravity method of gate construction should be considered.

ACKNOWLEDGMENTS
This project proceeded on a fast-track design-build basis from the outset. The authors would like to thank Horizontal Technologies, which had the only continuous pass trencher capable of working to 30-plus-foot (9-plus-meter) depths. They also recognize the Oregon Department of Environmental Quality's Voluntary Cleanup Division for giving innovative technology a chance.

REFERENCES
Vogan, J. L. 1997. "Monitoring of Permeable Iron Treatment Installations." Technical memorandum 31099.77. EnviroMetal Technologies Inc., Guelph, Ontario, Canada. October 15.

Vogan, J. L., B. J. Butler, M. S. Odzienkowski, G. Friday, and R. W. Gillham. 1998. "Inorganic and Biological Evaluation of Cores from Permeable Iron Reactive Barriers." Presented at First International Conference on Remediation of Chlorinated and Recalcitrant Compounds, Monterey, CA, May 18-21.

TECHNICAL UPDATE: THE FIRST COMMERCIAL SUBSURFACE PERMEABLE REACTIVE TREATMENT ZONE COMPOSED OF GRANULAR ZERO-VALENT IRON

Scott D. Warner, Carol L. Yamane, Nancy T. Bice and Frank S. Szerdy
(Geomatrix Consultants, San Francisco, California)
John Vogan (EnviroMetal Technologies, Inc., Guelph, Ontario)
David W. Major (Beak International, Ltd., Guelph, Ontario)
Deborah A. Hankins (Intersil, Inc., San Francisco, California)

ABSTRACT: The first commercial permeable reactive treatment zone (PRTZ) composed of granular iron and designed to treat chlorinated volatile organic compounds (VOCs) in groundwater has been operational since being installed during the winter of 1994-1995. The PRTZ was installed at a former semi-conductor manufacturing facility in the south San Francisco Bay area as the final remedy for shallow groundwater replacing a 7-year-old pump-and-treat remediation system. The final remedy was required to treat shallow groundwater that in 1991 contained greater than 2 milligrams per liter (mg/L) total VOCs (including greater than 0.5 mg/L vinyl chloride) to state maximum contaminant limits (MCLs). Since installation of the PRTZ, which consists of an approximately 40-foot long by 8-foot wide by 15-foot deep treatment cell, groundwater collected from performance wells within the PRTZ has met regulatory objectives. Assessment of the effectiveness of the PTZ has relied on the analysis of organic, inorganic, and biological constituents within groundwater samples collected from wells installed within and adjacent to the PRTZ.

INTRODUCTION

Construction of the first commercial full-scale subsurface permeable reactive treatment zone (PRTZ) composed of granular iron and designed to passively treat groundwater affected by chlorinated volatile organic compounds (VOCs) was completed in February 1995. Installed at a former semi-conductor manufacturing facility in Sunnyvale, California, the treatment system, consisting of an *in situ* "iron wall" and lateral low permeability "slurry" walls, was approved by the California Regional Water Quality Control Board (CRWQCB) as the final remedy for the site. This final remedy replaced a groundwater extraction and aboveground treatment system that had been operating since 1987. Although the first, and longest demonstration of a PRTZ composed of granular iron was conducted at the Canadian Forces Base, Borden, Ontario (Borden site) between 1991 and 1996 (O'Hannesin and Gillham, 1998), the Sunnyvale PRTZ provides performance data from the longest-running commercial installation of an *in situ* iron wall.

Several papers and conference proceedings documenting the design criteria and initial monitoring results from the Sunnyvale installation have previously been published (for example, Yamane, et al., 1995; and Szerdy, et al., 1996). The purpose of this paper is to update the monitoring results and summarize the long-term performance of the remedy at this site.

Site History. VOCs, chiefly trichloroethylene (TCE), cis-1,2-dichloroethylene (cis-1,2-DCE), and vinyl chloride (VC) were detected in soil and groundwater beneath the site in 1983. Remediation methods consisting of soil excavation and groundwater extraction with above-ground treatment by air-stripping were performed beginning in 1986 and 1987, respectively. The groundwater extraction and treatment system, acceptable as a final remedy for the site by the CRWQCB, incorporated three extraction wells pumping at an average total rate of 15 gallons per minute. Although the extraction system provided hydraulic capture of the affected groundwater, it was inefficient in removing VOC mass from the aquifer and was expected to require many decades to centuries to lower the concentrations of the target VOCs in groundwater to cleanup standards (0.005 milligrams per liter (mg/L) for TCE, 0.006 mg/L for cis-1,2-DCE, and 0.0005 mg/L for VC).

The affected aquifer lies between approximately 10 and 20 feet below ground surface, although the thickness of the aquifer is variable and ranges between approximately 2 and 10 feet. The aquifer is composed of interbedded silt, sand, and clay, generally is semi-confined, and is separated from the next deepest water-yielding zone by 55 to 65 feet of silty clay. The magnitude and direction of the shallow, horizontal hydraulic gradient is approximately 0.004 foot per foot northward, although shifts of up to 50 degrees from northwest to northeast have been observed. The ambient velocity of groundwater appears to range between 0.5 and 1 foot per day and the magnitude of the vertical gradient has been measured to be 0.005 foot/foot upward. The results of aquifer testing suggest an average transmissivity of approximately 375 square feet per day. Hydrochemically, the groundwater within the aquifer ranges from calcium bicarbonate to calcium sulfate type. The concentration of total dissolved solids (TDS) ranges from approximately 1500 mg/L at the southern end of the site to greater than 3000 mg/L immediately north of the site.

In late 1991, the granular iron technology was identified for the site as a possible remediation method that would treat the groundwater passively and *in situ*. Laboratory batch and column studies using site groundwater were performed by the University of Waterloo and demonstrated the efficacy of the treatment process. During 1992, a nine-month on-site pilot test was performed. The pilot test consisted of routing a side-stream of extracted groundwater through a 6-foot high by 2-foot diameter cylinder filled with a 4-foot-thick flow-through zone of granular iron that was sandwiched between sand and native soil. The pilot test successfully demonstrated the effectiveness of the treatment process and provided information, including flow rates and degradation rates, that would be used to design a full-scale system. A design for a full-scale application of the granular iron treatment technology was prepared in 1993. Regulatory approval for the system, which included removing the groundwater extraction and treatment system, was granted by the CRWQCB in November 1993.

Construction Details. Construction of the PRTZ began in November 1994 and the remedy became fully operational in February 1995. The system as described in Yamane, et al. (1995) and Szerdy, et al. (1996) consists of the design elements described below (Figure 1a).

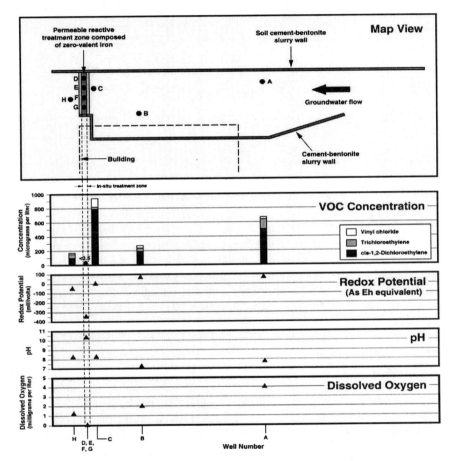

FIGURE 1. Diagram showing: (a) the schematic representation of the PRTZ; and graphical representations of (b) VOC concentrations and field parameter measurements.

- An approximately 40-foot long by 15-foot deep by 8-foot wide treatment cell, which consists of a 4-foot wide flow-through zone of 100 percent granular iron sandwiched between 2-foot wide upgradient and downgradient pea gravel sections. The thickness of the granular iron zone was designed to provide a minimum two-day residence time based on the anticipated influent concentration of VC (0.5 mg/L) and the pilot-test estimated half-life (42 hours) of dissolved VC in contact with granular iron.

- Lateral low-permeability barriers constructed upgradient and on both east and west sides of the treatment cell. The low permeability barriers were constructed using soil-cement-bentonite slurry and extend to lengths of approximately 300 feet and 225 feet, respectively.

- A lateral hydraulic barrier constructed of steel sheet piles and extending approximately 20 feet downgradient (north) of the treatment cell.

The treatment cell was constructed by driving sheet piles along the designated alignment of the treatment cell, excavating the soil within the sheet piled area, backfilling the trench with the granular iron and pea gravel, and extracting the sheet piles. The upgradient lateral barriers were emplaced using standard slurry wall construction methods. Both the treatment cell and lateral barriers extend at least 2 feet into the underlying silty clay aquitard material.

MONITORING RESULTS

Groundwater samples are collected from a network of monitoring wells located upgradient, downgradient, and within the PRTZ. Samples are collected quarterly and analyzed for VOCs by gas chromatography to meet regulatory objectives. Additional samples have been collected infrequently and analyzed for inorganic species, dissolved hydrocarbon gasses, and volatile fatty acids. The field water quality parameters including dissolved oxygen concentration (DO), pH, and reduction-oxidation potential (converted to equivalent Eh), are measured using a flow-through cell. Groundwater samples are collected using low-flow purging and sampling methods.

The general results of more than three years of sampling are summarized in the following tables for selected parameters. The results represent average conditions for the general regions — upgradient wells, downgradient wells, and performance wells located within the PRTZ— as indicated in Tables 1 through 4.

Table 1. Summary of field parameter measurements 1995-1998.

Location	pH	DO (mg/L)	Eh (millivolts)
Upgradient	7.5 to 8.0	1.5 to 2.5	+50 to +150
Performance Wells	10 to >11	<0.2	-300 to -400
Downgradient	7.7 to 8.3	1.0 to 2.5	-100 to +100

Table 2. Summary of VOC analytical results 1995-1998.

Location	TCE (mg/L)	cis-1,2-DCE (mg/L)	VC (mg/L)
Upgradient	0.03 to 0.230	0.010 to 0.500	0.030 to 0.210
Performance Wells	< 0.0005	< 0.0005	< 0.0005
Downgradient	0.012 to 0.460	0.010 to 0.640	0.001 to 0.080

Table 3. Summary of selected inorganic analytical results 1995-1998.

Location	Cation Ratio[1] (% meq/L)[2]	Bicarbonate Ratio[3] (% meq/L)	SO_4^{-2} (mg/L)	Fe (mg/L)
Upgradient	42.9	30.6	400	0.06
Performance Wells	3.3	1.5	6	<0.05
Downgradient	41.8	53.6	72	0.06

Table 4. Summary of selected dissolved hydrocarbon gases, volatile fatty acids, and bacteria count analytical results 1995-1998.

Location	Methane (parts per billion)	Ethane + Ethene (parts per billion)	Propionic Acid (mg/L)	Equivalent Cells (per mL)
Upgradient	870	38	175	113,000
Performance Wells	140	4 to 15	3.7	9140
Downgradient	5000	11	147	289,000

The results of monitoring indicate that the final remedy successfully meets regulatory objectives. Key conclusions of the results are as follows:

- Concentrations of target VOCs, including VC, are reduced to cleanup standards within the PRTZ while concentrations of VOCs detected in groundwater samples collected upgradient and downgradient of the PRTZ represent the residual chemistry of untreated groundwater (Figure 1b).

- Values of water quality parameters such as pH and Eh are consistent with the anticipated effect of iron oxidation and water reduction within the PRTZ (Figure 1b). The return of these parameters to near ambient conditions in groundwater samples collected downgradient of the PRTZ demonstrates the buffering capacity of the aquifer system.

---

[1] Cation ratio = $(Ca^{2+} + Mg^{2+}) / (Ca^{2+} + Mg^{2+} + Na^+ + K^+) * 100$ where concentrations are in units of milliequivalents per liter.

[2] meq/L = milliequivalents per liter.

[3] Bicarbonate Ratio = $(HCO_3^-) / (HCO_3^- + CO_3^{2-} + SO_4^{2-} + Cl^-) * 100$ where concentrations are in units of milliequivalents per liter.

- The depressed values of the cation ratio and bicarbonate ratio in groundwater samples collected from performance wells compared to results from upgradient and downgradient wells suggest that some secondary precipitation is occurring within the PRTZ. This is consistent with the anticipated effect from the higher pH conditions within the treatment zone. However, the effectiveness of the treatment process does not appear to have decreased as a result of the loss of inorganic species.

- The concentrations of dissolved hydrocarbon gases suggest that the dechlorination process is occurring both within and outside the PRTZ.

- Biological activity within the PRTZ is low relative to upgradient and downgradient groundwater as indicated by: (1) low concentrations of propionic acid, an anaerobic fermentation product; and (2) low biomass as indicated by low microbiological cell counts, within groundwater collected from PRTZ performance wells, compared to higher values of these parameters detected in groundwater collected from upgradient and downgradient wells.

ACKNOWLEDGEMENT

The authors acknowledge M. Ledesma, D. Sorel, and J. Gallinatti of Geomatrix Consultants, Inc. who have been instrumental in collecting, managing, and interpreting data collected for this project.

REFERENCES

O'Hannesin, S.F., and R. W. Gillham, 1998. "Long-term performance of an in situ 'iron wall' for remediation of VOCs." *Ground Water*, Vol. 36, No. 1, pp. 164-170.

Szerdy, F.S., J.D. Gallinatti, S.D. Warner, C.L. Yamane, D.A. Hankins, and J.L. Vogan, 1996. "In situ groundwater treatment by granular zero-valent iron: Design, construction and operation of an in situ treatment wall." American Society of Civil Engineers, National Meeting.

Yamane, C.L., S.D. Warner, J.D. Gallinatti, F.S. Szerdy, T.A. Delfino, D.A. Hankins, and J.L. Vogan, 1995. "Installation of a subsurface groundwater treatment wall composed of zero-valent iron." Division of Environmental Chemistry, American Chemical Society National Meeting, Anaheim, California, April 2-7, *in*, Preprints of Papers, Vol. 35, No. 1, pp. 792-795.

FIELD PERFORMANCE OF VERTICAL HYDRAULIC FRACTURE PLACED IRON REACTIVE WALLS

Grant Hocking, Samuel L. Wells and Rafael I. Ospina
Golder Sierra LLC, Atlanta, GA, USA.

ABSTRACT: Orientated vertical hydraulic fracturing technology has placed permeable iron reactive walls 3 to 4" thick in highly permeable sand and gravel formations down to moderate depths. Pre construction monitoring indicated that the site ground water was contaminated with chlorinated solvents primarily trichloroethene (TCE). The hydraulic fracture iron reactive wall was constructed by injecting through wells installed along the wall alignment. The field performance of the vertical hydraulic fracture placed permeable iron reactive wall is presented for both the installation and post construction phase. During installation, the fracture geometry is monitored in real time by electrically energizing the fracture and monitoring it's propagated geometry by down hole resistivity receivers. Hydraulic pulse interference tests verify the wall's continuity and hydraulic effectiveness. Up and down gradient groundwater monitoring wells are sampled and analyzed to assess the wall's degradation performance.

INTRODUCTION

In situ permeable iron reactive walls have been placed at a number of sites to abiotic degrade chlorinated solvents in groundwater. The conventional means of installing iron reactive walls for remediation of ground water contaminated with chlorinated solvents has been by shoring and excavation or continuous trenching. Recently orientated vertical hydraulic fracturing has been used as an alternate mode of placing permeable iron reactive walls in situ, at modest depth and with minimal site impact, (Hocking and Wells, 1997).

Zero valent metals have been known to abiotic degrade halogenated compounds as detailed in Gillham and O'Hannesin (1994). The abiotic degradation of halogenated aliphatics in the presence of iron can be approximated by a first order reduction process. The compounds are progressively degraded to daughter products and eventually broken down into ethanes and ethenes. In the hydraulic fracturing technology, the iron in the form of iron filings is injected into the subsurface by a viscous cross linked gel. The gel is a water based cross link gel, hydroxypropylguar (HPG), and has minimal impact on the iron's reactivity as determined by laboratory column tests. Cross linked, the gel is extremely viscous, ensuring the iron filings remain suspended during the installation of the wall. Once the gel degrades, a permeable iron reactive wall remains in situ with a wall thickness typically of 3" to 4" in sand and gravel formations.

CONSTRUCTION MONITORING

The hydraulic fracture place permeable reactive wall is constructed from wells installed along the wall alignment. A special down hole tool is inserted into each well and a controlled vertical fracture is cut and initiated at the required azimuth orientation and depth. Upon initiation of the controlled fracture, the tool is withdrawn and a packer is set in the well. Multiple well heads are then injected with the iron gel mixture to form a continuous permeable iron reactive wall.

The gel is dispersed and hydrated in 3000 gallon batches prior to injection. The hydrated gel is pumped to the mixing and blending unit at the required rate. The iron filings are pre-loaded into 250 cft hoppers and feed to the mixing and blending system via calibrated inclined screw conveyors. The feed of gel and iron filings is controlled to ensure the mix is of the correct consistency and density. Computerized instrumentation records and controls flow rates, volumes, iron feed rate, tonnages and mix density. The mix density is monitored by a precise in line mass flow meter. The gel is also batch tested for pH, conductivity and viscosity. The iron filings are sampled and tested for grain size prior to loading into the hoppers to ensure consistent quality. The iron gel mix is also batch tested for density and degree of cross link.

Following mixing and blending, the iron gel mix is pumped to a plunger hydraulic fracturing pump for injection into the prepared well heads. The gel and iron filings are feed to the pumping unit and cross linked in line, to form a highly viscous cross linked gel of a specific gravity typically around 2. During injection, the iron gel mix is electrically energized with a low voltage 100 Hz signal. Down hole resistivity receivers are monitored to record the in phase induced voltage by the propagating fracture, see Figure 1. From monitoring the fracture fluid induced voltages and utilizing an incremental inverse integral model, the fracture fluid geometry can be quantified during the installation process. The active resistivity method requires the fracture fluid to be at least thirty times (30x) more conductive than the ground to provide a high contrasting image. Shallow fracture injections can be geometrically traced from surface installed resistivity receivers, provided high conductive layers are not present between the fracture

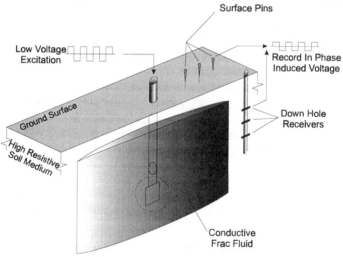

and the surface.

FIGURE 1. Fracture Mapping by Active Resistivity.

Active resistivity monitoring has the added benefit of determining when the individual fractures coalesce and thus become electrically connected. That is, by energizing each injected well head individually and in unison, the fracture electrical coalescence is clearly recorded. The imaging and inversion of the down hole resistivity data focuses on quantifying the continuity of the reactive wall and assessing if any holes or gaps are present. Such monitoring enables construction procedures to be modified to ensure the wall is installed as planned, and enable contingency measures to be implemented immediately, e.g. an additional fracture to patch any holes or additional injection volumes to ensure coalescence or sufficient overlap is attained.

The down hole resistivitiy receiver locations are shown in Figure 2 for the iron reactive wall constructed in a highly permeable sand, gravel and boulder sequence. The resistivity receivers are copper collars attached to a cable connected to the instrumentation data acquisition system. The wells B1 and B2, see Figure 3, were injected and the fracture geometry was determined by the measured induced voltages at the down hole receivers locations, see Figure 3. The cross hatched area on this figure is the geometrical outline of the vertical hydraulic fracture, which was calculated to be 3.5" in average thickness. All of the thirty injections at this site have been recorded and geometries delineated by active resistivity.

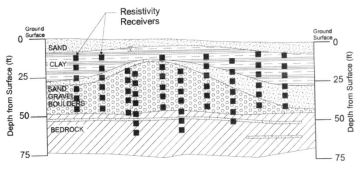

FIGURE 2. Down Hole Resistivity Receivers.

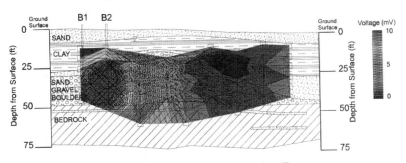

FIGURE 3. Induced Voltages from Propagating Fracture.

POST CONSTRUCTION MONITORING

The permeable reactive wall is tested for it's hydraulic effectiveness by pulse interference tests. Pulse interference tests, (Johnson et. al., 1966), involve a cyclic injection of fluid into the source well, and by high precision measurement of the pressure pulse in a neighboring well, detailed hydraulic characterization between wells can be made, see Figure 4.

The pulse interference test is highly sensitive to hydrogeological properties between the wells, and relatively insensitive to conditions outside of the wells. The time delay and attenuation of the hydraulic pulse enable the hydraulic effectiveness of the wall to be assessed. Before the gel cross link is broken, the wall acts as a temporary flow barrier, because the gel is an impermeable viscous fluid. If the wall is continuous, significant attenuation of the hydraulic pulse will occur. If holes are present, the time delay and lack of attenuation of the pulse enables the gross area and approximate location of any holes to be delineated. Following breaking of the gel, a permeable iron reactive wall remains, with minimal gel residue. Laboratory permeater tests have quantified that the gel residue is minimal and does not impact the permeability of the iron reactive wall.

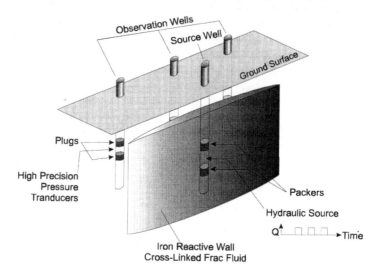

FIGURE 4. Hydraulic Pulse Interference Test.

The pulse monitoring wells are installed before the fracture wall is constructed. Prior to the wall construction, numerous hydraulic pulse tests of the ground between the wells is conducted to characterize the site's hydrogeological parameters. By systematic locations of pulse source and receiver arrays, the differences in the hydrogeological properties in the horizontal and vertical dimensions can be quantified.

To maximize the pulse test's resolution, a small section of the injector well is isolated by packers and monitored by a high precision pressure transducer. The receiver wells' pressure transducers are isolated by packers as well. Thus the pulse is basically a point source, and borehole storage effects are eliminated from both the injector and receiver wells. The injector well is pulsed for a set time, shut in for the same time period, and the cycle repeated.

The fracture injected wells have been mapped by borehole camera after the wells has been flushed of iron and gel. In most cases the borehole camera has been able to view induced hydraulic fractures down to moderate depths and confirm the fracture width at least at each well. The fracture width in the casing will be less than the fracture's width immediately inside the formation, because the casing is grouted in place and the iron gel mix injected needs to displace both the casing and surrounding grout. A down hole view of the fracture width inside of the well casing is shown in Figure 5. The left image in this figure is a view at the top of the initiated controlled fracture where the casing is intact and the fracture at its smallest width. The right hand view is some two feet deeper, where the iron filled fracture is approximately 2.5" thick. Immediately beneath this view the fracture is wider approximately 3.5" thick and only one half of the casing is within the camera view.

FIGURE 5. Down Hole Camera View of Iron Fracture Width.

The reactive wall's degradation performance is quantified from groundwater samples collected from upgradient and downgradient ground water monitoring wells. The anticipated degradation performance of a hydraulic fracture iron reactive wall of 3" in thickness is illustrated in Figure 6 for the respective contaminants for trichloroethene (TCE), cis-1,1-dichloroethene (cDCE) and vinyl chloride (VC). In order to fully evaluate the wall's performance, groundwater samples are analyzed for volatiles, semi-volatiles, metals and other parameters such as pH, ORP, alkalinity, carbonate, etc. From laboratory column desorption tests of the site's soils, it is clear that desorption of contaminants from the site soils will mask the early time history data of the wall's real performance.

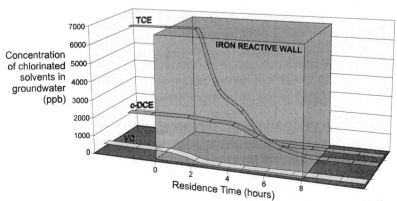

FIGURE 6. Anticipated Degradation Performance of Iron Reactive Wall.

CONCLUSIONS

Orientated hydraulic fracturing technology has placed permeable iron reactive walls of approximately 3.5" thick in highly permeable sands and gravel. The construction monitoring of the wall by active resistivity has provided clear images of the fracture geometry during injection. Fracture coalescence between wells has been observed from both active resistivity and physical evidence. The hydraulic pulse interference tests have confirm the presence of the hydraulic fracture placed iron wall and have assessed it's hydraulic effectiveness. The fracturing gel has been shown not to interfer with the wall' permeability nor impact the iron's reactivity. Borehole camera views of iron filled fracture widths have confirmed the computed wall's thickness at least in each injected well. Degradation performance of the wall will be quantified from analyses of upgradient and downgradient groundwater samples.

REFERENCES

Gillham, R. W., and S. F. O'Hannesin. 1994. "Enhanced Degradation of Halogenated Aliphatics by Zero-Valent Iron", *Ground Water*, Vol. 32, No. 6, pp958-967.

Hocking, G., and S. L. Wells. 1997. "Orientated Vertical Hydraulic Fracture Iron Reactive Walls", presented at the ASCE Env. Group Meeting, Atlanta, GA, Mar'97.

Johnson, C. R., R. A. Greenhorn and E. G. Woods. 1966. "Pulse-Testing: A New Method for Describing Reservoir Flow Properties Between Wells". *JPT*, Vol. 237, pp1599-1604, Trans., AIME.

HYDRAULIC EVALUATION OF A PERMEABLE BARRIER USING TRACER TESTS, VELOCITY MEASUREMENTS, AND MODELING

Neeraj Gupta, Bruce M. Sass, Arun R. Gavaskar, Joel R. Sminchak, Tad C. Fox, and Frances A. Snyder (Battelle, Columbus, Ohio, USA)
Deirdre O' Dwyer* (Tetra-Tech EMI Inc., Denver, Colorado, USA)
Charles Reeter (Naval Facilities Engineering Services Center, Port Hueneme, California, USA)

ABSTRACT: The monitoring and performance evaluation of a pilot-scale permeable barrier using granular iron and a funnel-and-gate configuration for treatment of a dissolved chlorinated solvent Volatile Organic Compounds (VOCs) plume at Moffett Airfield, California has been in progress since 1996. The project involves evaluation of the contaminant degradation as well as the associated inorganic geochemical and hydrogeologic issues. The hydrogeologic evaluation consists of using water-level measurements, down-hole velocity measurements, tracer tests, and groundwater flow modeling to determine the groundwater flow velocities and directions at this geologically heterogeneous site. The results of the investigations showed that the groundwater flow is moving through the permeable barrier at a reasonable flowrate. However, significant vertical and lateral complexities in flow exist, especially in the downgradient and upgradient pea gravel zones and in the aquifers.

INTRODUCTION

The use of permeable reactive barriers (PRB) for in-situ treatment of dissolved chlorinated solvents contamination in the groundwater is gaining increased popularity. A common version of this technology involves the use of granular iron reactive media and a funnel-and-gate configuration for reductive dechlorination of volatile organic compounds (VOCs) such as trichloroethene (Gavaskar et al., 1998). The objective of this paper is to verify the hydrologic behavior of a pilot-scale, funnel-and-gate PRB, installed in a regional plume of dissolved chlorinated solvents VOCs at Moffett Airfield, California in April 1996, by examining the groundwater velocities and flow patterns in the vicinity of the barrier. The performance evaluation consisted of groundwater flow modeling, down-hole groundwater velocity measurements, slug tests, quarterly and continuous water level monitoring, and two field tracer tests. Performance evaluation of the system is being conducted with a grant to the Naval Facilities Engineering Service Center (NFESC) from the Environmental Security Technology Certification Program (ESTCP). The detailed performance monitoring plan for the project is described in Battelle (1997a). The reactive performance and geochemistry are summarized in Sass et al. (1998).

* Present address: Gannett Fleming, Denver, Colorado

The Moffett PRB is about 22 feet deep and consists of a 10-foot-wide and 6-foot-long treatment zone with 2 feet of pea gravel along the upgradient and downgradient edges. The interlocking sheet-pile funnel walls are 20 feet long on either side. The soil at the PRB site has been characterized as a complex mixture of alluvial-fluvial clay, silt, sand, and gravel. The PRB is placed in a sand channel in the upper aquifer (A1) which is separated from the lower aquifer (A2) by a confining layer. To avoid breaching of the thin confining layer, the PRB is not keyed into it and there is a 1 to 2-ft-thick gap between the bottom of the PRB and the top of the confining layer which may result in some underflow.

GROUNDWATER MODELING

A three-dimensional, finite-difference model using MODFLOW was constructed to evaluate groundwater flow at the site (Battelle, 1996). Particle tracking was conducted using RWLK3D, a Battelle-developed contaminant transport and particle tracking code. Incorporation of the heterogeneities in the model resulted in a 7-layer model with the top 4 layers representing the upper aquifer and the PRB system. The simulated capture zones and flow paths for these layers (Figure 1) show that heterogeneities cause the capture zones and flow velocities to be variable. The particle movement in the upper two layers is very slow due to the clayey and silty soils at these depths. The regions upgradient of the barrier in layers 3 and 4 consist of the sand channel with high flow velocities. The simulated average particle velocity through the cell was determined to be about 3 ft/day with the groundwater flux through the gate at 2.3 gallons/minute.

FIELD TESTS

The hydrologic and chemical conditions were monitored at a quarterly interval from June 1996 to October 1997 using a network of 48 wells in the PRB and 49 wells in the surrounding aquifers. Water level measurements conducted at the site showed that at most times the groundwater through the PRB is moving as expected. However, there is some evidence that after heavy rainfall events, there is mounding in the PRB which may cause some backflow to occur over brief periods of time. In general there is a flattening in groundwater gradient in the PRB compared to the aquifer gradients.

Direct measurements of groundwater velocity in the wells were attempted in April 1997 to aid in the interpretation of hydraulics at the site. The down-hole velocity measurements were performed using the Geoflo Groundwater Flowmeter System (KVA Analytical Systems, Falmouth, MA). The observed flow velocities and directions (Figure 2) show that significant lateral and vertical variations in flow exist, especially in the pea gravel zones and the aquifer. The variations in the pea gravel zones are most likely created by sharp permeability contrasts between the pea gravel and the adjacent aquifers or reactive media and by mixing of flow. Within the reactive cell, the flow velocities ranged between 1.1 and 6.1 ft/day and the velocity and directions are within the range expected from design and modeling calculations.

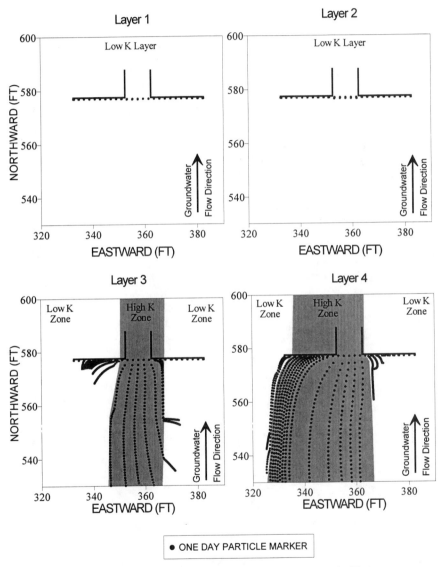

Figure 1. Backward Particle Tracking from Line Source in Four Model Layers in the Permeable Barrier Vicinity Showing Effects of Aquifer Heterogeneities on Capture Zones.

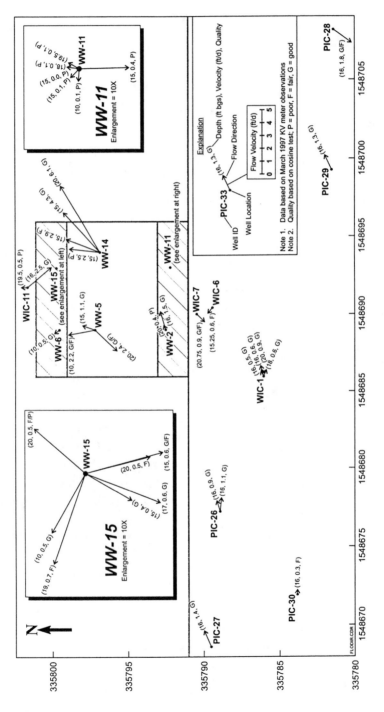

Figure 2.   Groundwater Flow Direction Test Results at Moffett Field

The tracer test involved injection of 12 liters of 3,000 mg/L bromide solution, a conservative tracer, in the upgradient pea gravel at a rate of 10 mL/min for two hours. The tracer was monitored with a combination of 17 specific ion electrodes with continuous data loggers and manual sampling. Rapid mixing and spread occurred within eight hours laterally within the upgradient pea gravel (Figure 3). Movement into the reactive cell was observed after about 4 days along a linear path. This tracer test confirmed that groundwater is flowing through the barrier. However, flow patterns and velocity variations are more complex than anticipated and the tracer appears to spread laterally in the pea gravel before moving into the reactive cell at a rate of about 0.5 ft/day. This is slower than was predicted from modeling or design calculations, from water level measurements, or from the velocity measurements. A second tracer test with injection in the upgradient aquifer was only partially successful due to scarcity of sampling points and showed that water is moving towards the PRB at a rate of about 0.5 ft/day.

SUMMARY

Hydraulic evaluation showed that groundwater is flowing through the reactive cell as expected. Within the range of uncertainty of each type of measurement – water levels, down-hole velocity probes, and tracer tests – the flow system in the reactive cell appears to match the design and model predicted behavior. Overall, complexity of the flow patterns developed in response to the permeable barrier placement points to need for more detailed hydraulic evaluations, modeling, and monitoring at these sites than is currently performed.

REFERENCES

Battelle. 1996. *Groundwater Modeling for Evaluation of a Permeable Barrier at Moffett Federal Airfield in Mountain View, California.* Prepared for NFESC. November.

Battelle. 1997a. *Performance Monitoring Plan for a Pilot-Scale Permeable Barrier at Moffett Federal Airfield in Mountain View, California.* Prepared for NFESC. July.

Battelle. 1997b. *Field Tracer Application to Evaluate the Hydraulic Performance of a Pilot-Scale Permeable Barrier at Moffett Federal Airfield.* Prepared for NFESC. October.

Battelle. 1998. *Field Tracer Application to Evaluate the Hydraulic Performance of a Pilot-Scale Permeable Barrier at Moffett Federal Airfield.* Prepared for NFESC. February.

Gavaskar, A. R., N. Gupta, B. M. Sass, R. J. Janosy, D. O'Sullivan. 1998. *Permeable Barriers for Groundwater Remediation: Design, Construction, and Monitoring.* Battelle Press, Columbus, OH.

Sass, B. M., A. R. Gavaskar, N. Gupta, W. Yoon, J. E. Hicks, D. O'Dwyer, and C. Reeter. 1998. *Evaluating the Moffett Field Permeable Barrier using Groundwater Monitoring and Geochemical Modeling.* The First International Conference on Remediation of Chlorinated and Recalcitrant Compounds. Monterey, California. May 18-21, 1998.

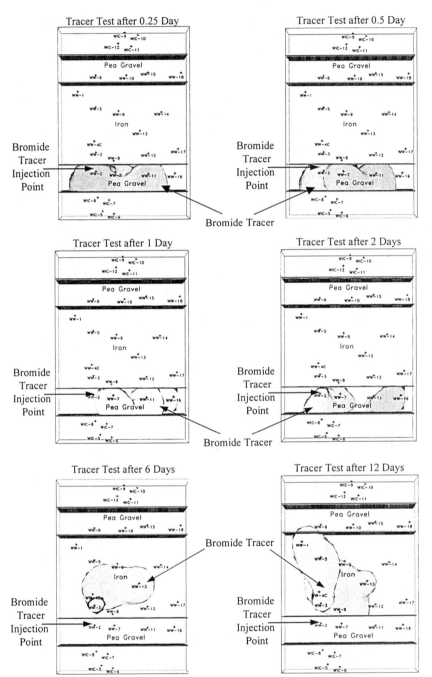

**Figure 3. Movement of Bromide Tracer Through Reactive Cell over a
Period of 12 Days Following Injection in Upgradient Pea Gravel.**

INORGANIC AND BIOLOGICAL EVALUATION OF CORES FROM PERMEABLE IRON REACTIVE BARRIERS

J.L. Vogan, EnviroMetal Technologies Inc., Guelph, Ontario, Canada
B.J. Butler, M.S. Odziemkowski, G. Friday and **R.W. Gillham**,
University of Waterloo, Waterloo, Ontario, Canada

ABSTRACT: Cores from two pilot-scale permeable iron reactive barriers were examined for mineral precipitates and indications of microbial activity. 100% iron was the treatment medium at both sites. Calcium carbonate precipitates predominated in the cores, however a variety of iron oxyhydroxide minerals, green rusts and magnetite were also identified. Estimated porosity losses due to carbonate precipitates of 6% over the first 0.3 m of iron were estimated at a Colorado site after 18 months of operation. At a New York site, after 2 years, the estimated porosity loss in the first few cm of iron was about 10%, declining sharply over the first 0.3 m to below 2%. Given the original porosity of these systems (about 0.5), groundwater flow patterns have not likely been affected. Lipid biomass analyses and microbial enumerations using both aerobic and anaerobic growth media were performed on core samples. Lipid biomass results in the order of 10^6 cells/g indicated microbial transport into the iron zone from the aquifer. Microbial enumerations, for the most part, did not exceed 10^4 CFU/g after 28 days, and several results were below detection limits. The results of the core analyses suggest that porosity loss due to mineral precipitates may contribute to periodic maintenance requirements of permeable barriers, rather than biofouling. The evidence suggests that most systems will perform for several years prior to requiring any maintenance.

INTRODUCTION

During the past few years, permeable reactive barriers have become a recognized option for the in-situ treatment of a variety of dissolved organic contaminants in groundwater (Gillham, 1996). Over twenty pilot-scale and full-scale permeable iron barriers have been installed since early 1995 for treatment of chlorinated aliphatic compounds such as trichloroethene, 1,1,1-trichloroethane, dichloroethene isomers and vinyl chloride. These systems have all been successful in meeting regulatory discharge criteria; however, their longevity remains in question. The concern revolves around two issues, the influence of mineral precipitates on iron reactivity and system hydraulics, and the potential for biofouling as a result of in-situ microbial activity.

In order to provide data to better evaluate these issues, cores were obtained from the iron zone of two pilot-scale permeable barriers which have been in operation for some time. These pilot-scale systems, located at a U.S. Air Force facility in Colorado and a private facility in New York, are similar in design. At both sites, 15 foot sheet pile wings (funnel sections) direct groundwater to a central treatment gate about 10 feet wide. Pea gravel on the upgradient and downgradient ends separates the 100% granular iron zone from the aquifer material. Both systems have been successful in degrading VOCs, as reported in Gallant (1997) and Focht et. al., (1996). Cores were collected from the Colorado site about 18 months after installation, and from the New York site about 24 months after installation. A variety of laboratory tests were performed on subsamples of the cores to investigate the effects of mineral precipitates and microbial activity on the iron materials.

Materials and Methods. The different core samples obtained at the New York and Colorado sites necessitated a slightly different approach to subdividing the core for laboratory analyses. Previous studies have indicated that significant inorganic changes in aqueous chemistry occur near the upgradient interface of in-situ granular iron walls. Consequently, efforts were made to collect continuous core samples that passed from the upgradient interface into the iron zone. At the New York site, cores were obtained with a drive point/piston sampler (Starr and Ingleton, 1992). A complete continuous angle core (Figure 1) was obtained from the interface, and subdivided into 5 cm sections for analysis. At the Colorado site, attempts to collect continuous angled cores using Geoprobe™ equipment were less successful, and therefore analyses at this site were completed on vertical cores obtained at various locations within the wall (Figure 1). In this case the cores were sectioned at 60 cm intervals and each section was homogenized to give a composite sample.

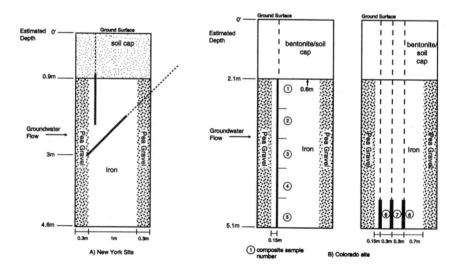

Figure 1: Core Sample Locations

Core samples were analyzed by the following methods:

(i)   total carbonate, using a method modified from Bundy and Bremner (1972). This method measures the CO_2 gas evolved when the sample is treated with a strong acid;

(ii) examination of surface coatings on the iron particles using Scanning Electron Microscopy (SEM), Raman micro-spectroscopy, and electron dispersive x-ray (EDX) techniques;

(iii) determination of lipid biomass using the method of Dobbs and Findlay (1991);

(iv) microbial enumerations (deep tube or plate counts), both anaerobically using anaerobic agar (Gibco Diagnostics) and aerobically using an R2A™ growth medium (Reasoner and Geldreich, 1985).

RESULTS AND DISCUSSION

Carbonate Content: Figure 2 shows the carbonate content versus distance for the New York site. The maximum carbonate content was 5.6% $CaCO_3$ (5.6 g/100 g solid), obtained from the sample nearest the interface. The profile shows a rapid decline in carbonate with distance with values of less than 1% beyond a distance of 15 cm. Based on the mass of iron placed in the gate during installation, the initial porosity of the material could be as high as 0.59. The percent carbonate content can be converted to a corresponding porosity loss by using the molar volume of calcium carbonate (36.8 cm^3/mol). Using an iron field bulk density of 180 lb/ft^3 (2.88 g/cm^3), 5.6% $CaCO_3$ represents 0.0016 mol and therefore would occupy 0.059 cm^3 per cm^3 of material. This represents a porosity loss of 10%, assuming an original porosity of 0.59.

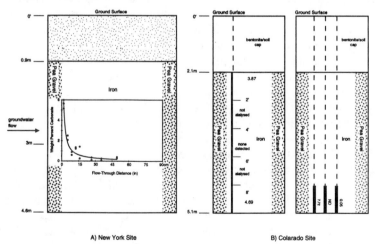

4.69 = weight percent carbonate

Figure 2: Carbonate Analysis Results

At the Colorado site, the maximum carbonate content was also detected in the cores closest to the upgradient interface, though, as indicated in Figure 2, samples from the midsection of the core showed little detectable carbonate. This suggests that little or no flow had occurred in this zone since the treatment system was installed. With the above noted exception, about 4% carbonate reported as $CaCO_3$ (4 g/100 g solid) was present in core samples nearest the upgradient interface. At this site, based on the mass of iron placed in the gate during installation, the initial porosity could be as high as 0.63. A carbonate content of 4% therefore represents a porosity loss of 6.2% assuming an original porosity of 0.63.

Raman Spectroscopy, SEM, and EDX Results: Raman spectroscopy, SEM, and to a lesser extent EDX, confirmed the results of carbonate analyses. Examination of samples near the upgradient interface showed discolouration and the presence of calcite and aragonite, with the exception of samples from the middle of core RW1 from the Colorado site. The colour of this sample approached that of fresh iron, and carbonates were not detected. Samples from sections of cores further downgradient also showed less carbonate precipitates. Several randomly distributed iron oxides and oxyhydroxides were also detected. Samples from sections of cores further downgradient showed less areas covered with Fe_2O_3 or FeOOH. Downgradient the predominant species were green rust compounds and magnetite. Magnetite or green-rusts compounds might form due to autoreduction of native Fe(III) films such as $a,y - Fe_2O_3$ (Odziemkowski and Gillham, 1997) or by precipitation from solution (Odziemkowski et al., 1998). Calcite and aragonite were the predominant carbonate species observed, with only minor amounts of siderite ($FeCO_3$). Forms of "green rust" were observed on iron samples from both sites. Green rusts are complex iron hydroxides containing both Cl^-, SO_4^{-2} and carbonate (CO_3^{-2}). No sulphide precipitates were observed in the core samples; however, declines in aqueous sulphate concentration at both sites indicate that sulphide or sulphate containing minerals may be forming. The molar volume of sulphide and hydroxide minerals are smaller than that of carbonate, so their effects on porosity loss would be less significant for a given mass of precipitate formed relative to carbonate mineral formation. From the results obtained in this study, it does not appear that oxide and/or sulphide formation is having an appreciable affect on the porosity of the system.

RESULTS OF MICROBIAL ANALYSES

The microbial enumerations and lipid biomass results show similar trends. Lipid biomass results from both sites were in the order of 10^6 cells/g dry weight. Microbial enumerations in both the aerobic and anaerobic cultures ranged from less than the detection limit to 10^4 Colony Forming Units (CFUs)/g wet weight after 28 days of growth. Samples closest to the upgradient interface show slightly higher populations than samples taken from within the wall. No microbial films were noted on the samples during microscopic examination.

It is suspected that the results reflect transport of microbial populations into the zone from the upgradient aquifer. The high pH, low Eh environment established in the iron zone may be relatively hostile to many types of bacteria (Major, 1997). Microbial populations in the order of 10^6/g are common in many aquifer and soil environments. Lipid biomass analyses performed on groundwater samples from the New York pilot after 6 months of operation showed microbial populations in the range of 10^3 to 10^4 cells/g; the higher values in core samples may reflect the commonly observed increase in biomass measurements in solid samples relative to groundwater samples.

CONCLUSIONS

Based on the analyses of core samples it appears that mineral precipitates represent the primary maintenance issue when dealing with the long-term performance of permeable barriers. At each of the sites examined, carbonate precipitates have apparently caused about a 6% to 10% loss of porosity in the upgradient few cm of the iron zone. These carbonate precipitates are mainly calcite and aragonite, with only minor evidence of siderite. The remainder of the wall appears unaffected by carbonate precipitate formation. There is no evidence of significant microbial populations, suggesting that biofouling is not an issue in long term performance. Based on these results, there is no reason to suspect that systems at these sites would not continue to perform for several more years prior to rejuvenation/refurbishing.

ACKNOWLEDGMENTS

The authors would like to acknowledge the Air Force Center for Environmental Excellence (AFCEE), Brooks AFB, and the private owner of the New York installation for providing access and funding for this evaluation. The assistance of Versar Inc. and Stearns and Wheler Inc., in collecting the core samples is also greatly appreciated.

REFERENCES

Bundy, L.G. and J.M. Bremner. 1972. "A Simple Titrimetic Method for Determination of Inorganic Carbon in Soils". *Soil Soc. Amer. Proc.* 36: 273-275.

Dobbs, F.C. and R.H. Findlay. 1991. "Analysis of Microbial Lipids to Determine Biomass and Detect the Response of Sedimentary Microcosms to Disturbance. Handbook of Microbial Ecology". Lewis Pubishers.

Focht, R.L., J.L. Vogan, and S.F. O'Hannesin. 1976. "Field Application of Reactive Iron Walls for In-Situ Degradation of Volatile Organic Compounds in Groundwater". *Remediation; Summer 1996. pp 81-94.*

Gallant, W.G. 1997. *Zero Valence Metal Reactive Wall Demonstration Project Results and Lessons Learned.* Presented at HazWaste World, Superfund XVIII Conference, Washington, D.C., Dec. 2-4.

Gillham. 1996. "In-Situ Treatment of Groundwater: Metal Enhanced Degradation of Chlorinated Organic Contaminants". *In Advances in Groundwater Pollution Control and Remediation* 249-274. Kluwer Academic Publishers.

Odziemkowski M.S., and R.W. Gillham. 1997. 213th ACS National Meeting, San Francisco, Ca. Extended Abstracts, Division of Environmental Chemistry, Vol.37, p.177.

Odziemkowski M.S., T.T. Schuhmacher, R.W. Gillham, and E.J. Reardon. 1998. Corrosion Science, in print.

Reasoner, D.J. and E.E. Geldreich. 1985. "A New Medium for the Enumeration and Subculture of Bacteria from Potable Water". *Appl. Environ. Microbiol.* 49:1-7.

Starr, R.C. and R.A. Ingleton. 1992. "A New Method for Collecting Core Samples Without a Drilling Rig". *Groundwater Monitoring Review*, Winter 1992 pp 91-95.

EVALUATING THE MOFFETT FIELD PERMEABLE BARRIER USING GROUNDWATER MONITORING AND GEOCHEMICAL MODELING

Bruce M. Sass, Arun R. Gavaskar, Neeraj Gupta, Woong-Sang Yoon,
and James E. Hicks (Battelle, Columbus, Ohio, USA)
Deirdre O'Dwyer (Tetra Tech EM Inc., Denver, Colorado, USA)
Charles Reeter (Naval Facilities Engineering Service Center,
Port Hueneme, California, USA)

Abstract: The pilot-scale permeable barrier at Moffett Field uses a funnel-and-gate design, where the funnel is made of interlocked sheet piles and the gate consists of a reactive cell with 100% granular iron. The results of groundwater monitoring, geochemical modeling, and core sample analysis are presented here in an effort to demonstrate the effectiveness of the barrier in remediating groundwater containing dissolved chlorinated compounds. Concentrations of inorganic species have been monitored as well for five consecutive quarters. These data have been used in a geochemical model to assess the possibility of precipitation in the reactive medium. Precipitation in the pore space of the reactive medium decreases the efficiency of the PRB by decreasing flow through the gate and decreasing residence time within the reactive cell. Results of analysis of core samples show little direct evidence for any kind of fouling of the reactive barrier at this time.

INTRODUCTION

The purpose of this study was to evaluate the performance of a pilot-scale permeable reactive barrier (PRB) that was constructed at the former Naval Air Station (NAS) Moffett Field in April 1996. The U.S. Navy's Engineering Field Activity West and the Naval Facilities Engineering Service Center funded research at Moffett Field to investigate alternative technologies that have potential technical and cost advantages over conventional pump-and-treat systems. The Moffett Field PRB is situated within a large regional plume of dissolved chlorinated volatile organic compounds (cVOCs). Historically, trichloroethene (TCE) and *cis*-1,2-dichloroethene (*cis*-1,2-DCE) are the predominant groundwater contaminants in the vicinity of the PRB. TCE is the most abundant contaminant in the groundwater, typically 2,000 µg/L, while concentrations of *cis*-1,2-DCE typically range from approximately to 200 to 300 µg/L. Perchloroethene (PCE), which also exists in the plume, is present in low concentrations near the PRB (10 to 20 µg/L).

The Moffett Field PRB is a funnel-and-gate system that consists of a reactive cell with adjoining pea gravel sections in the upgradient and downgradient flow directions (Figure 1). The pea gravel homogenizes heterogeneous flow from the aquifer before it passes into the reactive cell. The reactive cell contains 100% granular zero-valent iron (Peerless Metal Powders, Inc.) and is 6 ft thick, 10 ft wide, and 22 ft deep. The overall funnel-and-gate system is 50 ft wide. The wings of the funnel are interlocked sheet piles with grouted joints. The pilot barrier penetrates only the uppermost aquifer zone (designated A1), although contamination is present at deeper levels. Soil at the permeable barrier site has been characterized as a complex mixture of alluvial-fluvial clay, silt, sand, and gravel.

PERFORMANCE MONITORING ACTIVITIES

Monitoring hydrologic and chemical conditions in and around the PRB was conducted quarterly from June 1996 to October 1997, to determine whether the barrier was functioning as designed. Two tracer tests using bromide injections were completed in April and August 1997 at the site to verify the capture of the targeted groundwater, monitor its progress through the reactive cell, and determine residence time. Results of the tracer tests are reported elsewhere (Gupta et al., 1998). Several core samples were extracted from the reactive cell and adjacent aquifer in December 1997 to look for signs of iron encrustation, precipitate formation, and microbial population growth. These factors decrease the efficiency of the PRB by decreasing flow through the gate and decreasing residence time within the reactive cell. They also affect the longevity of the barrier and hence the operating costs.

The monitoring well locations are shown in Figure 2. All of the wells in the PRB and some wells in the aquifer were pre-installed during construction. Additional wells were installed in the aquifer during performance monitoring. Water samples from these monitoring wells provided essential information on water movement, organic contaminant levels, and inorganic chemistry needed to understand and model the performance of the

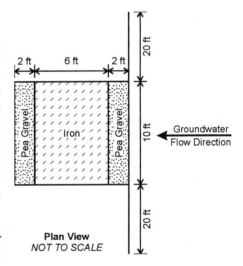

Plan View
NOT TO SCALE

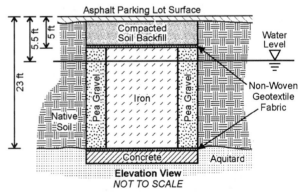

Elevation View
NOT TO SCALE

Figure 1. Plan and Elevation Views of Moffett Field Funnel-and-Gate System

permeable barrier. Samples were collected and prepared for laboratory chemical analysis; field parameters were analyzed on site for dissolved oxygen (DO), pH, redox potential, and temperature. The cVOCs were measured according to EPA Method 8260. Teflon™ tubing of ¼-inch outside diameter was used to sample each multilevel monitoring well. Micropurging and sampling at low pumping rates (typically 20-40 mL per min) were employed to minimize the disturbance to flow gradients in the PRB. Analysis of VOCs helped identify the distribution of contaminants in and around the permeable barrier, as well as potential byproducts of degradation.

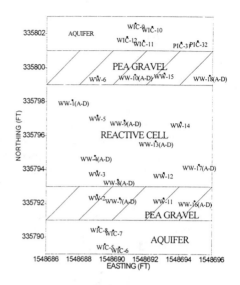

Figure 2. Map Showing Location of Monitoring Wells

Geochemical modeling was used to evaluate the possibility of precipitation due to reactions of inorganic groundwater constituents. Precipitation can directly affect the reactivity and hydraulic performance of the reactive cell. The U.S. Geological Survey code, PHREEQC (Parkhurst, 1995), was used to calculate saturation levels of possible precipitates based on measured inorganic and field parameters.

RESULTS AND CONCLUSIONS

Results of laboratory and field data were compiled after each quarterly monitoring event. Table 1 lists several parameters for selected monitoring wells from a sampling event in January 1997. The following paragraphs describe trends that have been observed within the past year.

Chlorinated VOCs. Concentrations of cVOCs range from approximately 1,000 to 2,000 µg/L in the upgradient A1 aquifer zone and adjoining pea gravel to < 0.5 µg/L in the reactive cell (see Table 1). TCE remains at 10 µg/L or below in the downgradient pea gravel. In the aquifer well cluster immediately downgradient of the cell, TCE concentrations range from 100 to 500 µg/L above the A1/A2 aquitard (WIC-9,-10,-11) and exceeds 3,000 µg/L below the A1/A2 aquitard (WIC-12). These results are consistent with an observed trend in decreasing aquifer contamination downgradient of the barrier over the course of this performance monitoring study. It appears that decontaminated water exiting the barrier is slowly causing desorption of contaminants in the soil. High concentrations of contaminants below the gap level may be caused by upward migration of groundwater from the more highly contaminated A2 aquifer zone.

The distribution of *cis*-1,2-DCE follows a similar trend, except that *cis*-1,2-DCE persists over a longer distance in the reactive cell, probably due to its longer degradation half life.

Field Parameters. Eh values are positive in all of the aquifer wells and generally negative in the reactive cell wells. Similarly, pH values are close to neutral in the aquifer wells and are alkaline (pH ~ 10 to 11) in the reactive cell wells. A decrease in Eh and an increase in pH are expected trends in the reactive cell due to chemical reactions involving zero-valent iron (Battelle, 1997). In the downgradient pea gravel and aquifer, Eh values increase somewhat and pH values decrease. This may result from mixing of treated and untreated water in the downgradient pea gravel. DO concentrations are generally <0.1 mg/L everywhere within the PRB and in the aquifer.

Inorganic Groundwater Constituents. The predominant ions in the A1 aquifer zone groundwater are calcium, magnesium, sodium, sulfate, bicarbonate, and chloride. The data in Table 1 indicate that the concentrations of major inorganic species decrease in the reactive cell. The decline in non-redox sensitive species

**TABLE 1. RESULTS OF GROUNDWATER ANALYSIS
AT MOFFETT FIELD IN JANUARY 1997**

Well ID	TCE	*cis*-1,2-DCE	Eh	pH	Alkalinity	Ca	Mg	Sulfate
	µg/L		mV	S.U.	mg/L			
Upgradient A1 Aquifer Zone Wells								
WIC-1	~1600	230	207	6.12	377	165	63.7	346
Upgradient Pea Gravel Wells								
WW-2	1400	240	98	7.04	407	158	62.4	365
WW-7C	750	210	NA	NA	329	157	63.1	335
WW-7D	NA	66	124	7.49	106	7.0	49.9	142
Reactive Cell Wells								
WW-3	11	220	-108	9.47	65	5.5	52.4	210
WW-4C	NA	74	-129	8.30	73	4.1	43.2	133
WW-4D	1400	280	-125	8.47	408	154	61.3	363
WW-5	NA	NA	-112	8.59	23	8.6	26.4	136
WW-8C	7	110	NA	NA	82	5.4	52.8	170
WW-8D	2	44	-167	10.44	50	16	29.4	154
WW-9C	NA	NA	-176	9.46	< 10	5.7	30.9	153
WW-9D	NA	3	-228	9.10	102	110	4.6	274
Downgradient Pea Gravel Wells								
WW-6	3	NA	94	7.00	< 10	63	16.5	314
WW-10C	3	NA	187	7.27	< 10	53	10.9	214
WW-10D	4	NA	-63	7.39	< 10	53	12.3	222
Downgradient A1 Aquifer Wells								
WIC-9	800	91	207	7.00	172	73	24.0	183
WIC-10	190	5	307	8.09	13	56	6.8	203
WIC-11	420	46	197	7.00	609	111	< 10	125
WIC-12	3200	NA	194	7.14	243	125	42.1	321

can be explained by precipitation of calcite or aragonite, based on geochemical modeling results described below. Redox sensitive ions such as nitrate and sulfate appear to be degraded along the flow path. Nitrate concentrations decrease from about 1 mg/L in the A1 aquifer zone to below detection in the initial part of the reactive cell. Sulfate concentrations decrease from approximately 300 to 400 mg/L in the upgradient wells to less than 200 mg/L in the reactive cell. Geochemical modeling predicts that reducing conditions bring about reduction of sulfate to sulfide, leading to precipitation of ferrous sulfide. Water samples were analyzed for sulfide, but none was detected.

Geochemical Modeling. Table 2 contains results of calculations of mineral saturation indices calculated from the data in Table 1. The mineral saturation index (SI) is defined by SI = log (IAP/K), where IAP is ion activity product and K is the thermodynamic equilibrium constant for a particular mineralogical reaction. When SI = 0, the mineral and groundwater are considered to be in equilibrium; negative values imply undersaturation of the mineral phase and positive values imply oversaturation. In practice, mineral equilibrium may be assumed when SI = $\pm$ 0.2.

TABLE 2. RESULTS OF PHREEQC CALCULATION OF GROUNDWATER SATURATION INDICES[a]

Well ID	Aragonite	Brucite	Calcite	Dolomite	Fe(OH)3	Goethite	Gypsum	Siderite
Upgradient A1 Aquifer Zone Wells								
WIC-1.	-1.14	-7.85	-1.00	-2.13	-5.83	**0.01**	-0.93	-2.22
Upgradient Pea Gravel Wells								
WW-2.	**-0.21**	-5.98	-0.06	-0.24	-5.49	**0.36**	-0.92	-1.85
WW-7C	-0.31	-5.93	-0.16	-0.41	**1.23**	**7.10**	-0.96	-1.83
WW-7D.	-1.55	-5.00	-1.41	-1.66	-3.76	**2.10**	-2.47	-2.03
Reactive Cell Wells								
WW-3	**-0.17**	-1.09	-0.02	**1.23**	-0.85	**5.01**	-2.44	**0.41**
WW-4C	-1.15	-3.47	-1.01	-0.69	-5.25	**0.61**	-2.70	-1.04
WW-4D	**1.17**	-3.08	**1.31**	**2.53**	-5.23	**0.63**	-0.94	-0.73
WW-5	-1.05	-3.11	-0.90	-1.02	-4.49	**1.37**	-2.33	-1.65
WW-8C	-0.37	-1.94	-0.22	**0.86**	-0.66	**5.21**	-2.52	**0.01**
WW-8D	**0.29**	**0.61**	0.43	**1.41**	-0.38	**5.48**	-2.04	-1.08
WW-9C	-1.02	-1.37	-0.87	-0.73	-2.71	**3.14**	-2.47	-1.16
WW-9D	**1.02**	-2.88	**1.17**	**1.26**	-5.27	**0.59**	-1.07	-0.98
Downgradient Pea Gravel Zone								
WW-10C	-1.96	-6.23	-1.82	-4.04	-2.79	**3.05**	-1.41	-2.70
WW-10D	-1.83	-5.86	-1.69	-3.71	-7.23	-1.37	-1.40	-3.16
Downgradient A1 Aquifer Wells								
WIC-3	-0.31	-6.12	**-0.16**	-0.47	-2.81	**3.04**	-0.94	-1.33
WIC-9	-0.87	-6.42	-0.72	-1.64	-3.99	**1.86**	-1.38	-2.46
WIC-10	-1.02	-4.80	-0.88	-2.39	**0.68**	**6.53**	-1.40	-2.80
WIC-12	-0.40	-5.90	-0.25	-0.68	-3.58	**2.28**	-1.02	-1.98

(a) Based on January 1997 sampling data. Bold values indicate that the water is close to saturation or oversaturated with respect to the referenced mineral.

The data in Table 2 indicate that saturation indices vary spatially for most minerals in the permeable barrier. Exceptions include the carbonate minerals, calcite, aragonite ($CaCO_3$), and dolomite [$MgCa(CO_3)_2$], which are close to equilibrium at all locations. Because calcium and alkalinity decline in the reactive cell and saturation indices for calcite and aragonite remain relatively constant, suggests that calcium carbonate may be precipitating in the reactive cell. The pore water chemistry is close to equilibrium with respect to dolomite in the upgradient pea gravel, then becomes oversaturated in the reactive cell, and undersaturated in the downgradient pea gravel. Oversaturation with respect to dolomite is not uncommon because precipitation kinetics are very slow. Siderite ($FeCO_3$), a ferrous carbonate mineral, is below saturation throughout most of the permeable barrier, but its saturation index is close to zero in some of the reactive cell wells, indicating that siderite may be precipitating. Ferric hydroxides, goethite ($FeOOH$) and $Fe(OH)_3$, seem to be near saturation. However, these species are not likely to be major precipitates because ferric ion concentrations would be exceedingly low under such strongly reducing conditions. Sulfate minerals (gypsum, anhydrite, and melanterite) are undersaturated at all locations. This suggests that the decline in sulfate levels in the reactive cell is not due to precipitation of sulfate minerals. A more likely explanation is that some of the sulfate is reduced to sulfide, due to the low redox potential, and precipitates as iron sulfide. Because ferrous sulfide has very low solubility may explain why sulfide ion could not be detected and the concentration of dissolved iron is very low (typically less than 0.1 mg/L).

Core Samples. Analysis of twelve core samples from the reactive cell showed that Ca concentrations range from approximately 6 to 260 mg/kg of iron and Mg concentrations range from approximately 25 to 250 mg/kg of iron. Unused iron from construction of the barrier contained much lower concentrations of Ca and Mg: 2.9 and 0.6 mg/kg, respectively. These results indicate that precipitation of Ca and Mg compounds has occurred in the reactive cell, but that the concentrations are relatively small (< 0.1%) and probably do not impact flow in the PRB.

Results of Raman spectroscopy, scanning electron microscopy, and X-ray diffraction analyses indicate that corrosion coatings predominantly consist of hematite (α-Fe_2O_3) and magnetite (Fe_3O_4). These species were present in all core samples and in the unused iron. Minor amounts of aragonite and marcasite (FeS_2) also were found. These minor species were not observed in the unused iron and thus must have formed during operation of the PRB. Although the analyses are qualitative, the abundance of aragonite and marcasite appears to be small.

Three core samples from the reactive cell were analyzed under anaerobic conditions for heterotrophic plate counts and GC-FAME (gas chromatograph-fatty acid methyl ester) of microbial strains. None of the iron samples showed any measurable colony forming units (CFU) after 48 hours of incubation. These results indicate that biofouling at the Moffett Field PRB is not presently a matter of concern.

Results of geochemical modeling and analysis of the core samples indicate early signs of iron-groundwater interactions. The processes involved potentially include oxidation of iron, precipitation of carbonate minerals and ferrous sulfide, and promotion of anaerobic microbial growth. Using these results as a baseline, coring could be repeated after five years of barrier operation to obtain further confirmation of such processes and their effect on barrier performance.

REFERENCES

Battelle. 1997. *Performance Monitoring Plan for a Pilot-Scale Permeable Barrier at Moffett Federal Airfield in Mountain View, California.* July.

Gupta, N., B.M. Sass, A.R. Gavaskar, J.R. Sminchak, T. Fox, F.A. Snyder, D. O'Dwyer, and C. Reeter. 1998. *Hydraulic Evaluation of a Permeable Barrier using Tracer Tests, Velocity Measurements, and Modeling.* The First International Conference on Remediation of Chlorinated and Recalcitrant Compounds. Monterey, California. May 18-21, 1998.

Parkhurst, D.L. 1995. *Users Guide to PHREEQC–A Computer Program for Speciation, Reaction-Path, Advective-Transport, and Inverse Geochemical Calculation.* Water-Resources Investigations Report 95-4227. U.S. Geological Survey.

DEGRADATION OF PESTICIDES BY FREE *Pseudomonas fluorescens* CELL CULTURES.

Erick R. Bandala, Juan A. Octaviano
(Instituto Mexicano de Tecnología del Agua. Mexico),
Veronica Albiter, and Luis G. Torres (Instituto de Ingeniería/UNAM. Mexico)

ABSTRACT: Degradation of four pesticides (heptachlor, aldrin, dieldrin and heptachlor epoxide) was tested using free cultures of *Pseudomonas fluorescens* under controled conditions. Pesticide concentrations were monitored by gas chromatography during 120 hours. Percentage of degradation and biodegradation rates (BDR) were calculated. Data show a trend suggesting a relation between chemical structure and degradability. Degradation kinetics for each pesticide tested showed that the higher degradation rates were found in the first 24 hours. GC/MS analysis of the final extracts allows the identification of chlordene and monochlorodieldrin which had been reported as final metabolite produced in the biodegradation of this kind of compounds.

INTRODUCTION

The application of synthetic pesticides to control weeds, insect pest, and fungi diseases has been routinely in agriculture for the past half century (Lawrence *et al.*, 1996). It had been showed that residues of this xenobiotics are the precursors of many health effects and that many kinds of pesticides have been used in indiscriminated way in the past (Bandala *et al.*, 1995).

Due to their inherent persistence and bioaccumulation, pesticides have become a serious environmental issue in the last decades (Nair *et al.*, 1996). Many papers dealing with the ubiquity of these chemicals in the environment have been published in order to demonstrate the risk involved on their irrational use (Sitarska *et al.*, 1995). Despite this undesirable characteristics, developing countries contribute with more than 70 percent of the total pesticide consumption (Nair *et al.*, 1991).

In Mexico, presence of pesticide in environmental samples have been detected in many reservoirs all around the country at Nazas, Lerma, Conchos, Colorado, Coatzacoalcos, Pánuco and San Juan rivers. In the same way, pesticides have been detected in underground water from the Yaqui and Culiacán Valleys, as well as in human milk from women in the Yaqui tribe (Bandala *et al.*, 1995).

Although there are many reports dealing with the removal of pesticides from water and other environmental matrix, relatively few studies had been reported about the removal of pesticides using biological processes.

Objective. The aim of this paper is to demonstrate the feasibility of *Pseudomonas fluorescens* free cell cultures in the biodegradation of pesticides in water.

MATERIALS AND METHODS

Pesticide solution preparation. Two pesticides and their major metabolites: aldrin, dieldrin, heptachlor, and heptachlor epoxide were chosen for this study mainly because of their significative presence in superficial waters from Mexico. Pesticide standard solutions were prepared as stock solutions in methanol and spiked in the cell cultures in order to obtain the work concentrations.

Bacteria culture. The bacteria *Pseudomonas fluorescens* was grown in liquid YPG (in mg/L: yeast extract, 10; peptone of casein, 10, and glucose 10) media for 24 hours at 35° C. The minimum medium employed was that reported by Dapaah and Hill (1992). Media were adjusted at pH=7.0 and cell growth was registered as the increment on optical density (O.D.) at 610 nm (1:10 dilution). This strain was previously used in the biodegradation of chlorophenols and different chlorophenols mixtures (Torres *et al.*, 1997).

Biodegradation test. The biodegradation of pesticides by *Pseudomonas fluorescens* free cells was evaluated as follows. Sterile minimum media (45 ml) were inoculated with 5 ml of a 24 hours YPG culture with a fixed biomass content (optical density *ca.* 0.20). The culture was grown at 32° C in a shaker. After 24 hours, the optical density was evaluated and pesticide solution was added until reaching 10 mg/L. Pesticide concentrations were monitored during 120 hours by gas chromatography.

Analytical methods. Pesticide concentrations were monitored as follows. The sample extractions were carried out using methilene chloride nanograde. The organic phases were dried with anhydride sodium sulfate. Once dried, the extracts were concentrated until reaching 0.5 ml volume at reduced pressure and temperature not higher than 40° C. Approximately one microliter of the concentrate sample was injected to a gas chromatograph Hewlett-Packard 5890 series II coupled with an electron capture detector. A 25 m x 0.2 mm x 0.33 μm capillary Ultra 2 column (5% methylphenyl polisiloxane) was employed. The operational conditions were as follows: carrier flow, 80 mL/min; split/splitless ratio, 1:25; initial temperature, 70° C; final temperature, 250° C; increment, 15° C/min.

End products determination. The end products determination was evaluated using a gas chromatograph Hewlett-Packard 5890 series II coupled with a mass spectrometer Hewlett-Packard 5971 series. Operational conditions and sample extraction techniques are described above.

RESULTS AND DISCUSSION

The biodegradation typical profiles for aldrin and dieldrin are shown on figures 1a and 1b. As shown, degradation kinetics for each pesticide tested showed that the higher degradation rates were found in the first 24 hours. Error bars indicate the reproducibility of the experiment, which is fair. Profiles for the heptachlor and heptachlor epoxide are very similar (figure not shown). The final pesticide concentrations were as low as 0.51, 2.26, 0.30, and 1.66 mg/L for the aldrin, dieldrin, heptachlor and heptachlor epoxide experiments, respectively.

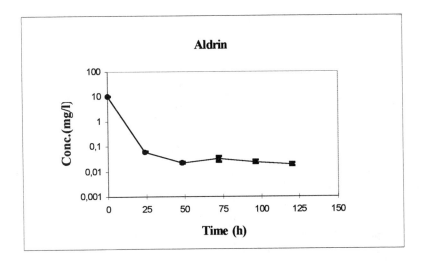

FIGURE 1a. Degradation kinetics for aldrin.

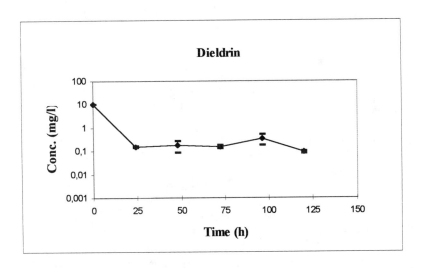

FIGURE 1b. Degradation kinetics for dieldrin.

The profiles are very similar to those reported by Lee *et al.*, 1997, in the biodegradation of antrazine and EPTC by a fixed biofilm sequencing batch reactor inoculated with *Agrobacterium radiobacter* J14a. For most of the cases, BDR values describe accurately the behavior of the pesticides in the first 24 hours.

Percentage of degradation and biodegradation rates (BDR) for each tested pesticide are shown on table 1.

Table 1. Percentage of degradation and biodegradation rates (BDR) for the tested pesticides.

Pesticide	Initial Conc. (mg/L)	Final Conc. (mg/L)	Percentage of degradation	BDR kg/m^3.d ($\times 10^{-3}$)
Aldrin	10	0.51	94.8	77.5
Dieldrin	10	2.26	77.3	13.29
Heptachlor	10	0.30	96.9	41.4
Heptachlor epoxide	10	1.66	83.4	18.1

Significative differences between chemical structure for pesticides and their metabolites can be observed in figure 2. The introduction of an epoxide group in the chemical structure produced a reduction in the degradation percentage up to 86% for the heptachlor-heptachlor epoxide pair, and 81% for the aldrin-dieldrin pair. This behavior could be explained by the increase in the oxidation state and its relation with the toxicity for the tested compounds. Apparently, toxicity of the pesticides is related to its oxidation state. The higher oxidation state, the higher toxicity. Thus, conversion of aldrin (LD50=70 mg/kg) to its high oxidized epoxide dieldrin (LD50= 38.5 mg/kg) reduced the LD50 value in 55%. In the same way, converting heptachlor (LD50= 100mg/kg) to its epoxide (LD50= 47 mg/kg) reduced in 47% the LD50 value. The increase in the toxicity is reflected in the decrease of the biodegradability. On each pesticide-metabolite pair, the compound with the lower toxicity is also the most easily biodegraded one. In the heptachlor-heptachlor epoxide pair, the first one, with the lower toxicity, is easier and faster to degrade (97% of degradation and BDR equal to 41.4 g/m^3.d) than heptachlor epoxide (83.4% of degradation and 18.1 g/m^3.d of BDR value). In the same way, for the aldrin-dieldrin pair, dieldrin, with the higher toxicity, has the lower percentage of degradation, and BDR value (77% and 13 g/m^3.d respectively).

This trend was also observed in the whole set of pesticides. Heptachlor, with the lowest toxicity, is more easily degraded than aldrin. Aldrin is less toxic than heptachlor epoxide and, therefore, easier to degrade. Finally, dieldrin, with the highest toxicity of the set, has the lowest degradability rate of the whole pesticide set tested.

Attempts to observe this trend in BDR values is easy for each pesticide-metabolite pair, but failed for the whole pesticide set. In the first case, the higher percent of degradation value, the higher BDR value. Nevertheless, in the second case, not always high biodegradability means high rates of degradation: heptachlor, with a higher percent of degradation (96.9%) showed a lower BDR value (41.4 g/m^3.d) than aldrin (77.5 g/m^3.d), with a lower percentage of biodegradation (95%). In this particular case, maybe thermodynamic aspects such as substrate-enzyme complex formation rate, or inhibitor/catalytic effects caused by products or structural effect, such as steric hindrance or solubility changes should be considered in order to explain this trend.

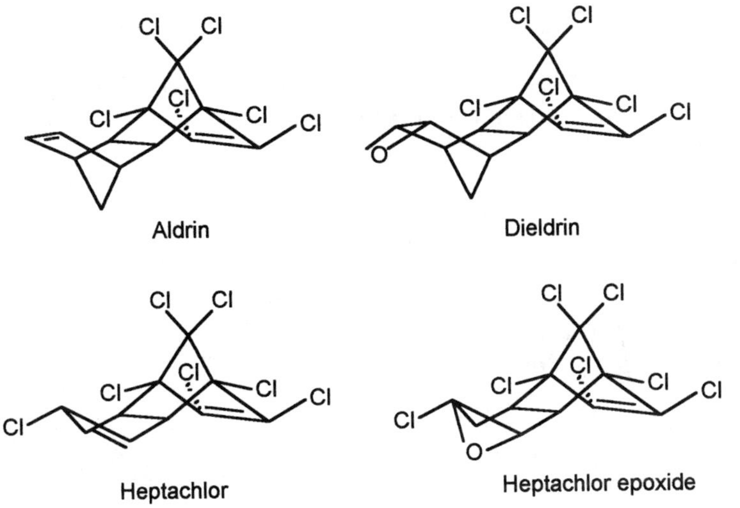

Figure 2. Chemical structure for pesticides and its metabolites.

The analysis by gas chromatography/mass spectrometry of the extract from final growth, allows us to determine, in the case of heptachlor and heptachlor epoxide runs, the presence of chlordene. For aldrin and dieldrin runs, presence of monodechlorodieldrin was detected. Both chlordene and monodechlorodieldrin have been reported as metabolites present in the degradation route of these compounds carried out by many soil microorganisms (Montgomery, 1993).

CONCLUSIONS

Degradation of heptachlor, heptachlor epoxide, aldrin and dieldrin was carried out in free cell cultures of *P. fluorescens* reaching degradation percentages up to 97% for pesticide concentration of 10 mg/L. Trends showed by the studied pesticides suggest that there could be a relationship between the chemical structure and the biodegradability of the compounds. Degradation kinetics suggest that the higher degradation rates occur during the first 24 hours. GC/MS analysis of the final extracts allows the identification of chlordene and monochlorodieldrin, which had been reported as the final metabolite produced in the biodegradation of this kind of compounds. The use of *P. fluorescens* cell showed a great potential in the degradation of heptachlor, heptachlor epoxide, aldrin and dieldrin highly contaminated groundwaters, and studies regarding the immobilization of the bacteria and the scaling up of the process are currently in progress.

ACKNOWLEDGEMENTS

The authors acknowledge with thanks the text revision by Manuel de la Torre.

REFERENCES

Bandala, E.R., P. Pastrana and L.G. Torres. 1995. "Pesticide adsorption in granular activated carbon".(In Spanish). Proceedings of the IX Environmental Engineering Conference. Maracaibo, Venezuela.

Dapaah, S.Y. and G.A. Hill. 1992. "Biodegradation of chlorophenols mixtures by Pseudomonas putida". *Biotech. Bioeng.* 40: 1353-1358.

Lawrence, A.B., B. Williams and A. Fairbrother. 1996. "Evaluating risk predictions at population community levels in pesticide registration: hypotheses to be tested". *Environ. Toxicol. Chem.* 15(4):427-431.

Lee P.H., R. Protzman and S.K. Ong. (1997). "Pesticide wastewater treatment using biomineralization". *In situ* and on-site bioremediation. B. Alleman and A. Leeson (Edit.) Volume 2. United States of America: 281-286.

Montgomery, J. 1993. *Agrochemicals desk reference: Environmental data.* Lewis Publishers. USA.

Nair, A., R. Mandapati, P. Dureja and M. Pallai. 1996. "DDT and HCH load in mothers and their infants in Delhi, India". *Bull. Environ. Contam. Toxicol.* 56(1):58-64.

Sitarska, E., W. Klucinski, R. Faaundez, A. Duszewska, A. Winnicka and K. Goralezyk. 1993. "Concentration of PCB's, HCB, DDT and HCH isomers in the ovaries, mammary gland and liver of cows". *Bull. Environ. Contam. Toxicol.* 55(6): 858-869.

Torres L.G., V. Albíter and B. Jiménez. (1997). "Removal of chlorophenols including pentachlorophenol at high concentrations from contaminated water". *In situ* and on-site bioremediation. B. Alleman and A. Leeson (Edit.) Volume 2. United States of America: 447-452.

BIOREMEDIATION OF SOILS CONTAINING 2,4-D, 2,4,5-T, DICHLORPROP, AND SILVEX

Jack T. Spadaro and Eileen L. Webb (AGRA Earth & Environmental, Inc., Portland, Oregon), Henry Schmid (Oregon Department of Transportation, Portland, Oregon), Alexandra Degher and Kenneth J. Williamson (Oregon State University, Corvallis, Oregon)

ABSTRACT: The chlorinated phenoxy herbicides 2,4-dichlorophenoxyacetic acid (2,4-D), 2,4,5-trichlorophenoxyacetic acid (2,4,5-T), 2-(2,4-dichlorophenoxy)-propionic acid (Dichlorprop), and 2-(2,4,5-trichlorophenoxy)-propionic acid (Silvex) were used extensively in the past because of their effectiveness and recalcitrance. They have been shown to be toxic, bioaccumulative carcinogens and mutagens. A field-scale bioremediation system for soils contaminated with these herbicides has been in operation since late 1996. The goal of the project is to lower herbicide concentrations to levels acceptable for landfilling. Significant reductions in herbicide concentrations have been observed in the soils and saturation pore waters.

INTRODUCTION AND BACKGROUND

On behalf of the Oregon Department of Transportation (ODOT), AGRA Earth & Environmental, Inc. (AEE) and Oregon State University (OSU) have tested, designed, installed, and operated a bioremediation to treat herbicides in soils at the ODOT Baldock site, a highway maintenance facility located in Portland, Oregon. In 1988, a release of herbicide solutions was discovered; an estimated 160 yd^3 (122 $\dot{m}$) of soil containing the chlorinated herbicides 2,4-D, 2,4,5-T, Dichlorprop, and Silvex were excavated. The soils were placed into 26 plastic-lined, 6-yd^3 (4.6-m^3), steel storage bins (bins) in an on-site building pending development of future treatment and disposal options. The soil consistencies range from clayey silty sands to cobble gravels, inconsistent between bins.

From 1993 to 1995, OSU conducted laboratory and field studies to optimize methods for herbicide bioremediation in the soils. 2,4-D and 2,4,5-T (phenoxyacetates) degraded quickly in a prototype anaerobic field bioreactor that used methanol and ethanol as electron donors (Degher et al., 1997; Williamson & Woods, 1995; Williamson et al., 1995). Dichlorprop and Silvex (phenoxypropionates) were not degraded anaerobically but degraded significantly during a 3-month aerobic laboratory study using methanol and ethanol as primary substrates.

The Oregon Department of Environmental Quality directed ODOT to achieve herbicide concentrations less than applicable Land Disposal Restriction (LDR) concentrations in the 17 bins in which concentrations exceeded LDR values to allow Subtitle C landfill disposal. The applicable LDR concentrations are:

Dichlorprop - 0.435 mg/kg; 2,4,5-T and Silvex - 7.4 mg/kg; and 2,4-D - 10 mg/kg.   AEE's June 1996 work plan details bioremediation preparations; remediation commenced in September 1996 (Webb and Kuiper, 1996).

MATERIALS AND METHODS

The biotreatment system initially was designed to operate under either aerobic or anaerobic conditions and included saturating the soil with approximately 200 to 250 gal. (750 to 950 L) of water per bin. AEE designed an air-lift system for mixing, aerating, and recirculating pore water (Figure 1) using a slotted, 1-in.-diameter, Sch. 40 polyvinylchloride (PVC) air-lift well located centrally in each bin. Peripheral, sampling wells were installed in the bin corners. A regenerative blower, control manifold, and 0.5-in.-diameter, polyethylene tubing were integrated to deliver air to the wells at approximately 25 to 30 ft^3 (0.7 to 0.8 m^3) per hour. The control manifold includes a 1.5-in.-diameter Sch. 40 PVC header pipe connected to individual air flow meters and valves by reducer tees. Plastic cones, constructed of 0.062-in.-thick, high-density, polyethylene sheeting, were installed around the air-lift wellheads to promote aerated water dispersal.

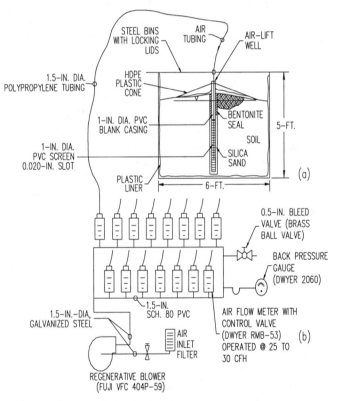

FIGURE 1. Field-scale bioreactor design. (a) Cross section showing air-lift well design and placement. (b) Schematic of air delivery system.

The remediation process included adding to each treatment bin a nutritive amendment of: 800 mL methanol, 600 L ethanol, and 25 g ethyl vanillin or sodium benzoate as carbon sources; 80 g calcium hydroxide and 50 g sodium bicarbonate as buffers; and 250 g ammonium chloride and 25 g monobasic sodium phosphate as nitrogen and phosphorus sources. The nutritive amendment was added to each bin on a periodic basis, generally every four to six weeks. Approximately 16 gal. (60 L) of pore water were pumped from a sampling in each bin into a plastic container, nutrients were mixed with the water, and the amended water then was applied onto the soil surface. Approximately 25 gal. (95 L) of anaerobic digest supernatant from a sewage treatment plant were initially added to each treatment bin as an inoculum. At the start of treatment, an amendment containing 100 g guaiacol, 35 g ethyl vanillin, 25 g sodium benzoate and 210 g propylene glycol as carbon sources was used in some bins, but its use was abandoned in January 1997 upon finding no clear degradation advantage when compared to using methanol and ethanol.

Pore Water Monitoring, Sampling, and Analysis. Pore water monitoring was performed approximately every two to four weeks to ensure maintenance of proper bacterial growth conditions. The parameters monitored included pH, temperature, dissolved oxygen (DO), and conductivity. Pore water was pumped peristaltically from sampling wells at approximately 500 to 1000 mL per minute into a jar containing a water analyzer multiprobe.

Pore water samples also were periodically collected from sampling wells for analysis of herbicides, total organic carbon (TOC), ammonia-nitrogen, orthophosphate-phosphorus, and heterotrophic microbial counts. Samples were sealed in appropriately preserved bottles and stored at 4°C prior to analysis. Herbicides were analyzed by EPA Method 8151 (United States Environmental Protection Agency, 1986) by gas chromatography with an electron capture detector (October through December 1996) and by high pressure liquid chromatography (February/March 1997).

Soil Sampling and Analysis. Soils were sampled periodically using either a stainless-steel, hand auger (2- or 3-in.-diameter bit) or a 3-ft-long power auger (6-in.-diameter, steel bit). The May 1996 samples were collected at one point per bin at 1.5 ft. (0.46 m) below soil surface (bss). In December 1996, soils were sampled at 1 and 3 ft. (0.3 and 0.9 m) bss from two borings in each bin and were composited. During the March 1997 sampling event, six 4-ft. (1.2-m) top-to-bottom soil cores from each bin were composited. Soil samples were placed into glass jars and preserved at 4°C prior to herbicide analysis by EPA Method 8151.

RESULTS AND DISCUSSION
The following summary of results presents data from four bins (3, 16, 18, and 23) containing the greatest soil herbicide concentrations.

Pore Water Monitoring. The bin treatment pore water pH values varied significantly at treatment startup, with a tendency to drop to the range of 4 to 5; periodic addition of sodium bicarbonate and calcium hydroxide raised the pH to 6 to 6.5. The soil treatment bins were maintained in the temperature range of 18 to 35°C; initial temperatures were in the high end of the range. Building heaters were set to maintain an air temperature ≥25°C. Conductivity varied considerably between bins and ranged from 500 to 8,000 milliSiemens, with an increase during treatment because of addition of nutritive salts and buffers. The DO concentration varied significantly in the bins with extremes of 0.0 and 5.9 mg/L during monitoring events (Table 1).

TABLE 1. Aqueous herbicide and dissolved oxygen concentrations.

Bin	Date	2,4-D (mg/L)	2,4,5-T (mg/L)	Dichlorprop (mg/L)	Silvex (mg/L)	DO (mg/L)
3	10/14/96	57.00	1.10	67.00	0.45	0.1
	11/12/96	8.10	0.19	<24.00	0.20	3.4
	12/10/96	1.00	0.023	34.00	0.100	0.8
	2/19/97	<0.01	<0.01	5.30	<0.01	0.1
16	10/14/96	21.00	3.80	14.00	0.210	0.1
	11/12/96	6.40	<15.00	<30.00	0.510	5.9
	12/10/96	4.10	52.00	67.00	1.30	0.5
	2/19/97	<0.01	34.30	4.00	<0.01	0.2
18	10/14/96	8.60	13.00	54.00	0.510	1.3
	11/12/96	<1.00	<7.00	<5.50	0.500	0.9
	12/10/96	0.100	0.058	3.50	0.110	3.5
	3/3/97	<0.01	<0.01	<0.01	<0.01	3.4
23	10/14/96	2.80	0.090	7.10	0.190	0.2
	11/12/96	<0.200	<0.005	<0.200	0.030	2.7
	12/10/96	0.150	<0.010	2.70	0.038	1.1
	2/19/97	<0.01	<0.01	0.220	<0.01	4.0

No clear pattern in DO behavior was apparent. A rapid decline in TOC in many bins indicated a high microbial metabolism rate. Nitrogen and phosphorus concentrations consistently were depleted. Heterotrophic microbial population counts demonstrated that bacterial populations essentially remained stable throughout the treatment process (Table 3).

TABLE 2. Heterotrophic plate count results for pore water.

	Plate Count (CFU/mL)			
Bin	December 1996	May 1997	August 1997	October 1997
3	NT	NT	1.28×10^6	5.5×10^5
16	1.3×10^6	7.8×10^4	8.0×10^4	7.6×10^4

CFU/mL - Colony Forming Units per mL of solution.
NT - Not tested by this method.

Chlorinated Herbicides. Treatment progress initially was measured by herbicide concentrations in pore water (October through December 1996, and February/March 1997) (Webb and Spadaro, 1997). Aqueous herbicide concentrations declined significantly in Bins 3, 16, 18, and 23 (Table 3). Aqueous concentrations of 2,4,5-T, Dichlorprop, and Silvex in bin 16 contradicted the general trend during the first two months of monitoring, probably due to inadequate initial mixing of the pore water and soil.

TABLE 3. Herbicide concentrations in soil.

Bin	Date (a), (b), (c)	2,4-D (mg/kg)	2,4,5-T (mg/kg)	Dichlorprop (mg/kg)	Silvex (mg/kg)
3	5/8/96	0.3	0.14	0.5	<0.05
	12/10/96	1.1	<0.25	1.4	<0.25
	3/5/97	<0.93	<0.35	2.96	<0.069
16	5/8/96	1,100	240	370	2.9
	12/10/976	18	140	22	1.3
	3/5/97	<6.39	46	9.8	<0.48
18	5/8/96	<0.2	0.7	0.1	<0.05
	12/10/96	<1	0.3	<0.5	<0.25
	3/5/97	<0.916	<1.96	1.48	<0.055
23	5/8/97	1.1	0.09	<2.6	0.14
	12/10/96	11	<0.25	10.0	<0.25
	3/5/97	<0.967	<0.087	1.54	<0.049

(a) Samples on 5/8/96 were from one hole per bin at 1.5 ft. (0.46 m) below soil surface (bss).
(b) 12/10/96 composite from two borings sampled 1 and 3 ft. (0.3 and 0.9 m) bss.

(c) Composite on 3/5/97 from six 4-ft. (1.2-m) top-to-bottom cores per bin.

Declines in herbicide concentrations were less apparent in the soils, except in Bins 16 and 23 where degradation was quite evident. By March 1997, LDR-exceeding concentrations of Dichlorprop remained in soil in Bins 3, 16, 18, and 23; an excessive 2,4,5-T concentration remained in bin 16; and 2,4-D and Silvex essentially had been removed. No significant chlorophenolic species concentrations were detected during subsequent analysis.

Both the phenoxyacetate and phenoxypropionate herbicides have been removed, indicative of a mixed aerobic-anaerobic environment in the treatment bins. By March 1997, soils in five treated bins met the LDR disposal requirements. In October 1997, soils in 14 bins (treated and untreated) that met the LDR disposal goals for herbicides, chlorinated phenols, dioxins, and furans were transported for disposal to a Subtitle C landfill. Twelve bins of soil are undergoing continued treatment to target Dichlorprop removal. Bin 16 may require further anaerobic treatment to remove 2,4,5-T.

The results demonstrate the successful bioremediation of relatively small quantities of soil, containing soluble chlorinated compounds, using a saturated soil bioreactor technology. The reactors employ recirculation of pore waters and additions of electron donors and acceptors. The technology should not necessarily be limited to soluble or chlorinated compounds; more development is necessary to define the limits of this technology with respect to the variety of soils and contaminants that can be treated. In this case study, application of this technology is resulting in a net savings of approximately 70 percent of the $2 million cost associated with incineration, the regulatory Best Demonstrated Available Technology (BDAT) treatment standard for the compounds in question.

ACKNOWLEDGEMENTS

The authors would like to thank Columbia Analytical Services (Kelso, Washington) for their assistance in the analysis of the soils and waters for herbicides, North Creek Analytical (Beaverton, Oregon) for analysis of nutrients, and Air, Food, and Water Analysis, Inc. (Portland, Oregon) for assistance with heterotrophic plate counts.

REFERENCES

Degher, A. B., K. G. Wang, K. J. Williamson, S. L. Woods, and R. Strauss. 1997. "Bioremediation of Soils Contaminated with 2,4-D, 2,4,5-T, Dichlorprop, and Silvex." In B.C. Alleman and A. Leeson (Eds.), *In Situ and On-Site Bioremediation: Volume 2*, p. 279. Battelle Press, Columbus, OH.

Spadaro, J. T., E. L. Webb, and J. L. Kuiper. 1997. "Quarterly Report Number 1, 4th Quarter 1996, Baldock Station Maintenance Facility, Oregon Department of Transportation, 9637 Southwest 35th Drive, Portland, Oregon, ODEQ Facility ID #1226." March 5, 1997. AGRA Earth & Environmental, Inc., Portland, OR.

United States Environmental Protection Agency. 1986. *Test Methods for Evaluating Solid Waste, SW-846, 3rd Edition, et. sequens.* November 1986.

Webb, E. L. and J. L. Kuiper, 1996. "Project Plan, Baldock Station Maintenance Facility, 9637 Southwest 35th Drive, Portland, Oregon, ODEQ Facility ID #1226." June 5, 1996. AGRA Earth & Environmental, Inc., Portland, OR.

Williamson, K. J., and S. L. Woods. 1995. "Proposal, Phase 3: Remediation of Contaminated Soil from the Baldock Station Maintenance Facility." July 1995. Western Region Hazardous Substance Research Center, Oregon State University, Corvallis, OR.

Williamson, K. J., S. L. Woods, R. S. Strauss, and K. G. Wang. 1995. "Progress Report, Remediation of the Contaminated Soils from the Baldock Station Maintenance Facility." May 1995. Western Region Hazardous Substance Research Center, Oregon State University, Corvallis, OR.

TREATMENT OF PCBS AND CHLORINATED SOLVENTS BY ELECTROCHEMICAL PEROXIDATION

Michele Wunderlich, Ronald Scrudato and Jeff Chiarenzelli (Environmental Research Center, SUNY at Oswego, Oswego, New York)

ABSTRACT: Laboratory and pilot-scale testing of the Electrochemical Peroxidation Process (ECP) have been carried out by the Environmental Research Center (ERC) on various aqueous systems and slurries contaminated with polychlorinated biphenyls (PCBs) and chlorinated solvents. These include water and slurries from a subsurface storage tank (SST) contaminated with chlorinated solvents and PCBs, condensate derived from PCB contaminated soils and sediments, and groundwater contaminated with chlorinated solvents. The ECP process utilizes a small electric current to enhance Fenton's Reagent reactions. Laboratory testing of ECP on 500 mls of the SST water and slurry containing a small amount (<1%) of organic sludge resulted in PCB reductions of 97.2 and 68.2% respectively. Similar results were achieved for solvents with greater than 94% degradation for chloroethane, dichloromethane, 1,1-dichloroethane, 1,1,1-trichloroethane, and acetone in both water and slurry. Pilot tests conducted on 200 liters of SST water resulted in 87.9 and 85.2% degradation of PCBs in duplicate treatments. Laboratory testing of ECP on condensate derived from sediment contaminated with biodegraded Aroclor 1248 and soil contaminated with Aroclors 1254 and 1260 resulted in 95 and 65.6% degradation respectively. Concentrations of trichloroethane (TCE), perchloroethane (PCE), and 1,1,2,2 tetrachloroethane (TCEA) in contaminated groundwater from the Massachusetts Military Reservation were reduced by 92%, 97.9%, and 65.3%, respectively.

INTRODUCTION

The ERC at SUNY Oswego, New York has developed an advanced oxidative process known as ECP which has been used to degrade organic contaminants including PCBs, BTEX, chlorinated solvents, and volatile organic compounds (VOCs). This process utilizes steel electrodes, an electrical current and peroxide to form free radicals; the process does not require the addition of soluble iron salt with Cl^- or SO_4^{2-} ions. The most significant mechanism for free radical formation is Fenton's Reagent. Fenton (1894) first reported the accelerated decomposition of hydrogen peroxide in the presence of ferrous iron. Fenton's Reagent forms ferric iron and the hydroxide free radical (equation 1) which can destroy organic matter.

$$Fe^{+2} + H_2O_2 \rightarrow Fe^{+3} + OH^- + OH° \qquad \text{(Walling 1975)} \qquad (1)$$

In the ECP process, iron is utilized as a true catalyst by the electrical reduction of ferric iron (equation 2).

$$Fe^{+3} + e- \rightarrow Fe^{+2} \tag{2}$$

If the polarity of the current is reversed, both electrodes act as anodes accelerating the recycling of ferrous iron.

Other pathways believed to be responsible for organic degradation by ECP are solvated electrons, zero valent iron from the surface of the electrode, cathode oxidation, and anode reduction.

Site Descriptions. Water and slurry samples were taken from a 3000 gallon SST located on a New York State Superfund Site in Oswego, N.Y. used for septic purposes and contaminated with PCBs and chlorinated solvents.

A condensate sample was obtained by steam distillation of a sediment collected from a small embayment of the St. Lawrence River adjacent to the General Motors Federal Superfund Site and Akwesasne Mohawk Nation. This sediment was contaminated by anaerobically degraded Aroclor 1248. A second condensate was obtained by steam distillation of a soil collected from the Mare Island Naval base in California. This soil was contaminated by a mixture of Aroclors 1254 and 1260 with no evidence of biodegradation.

The groundwater sample came from a well located near the Otis Air Force Base at the Massachusetts Military Reservation. The water is contaminated with TCE, PCE, and TCEA at a combined concentration of ~10 mg/L.

MATERIALS AND METHODS

Sample Collection. For the SST water bench scale work, water was collected with a peristaltic pump and teflon tubing. The water was stored in amber glass at $4\,^\circ$C until used. The slurry was created by mechanical agitation between the tank water and bottom sediment (sludge). A peristaltic pump and teflon tubing was used to collect the samples. The resulting slurry samples contained an organic black floc that settled out and comprised <1% by weight of the sample.

For the pilot scale work, the SST water was pumped with a peristaltic pump into a steel 220 liter drum and treated on-site.

To obtain the condensates from the St. Lawrence sediment and the California soil, a modified Nielsen-Kryger Apparatus (Veith and Kiwus 1977) was used to remove the PCBs. 500 grams of St. Lawrence sediment or 738 grams of California soil were placed into the apparatus with double deionized water. The water was boiled off and the condensate collected. Two liters of condensate was collected for the St. Lawrence sediment (~182 μg/L) and 3.02 liters from the California soil (122 μg/L). The condensates were stored in amber bottles at $4\,^\circ$C until used.

Groundwater was collected from monitoring well, 5DD, located adjacent to the Otis Air Force Base near Sandwich, Massachusetts. Three to five well volumes were purged before samples were collected in amber four liter bottles and headspace was eliminated. Samples were stored at $4\,^\circ$C until used.

ECP Experiments. Bench scale experiments were conducted for the SST water and slurry, condensates, and groundwater. Pilot scale experiments were conducted on SST water only. The solution of interest was placed into a precleaned container, and stirred with a teflon coated stir bar or in the case of the pilot scale experiment, an electric mixer. The pH was adjusted with the addition of either sulfuric (H_2SO_4) or nitric (HNO_3) acid and the temperature was either maintained or heated to 70°C with a hot plate. Control samples were taken. Steel electrodes, either parallel plates (6 by 8 cm or 30 by 90 cm), or nested pipes (6.3 cm and 3.8 cm diameters) were immersed into the containers to induce a current that ranged from 500 mA to 20000 mA with a polarity reversal every 5 seconds for the bench scale and 10 seconds for the pilot scale. Ferrous sulfate was added to the SST samples for VOC degradation and in the California condensate to increase iron concentrations due to elevated target compound concentrations. Hydrogen peroxide, either heated or at room temperature, was added to the solutions being treated. After treatment, duplicate samples were taken and stored at 4°C until extracted. Some solutions received additional treatments (Table 1).

TABLE 1. **Experimental parameters for Bench Scale and Pilot Scale ECP treatments of PCB contaminated samples.**

	SST Water & Slurry	SST Water & Slurry	SST Water Pilot	St. Lawrence Cond.	CA Condensate
Contaminant Treated	PCBs	VOCs	PCBs	PCBs	PCBs
Amount (mls)	500	580	200000	1000	1200
Temp. (°C)	70	5	15	16.8	20
pH/Acid Used	2/H_2SO_4	2.5/H_2SO_4	2/H_2SO_4	5/HNO_3	2/HNO_3
Electrodes	plates(6 by 8 cm)	plates(6 by 8 cm)	plates(30 by 90 cm)	nested pipe	nested pipe
Treatments	1	3	1	3	2
Ferrous Sulfate (mg)	0	24	0	0	280
H_2O_2 (mg/L)	866	1680	87.5	43 (70°C)	43 (70°C)
Current (mA)	500	500	20000	1000	1200
Reaction Time (min.)	5	1	5	~10	~10

PCB Analysis.   Congener specific PCB analyses were conducted by the Environmental Research Center in Oswego, N.Y., and were congener specific. All reagents, including hexane, sodium sulfate, florisil, TBA, and sulfuric acid were reagent grade and checked for contamination before use. The water used was double deionized; all samples had Decachlorobiphenyl (DCB) added as a surrogate standard to monitor PCB recovery.

Liquid-liquid extraction procedures were used and all samples were extracted in a separatory funnel using three sequential extractions with 50 ml of hexane. The hexane extracts were combined, dried with sodium sulfate, and oxidized with sulfuric acid. The extracts were treated with TBA to remove elemental sulfur, 4% deactivated florisil to remove polar compounds, and then condensed. The condensed hexane was stored in the dark at 4°C until analyzed on a HP5890 gas chromatograph (GC) with an electron capture detector and Ultra DB-S column calibrated after every fifth sample with a mixed Aroclor standard from the New York State Department of Health Wadsworth Laboratory.

VOC Analysis. The VOC analyses for the SST samples were conducted by O'Brien and Gere Laboratories, Inc. in Syracuse, NY, a state certified laboratory. The 40 ml glass vials with teflon lids were delivered within 2 hours of the treatment. Samples were analyzed within 5 days by the GC-EPA method 601/602 for purgeable organics modified to allow for acetone quantification.

The VOC analyses for the groundwater samples were performed by the Barnstable Department of Health Laboratory at Barnstable, MA. The samples were delivered directly after treatment and stored in the dark at 4°C until analyzed by EPA Methods 504 and 524.

RESULTS AND DISCUSSION

Results of ECP on PCB contaminated matrices are shown in Table 2. In the SST samples, there was near complete destruction of all congeners except for those congeners with retention times over thirty minutes (85% degradation for the water and ~ 40% degradation for the slurry), suggesting solubility imposed constraints in slurry systems. Those congeners with a lower retention time enter the aqueous phase

TABLE 2. Results of ECP experiments on SST water, SST slurry, and condensates contaminated with PCBs.

	SST Water	SST Slurry	Pilot Scale SST water	St. Lawrence Condensate	California Condensate
Control PCBs (μg/L)	19.8	99	16.3 13.65	182	122
ECP Treated PCBs (μg/L)	0.3 0.8	34.5 28.5	2 2	9.1	42.1
% Degradation	98.5 96	65.2 71	88 85	95	65.6

more readily, while those of less solubility and higher retention times sequester to solids. Free radicals can also be scavenged by other organic and inorganic compounds. Up-scaling of the process from bench-scale to pilot-scale (400 times) resulted in minor losses in efficiency. This is most likely due to proportional differences in electrode surface areas and/or spacing.

Both of the condensates showed preferential destruction of lower and orthochlorinated congeners. The St. Lawrence sediment was enriched in these congeners through anaerobic degradation, and was therefore more susceptible to ECP than the California condensate.

ECP treatment of the Otis Air Force Base groundwater degraded most of the chlorinated solvents with no measurable byproducts detected (Table 3).

TABLE 3. Results of ECP treatment of chlorinated solvent contaminated groundwater.

	trichloroethane (μg/L)	perchloroethane (μg/L)	1,1,2,2 tetra-chloroethane(μg/L)
Control	420	38	9500
Treatment #1	46	6.7	5100
Treatment #2	33	0.8	3300
% Total Degradation	92.1	97.9	65.3

There was a progressive destruction of the VOCs with the incremental additions of reagents (Table 4). The extent of contaminant degradation was nearly linear on log/linear plots of concentration versus peroxide dosage for all VOC compounds detected. Several compounds including chloroethane, 1,1-dichloroethane and dichloromethane were reduced to non-detectable levels.

Conclusion. ECP has been used to rapidly degrade chlorinated organic contaminants including PCBs and VOCs in water and slurry matrices. Future work at the ERC will focus on pilot scale applications, including a continuous flow ECP treatment system to be used in-situ to treat contaminated liquids.

TABLE 4. Results of ECP bench-scale experiments on VOC contaminated SST water (W) and slurry (S).

	Control (μg/L)	Treatment #1 (μg/L)	Treatment #2 (μg/L)	Treatment #3 (μg/L)	% Degrade
Chloroethane	W: 1750 S: 2150	W: 80 S: <5	W: 33 S: <1	W: <10 S: <1	W: >99.4 S: >99.9
Dichloromethane	W: <100 S: <130	W: 10 S: <5	W: 15 S: <1	W: 11 S: <1	W: - S: >99.2
1,1-Dichloroethane	W: 170 S: 185	W: 51 S: 6	W: <10 S: <1	W: <10 S: <1	W: >94.1 S: >99.4
1,1,1-Trichloroethane	W:<100 S: <100	W: 47 S: 14	W: 18 S: 5	W: 12 S: 6	W: - S: -
Acetone	W:73000 S: <9999	W: 29000 S: <500	W: 10000 S: <100	W: 3300 S: 659	W: 95.5 S: -

ACKNOWLEDGMENTS

We appreciate the assistance of James Pagano and Gideon Oenga of the ERC.   Tom Bourne and the Barnstable County Health Department provided analytical support and consultation.   Thomas Alexander and O'Brien and Gere Laboratories, also provided analytical support and consultation. Support for the research was provided by the National Institute of Environmental Health Science Basic Research Program, the US Environmental Protection Agency and the New York Sea Grant Institute.

REFERENCES

Fenton, H.J. 1894. *Journal of the Chemical Society. 65:* 899.

Veith, G. and L. Kiwus. 1977. "An exhaustive steam-distillation and solvent extraction unit for pesticides and industrial chemicals." *Bulletin of Environmental Contamination and Toxicology. 17*: 631-636.

Walling, C. 1975. "Fenton's Reagent Revisited." *Accounts of Chemical Research. 8*: 125-131.

SUCCESSFUL REMEDIATION OF PCDF CONTAMINATED SOIL USING A LABORATORY REACTOR

Christoffer Rappe (Umeå University, Umeå, Sweden)
Rolf Andersson, Lars Öberg and Bert van Bavel
(Umeå University, Umeå, Sweden)
Ryuji Uchida (EBARA Research, Fujisawa, Japan)
Shin Taniguchi, Akira Miyamura and Hisayuki Toda
(EBARA Corporation, Tokyo, Japan)

ABSTRACT: Soil contaminated by polychlorinated dibenzofurans (PCDFs) and mercury from a chloralkali plant was successfully decontaminated by treatment with sodium bicarbonate at 350°C and 400°C. Counted as toxic equivalents the decontamination exceeded 99.9%. The reduction of mercury in the soil exceeded 98%. Tetra- and penta CDFs were identified and quantified in the rinse and XAD cartridge. Inspections of the fragmentograms indicate that these more volatile compounds are released from the soil before the destruction starts. These constituents can be degraded in a secondary treatment.

INTRODUCTION

Dioxin is the generic name for 75 polychlorinated dioxins (PCDDs) and 135 polychlorinated dibenzofurans (PCDFs). Seventeen of these are included in the Toxic Equivalent Factor (TEF) System, by which Toxic Equivalents (TEQ) can be calculated. The 2,3,7,8-tetraCDD has been classified as a Group 1 human carcinogen, but all others as well as TEQ are not classifiable as to their carcinogenic effects (IARC, 1997).

The PCDDs and PCDFs are identified as toxic and very recalcitrant contaminants. Soil pollution can occur during the production and use of chlorinated pesticides and industrial products like chlorinated phenols, phenoxy herbicides and PCBs. In 1989 it was reported that the electrode sludge from the chloralkali process was highly contaminated by a series of PCDFs, in addition to the expected mercury contamination (Rappe et al., 1989).

Recently the Swedish EPA (1996) published target values for the remediation of soil contaminated by PCDDs and PCDFs. For soils without any restrictions in use a value of 10 pg TEQ/g d.m. was suggested. For industrial soil the suggested value was 250 pg TEQ/g d.m. In Germany the practical guideline for agricultural use of soil is 40 pg TEQ/g d.m. (Basler, 1994).

In a series of publications it has been reported that PCBs can be successfully dechlorinated (> 99.9%) by the base catalyzed decomposition process, BCD (Takada et al, 1997). The resulting products are dechlorinated organic compounds, alkali halides and water. In a pilot reactor the feeding rate was 120 kg per hour, in the laboratory scale reactor the feeding rate was 70 g per hour.

Objective. The objective of the study was to determine the feasibility of the base catalyzed dechlorination process (BCD) for the remediation of soil contaminated from the chloralkali process. A 30% reduction of a tetraCDD has been reported in a river sediment by Barkowskii and Adriaens (1995) in active microcosms during seven months, but the concentration remained unchanged in pasteurized microcosms. It is doubtful whether bioremediation can be used to reach the low guideline values suggested by the Swedish EPA. Another technique, which has been used for soil remediation, is a rotary kiln followed by an afterburner, operating at temperatures exceeding 900 - 1000°C. However, such incinerators need investments exceeding US $ 10 M.

Site description. The soil originated from the Eka Chemicals chloralkali plant situated at Göta River close to the west coast of Sweden. The production of chlorine at the site started in 1924, and the maximum production volume was approximately 100 000 tonnes/year. During the period up to the late 1970s the process included the use of grafite electrodes. The amount of contaminated soil in the landfill at the site is approximately 13 000 m^3.

MATERIALS AND METHODS

Soil sampling

From the landfill samples were collected in barrels. The content of one barrel was homogenized by mixing in a cement mixer and transferred into 10 kg portions. About 70 kg of this soil was sieved (0.4 cm), mixed and stored at +4°C before grinding. After grinding the soil was analyzed in duplicate and used for the experiments. It was found to contain as much as 12 000 and 13 000 pg TEQ/g d.m., see Table 1.

Soil treatment

Figure 1 shows the set-up used for the experiments: a flask with a heating mantle, a ribbon heater, a condenser with a flask and an XAD-2 trap. An amount of 2.1 g of NaHCO$_3$ was added to 70 g of the contaminated soil. After careful mixing the soil was treated at 350°C or 400°C for 60 min using the heating mantle. The ribbon heater was used to keep the upper part of the flask at around 200°C. A condensate was collected and the remaining constituents were collected at a XAD-2 trap. The condenser was rinsed with methanol and then toluene, and the rinse was combined with the condensate. The following products were collected: soil, the rinse with the condensate and XAD-2.

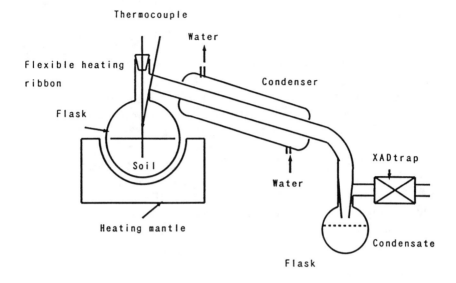

Figure 1. The experimental set-up.

Chemical analyses

After addition of 15 $^{13}C_{12}$-labelled PCDD and PCDF standards the un-treated and treated soil (3 g) samples were extracted in a Soxhlet extractor with toluene for 18 hours. The XAD-2 trap was Soxhlet extracted with toluene for 15 hours, and the combined wash and rinse was allowed to evaporate in a hood. The multistep clean-up procedure (three micro-columns) and the congener specific analysis (HRGC/HRMS) of PCDDs and PCDFs has earlier been described by Rappe et al. (1995). The reco-veries of the $^{13}C_{12}$-labelled standards were 50-90% for the treated soil but only 20-95% for the untreated soil.

RESULTS

Two campaigns have been performed. In both campaigns the soil was treated at 350°C and 400°C. In the first campaign only the treated soil samples were analyzed. In the second campaign the soil, the rinse and the XAD-2 have been collected and analyzed.

The analytical results of PCDDs and PCDFs from the homogenized untreated soil sample (duplicate) as well as results from the analyses of soil from the two campaigns are given in the table. In addition, results of the analyses of the XAD-2 trap and the combined rinse and condensates are also given in the table.

For campaign 1, where only the soil samples were analyzed, the most dramatic reduction was noticed for the 2,3,7,8-tetra CDF and the 2,3,4,7,8-penta CDF. Here we observed a reduction of > 99.99%. For the TEQ values the degradation was 99.997% which is a remarkably high reduction. The mercury content of the untreated soil was 46.7 $\mu g/g$, and in the two samples from the first campaign the content of mercury was found to be 0.6 and 0.7 $\mu g/g$, which is a reduction greater than 98%.

From campaign 2 the rinse and the XAD-2 were found to contain large amounts of PCDFs, especially the lower chlorinated congeners. In the table we have also given the values for the total destruction in these two experiments, which ranged from 32% (sum of tetra CDFs) to >99.9% (hepta- and octa chlorinated compounds). For the hexaCDDs the detection levels in the untreated soil were higher than the amount found after the treatment, so no figures are given for the degradation of these isomers.

CONCLUSIONS

The treated soil was found to contain 2 pg TEQ/g d.m. (350°C) or 0.29 pg TEQ/g d.m. (400°C). This is far below the generic guideline value for PCDDs/PCDFs (as TEQ) for contaminated soil in Sweden. The value for mercury is also below the guideline in Sweden, which is 2$\mu g/g$ d.m. (Swedish EPA, 1996).

The combined condensate and wash (rinse) as well as the XAD-2 were found to contain much larger amounts of Cl_4., Cl_5- and Cl_6 than the treated soil. These collected products could be treated in a second step using the same process or burned under controlled conditions in a small incinerator for liquids.

In order to study the origin of the remaining PCDFs found in the rinse and XAD we have studied the detailed pattern of these compounds in the untreated soil and in the products, see Figure 2. The great similarity found here supports the conclusion that these lower chlorinated and more volatile compounds are distilled or steam distilled with the remaining water from the soil before the degradation starts. A stepwise dechlorination of the Cl_6 - Cl_8 PCDFs to the Cl_4 and Cl_5 PCDFs could be expected to result in different patterns.

ACKNOWLEDGEMENTS

A grant from MISTRA (Foundation for Strategic Environmental Research) within the national research program COLDREM (Soil remediation in a cold climate) is gratefully acknowledged.

	Untreated soil 1	Untreated soil 2	Campaign 1 350°C	Campaign 1 400°C	Campaign 2, 350°C Soil	Wash	XAD	Efficiency	Campaign 2, 400°C Soil	Wash	XAD	Efficiency
	pg/g	pg/g	pg/g	pg/g	pg/g	pg/g	pg/g	%	pg/g	pg/g	pg/g	%
2378 TCDF	21000	23000	3.4	0.36	20	2400	730	85.5	0.58	3400	1400	78.2
SUM TCDF	42000	48000	10	0.86	46	7000	1900	59.5	1.5	11000	4100	31.8
2378 TCDD	15	11	<0.044	<0.042	<0.057				<0.056			
SUM TCDD	62	83	0.13									
12378 PeCDF	7700	8100	1.3	0.056	6.1	370	150	97.6	0.11	480	190	97.0
23478 PeCDF	6900	7200	0.78	0.065	4.0	380	110	97.8	0.11	490	140	97.1
SUM PeCDF	27000	29000	5.6	0.51	17	2000	540	88.2	0.84	2200	700	86.8
12378 PeCDD	18	23	0.12	0.10	<0.070	2.7	1.0	100	<0.11	2.4	1.4	100
SUM PeCDD	100	100	0.45	0.10		50	6.1	99.7		47	7.0	99.8
123478 HxCDF	6100	6500	0.64	0.085	1.9	190	40	99.0	<0.13	180	62	98.9
123678 HxCDF	1300	1300	0.28	0.046	0.46	48	10	99.7	<0.12	49	15	99.7
234678-HxCDF	570	650	0.13	<0.039	0.21	17	4.2	99.9	<0.082	24	6.4	99.9
123789-HxCDF	140	130	0.22	0.17	0.25	1.6	0.30	100	<0.063	2.1	0.59	100
SUM HxCDF	6600	7800	2.1	0.52	2.5	370	52	98.1	0.072	390	80	97.9
123478 HxCDD	<11	<14	0.067	<0.057	<0.12	0.39	0.19		<0.20	0.65	0.24	
123678 HxCDD	<9.9	<13	0.089	<0.055	<0.10	0.49	0.12		<0.18	1.6	0.58	
123789 HxCDD	<5.5	<8.0	0.063	<0.049	<0.074	1.6	0.37		<0.10	2.6	0.91	
SUM HxCDD			0.30			17	6.0			23	8.4	
1234678 HpCDF	1400	1600	0.85	0.57	0.57	21	4.3	99.9	<0.17	22	7.3	99.9
1234789 HpCDF	760	840	0.071	<0.049	<0.14	3.4	0.73	100	<0.21	3.9	1.5	100
SUM HpCDF	3100	3400	1.5	1.0	0.72	32	6.8	99.8		36	12	99.8
1234678 HpCDD	47	80	0.17	0.14	0.19	4.8	1.3	100	<0.17	5.8	1.9	100
SUM HpCDD	83	130	0.27	0.2	0.19	9.1	2.6	99.9		10	3.9	99.9
OCDF	3300	3600	1.7	1.2	0.69	3.7	0.42	100	0.58	4.9	1.0	100
OCDD	390	480	1.2	0.95	1.5	8.7	1.6	99.9	0.83	10	2.6	99.9
TEQ	6800	7200	1,0	0.19	4.7	480	140		0.22	640	230	

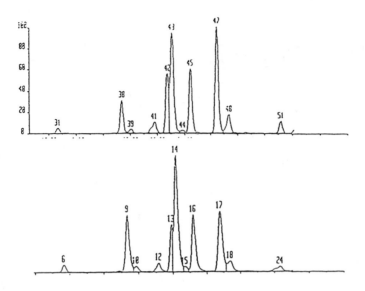

Figure 2. Fragmentograms of pentaCDFs in soil (upper) and XAD-2 trap (lower).

REFERENCES

Basler, A. 1994. Regulatory measures in the Federal Republic of Germany to reduce the exposure of man and the environment. *Organohalogen Compds.* 20: 567-570.

Barkowskii, A.L. and P. Adriaens, 1995. Reductive dechlorination of tetra-chlorodibenzo-*p*-dioxin partitioned from Passaic River sediments in an autochthonous microbial community. *Organohalogen Compds.* 24: 17-21.

IARC Monograph Vol: 69: Polychlorinated dibenzo-*para*-dioxins and dibenzofurans. Lyon, 1997.

Rappe C., L.-O. Kjeller, S.-E. Kulp, C. de Wit and O. Palm, 1990. Levels, patterns and profiles of PCDDs and PCDFs in samples related to the production and use of chlorine. *Organohalogen Compds.* 3: 311-314.

Rappe, C., R. Andersson, M. Bonner, K. Cooper, H. Fiedler, F. Howell, S.E. Kulp and C. Lau, 1995. PCDDs and pCDFs in soil and river sediment samples from a rural area in the United States of America. *Chemosphere* 34: 1297-1314.

Swedish EPA,1996. Development of generic guideline values. Report 4639.

Takada, M., R. Uchida, S. Taniguchi and M. Hisomi. 1997. Chemical dechlorination of PCDDs by the base catalyzed decomposition process. *Organohalogen Compds.* 31: 435-440.

REMEDIATION OF A LANDFILL CONTAINING DIOXINES, ORGANOCHLORINE PESTICIDES, CHLOROBENZENES AND CHLOROPHENOLS

Almar Otten and Ad Bonneur (Tauw Milieu bv, Deventer, The Netherlands)
Meindert Keizer and Jeroen van Laarhoven (Wageningen Agricultural University, Wageningen, The Netherlands)

ABSTRACT: This article describes the method and execution of the soil remediation of a former 1 hectare (illegal) landfill contaminated with chlorinated organic compounds. The remediation was carried out in the period of November 1992 until April 1994. The landfill was contaminated with the organic compounds: organochlorine pesticides, chlorobenzenes, chlorophenols and dioxins. These contaminants infiltrated in the underlying soil, mainly consisting of clay and peat layers. The contaminants were found at depths up to 8 meters in the soil. The remediation had been carried out by removal of the toxic waste material and the contaminated soil. After excavation the contaminated soil was transported to a special for this occasion designed disposal. The toxic waste was processed by combustion in an incinerator. The groundwater was extracted, purified in a water treatment plant and discharged at the surface water. Occupational hygiene and safety measures were of great importance during the remediation.

INTRODUCTION

The last decades soil contamination by chlorinated organic compounds has reached significant proportions. This contamination frequently occurs as a result of e.g. surface spills, improper waste disposal and organic liquid spills due to leakage from underground tanks.

In the early sixties a 1 hectare lake, the site Diemen-Noord near Amsterdam in The Netherlands, had been used for illegal disposal of toxic waste. In 1985 some rusty drums were discovered sticking out of the surface. The drums appeared to contain residues of the production of a herbicide, the so-called 2,4,5-T (Trichlorophenoxy-acetic acid). These residues consisted of the following compounds: organochlorine pesticides, chlorophenols, chlorobenzenes and dioxines. The soil at the site mainly consisted of clay and peat layers.

The remediation consisted of removal of the toxic waste material and the contaminated soil. After removal of the waste material the excavation of the contaminated soil took place in sections of 10 meters by 10 meters and in layers of 0.5 meter thickness. After excavation of one layer, samples from the underlying soil layer were taken and analyzed on contaminants. The excavation continued until all contaminated soil had been removed. In this way a lot of data had been obtained about the transport of organic contaminants through clay and peat layers in the soil.

SITE DESCRIPTION

The 1 hectare site Diemen-Noord is situated near Amsterdam and the Amsterdam-Rijncanal in the northwest of The Netherlands. The soil description at the site corresponds to the situation as indicated in Table 1.

TABLE 1. Soil description at the site Diemen-Noord

Depth (m)	Description
0.00 - 5.70	Clay and peat layers
5.70 - 7.95	Sandy clay layers
7.95 - 8.50	Peat layer
8.50 - 14 to 15	Sand layer
14 to 15 - 25	Clay layer

The groundwater level at the site varies between groundsurface to 0.50 meter below ground surface (bgs). The groundwater quality in the immediate surroundings of the site had not been affected by the contaminants. In the sand layer (8.50 - 14 meters bgs) in-between the peat and clay layer underneath the landfill, the groundwater quality had not been contaminated as well.

Several soil examinations showed that the waste area corresponds to an area of approximately 5,500 m^2 and the depth of the waste material varies from 1.00 meter to 3.50 meters bgs. In total the site occupied a maximal remediation area of 9,430 m^2. Figure 1 shows the remediation site Diemen-Noord including the sections used for excavation.

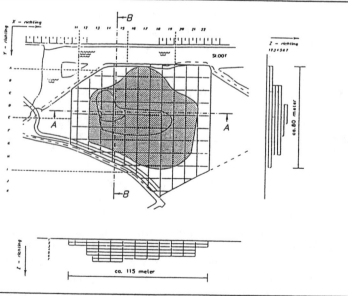

FIGURE 1. Contour of the site Diemen-Noord including the sections used for excavation

Figure 1 shows cross-sections of the site, A-A and B-B, that indicates up to which depth the toxic waste had been removed before the start of the soil excavation. Totally 7 layers of toxic waste, with a thickness of 0,5 meter each (z-direction) were excavated during the remediation. After removal of the toxic waste the underlying soil was excavated, also layer by layer. The diameter of the site in the x-direction (A-A) and y-direction (B-B) amounts respectively circa 80 meters and 115 meters.

Contaminants. The chlorinated organic compounds present in the waste material include: organochlorine pesticides, chlorobenzenes, chlorophenols and dioxines. Maximum concentrations in the soil varied between 3.9 mg/kg dry soil (d.s.) for chlorophenol and 520 mg/kg d.s. for chlorobenzenes.

The groundwater in the landfill was contaminated with organochlorine pesticides and less with aromatic solvents.

SOIL REMEDIATION

Excavation Strategy. The toxic waste present at the remediation site was removed by excavation. Next the first 0.5 meter of the original soil underneath and near the landfill was excavated.

Before the excavation of the remediation site started the groundwater level was lowered first. The remediation site had been excavated in sections of 10 meters by 10 meters and in layers of 0.5 meter. Every time a layer was removed, samples were taken from the underlying soil. In this way the site had been excavated to a maximum depth of 5.80 meter bgs. In total approximately 35,000 tonnes of soil and waste had been removed during the remediation. Figure 2 shows the excavation depths at the remediation site.

Excavation depths of the remediation site

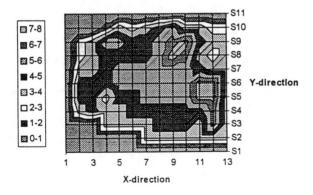

FIGURE 2. Excavation depths at the site Diemen-Noord

The maximum excavation depth at this site amounts approximately 8 meters. The depth of the toxic waste varied between 1.0 and 3.50 bgs. So the organic contaminants were transported in the soil for approximately 5 meters.

Contaminated Soil Processing. Both contaminated soil and non toxic waste material released during the remediation was stored at a special for this occasion developed disposal. The precautions taken for soil protection at this disposal consisted of a double confined layer of sand-bentonite and a HDPE-foil (high density polyethylene) of 2 mm thickness. The percolationwater, rainwater and compressionwater from the organic waste was collected and buffered by a separated drainage system.

In foundation from the taken samples and the executed analysis of the processed waste in this disposal, a rough estimate of the contaminant load was made. In Table 2 a view is given from the estimated amounts of removed organic contaminants brought to the disposal.

TABLE 2. Removed amounts of organic contaminants

Organic contaminant	Amount (kg)
Dioxines and Furanes	0.7
Chlorobenzenes and Chlorophenols	2,800
Pesticides	3,100

Toxic Waste Processing. After the removal of the drums and residues of the drums (including attached soil), this material was transported in a closed liquidtight container to a drumprocessor. Then the toxic waste material was combusted in an incinerator.

Water Treatment Plant. The purpose of the water treatment plant was to purify the water originating from the open drainage. Because of the presence of suspended solids with adsorbed contaminants, the first purpose of the water treatment was to remove these suspended solids. The second purpose was to remove the dissolved organic contaminants. The water treatment plant was dimensioned for a 5 m^3/hr flow rate. After purification the water was discharged at the surface water (Amsterdam-Rijncanal).

The water treatment plant built at the remediation site consisted of the following four compartments:

- flocculation-unit for regulating the pH of the water and addition of $FeCl_3$ (ferrichlorine) to stimulate sedimentation;
- sedimentation-unit; due to the utilization of sedimentation basins with a volume of 250 m^3 with a minimum retention time of 12 hours;
- sand filtration over 1.0 m^3 filtration sand with a grain size between 0.8 and 1.5 mm;
- bag filtration over two filters in principle with a retention of respectively 50 and 1 μm.

OCCUPATIONAL HYGIENE AND SAFETY MEASURES

Personnel. A decontamination-unit was build to be used by the personnel working at the remediation site and consisted of four compartments, all with different functions:
- dirt compartment;
- gas mask compartment with toilets;
- shower compartment;
- clean compartment.

The personnel working at the remediation site used liquidtight overalls and breathprotection by a blow-unit with dust filter and gas mask. During the remediation activities precautions were taken to prohibit dust. Never-the-less dust filters were changed every day.

Material. For the material a distinguish was made between trucks and other material like tow trucks and shovels. Both had different requirements mainly stem from the fact that tow trucks and shovels in principle did not leave the contaminated remediation site and other trucks drive to and from the remediation site.

The machinery used at the remediation site were equipped with overpressure cabins and dust filters for the airsupply. Beside the overpressure cabins with filters the tow trucks and shovels were equipped with a sparkle catcher at the funnel. Truck drivers were prohibited to leave there cabins at the remediation site. Windows and doors had to be kept closed at the contaminated area and the wash place.

Trucks working at the remediation site were cleaned from attached toxic waste and contaminated soil in special constructed washplaces before leaving the site. In this way contamination of the environment was prevented. The washplace was connected to a water treatment plant for the purification of the extracted groundwater.

Environment. The chlorinated organic contaminants found at the remediation site adsorb strongly to organic material of the soil. The biggest risk during the remediation was windblowing of contaminated soil particles. To reduce this risk, specific precautions were taken and dust concentrations in the air were measured continuously.

To prevent formation of dust during the remediation a sprinkler installation kept working. This sprinkler installation kept the remediation site wet at places were the groundwater level was lowered. At the other excavation frontages where no remediation activities took place were covered with agricultural foil to prevent formation of dust at these places.

In spite of these dust preventive measures dust measurements were performed during the remediation. These measurements took place by two methods, both windward and leeward. In case of the first method, the so-called HVS-method (High Volume Sampling), air was taken in during the day (70 m³/hr) which made an estimation of the dust concentration in the in-door air possible. A second method, the continuous dust analyzer, measered continue and gave information of actual dust concentrations at that moment in the air.

Besides the prevention of dust and controlling dust formation, activities at the remediation site were prohibited at wind speeds over 15 m/sec. To control this a continuous measuring weather station was placed at the remediation site.

ANAEROBIC/AEROBIC MICROBIAL COUPLING FOR THE BIODEGRADATION OF POLYCHLORINATED BIPHENYLS

Serge R. Guiot, Boris Tartakovsky, Jalal Hawari, Peter C.K. Lau
(Biotechnology Research Institute, NRCC, Montreal, Canada)

ABSTRACT: A consortium of aerobic and anaerobic microorganisms was developed and tested for degradation of polychlorinated biphenyls (PCBs) in a coupled anaerobic-aerobic reactor system (CANOXIS). The reactor was inoculated with anaerobic granular sludge augmented with biphenyl and chlorobiphenyl aerobic degraders, *Rhodococcus* sp. strain M5. The reactor system was fed, first, with a biphenyl-isobutanol mixture, and then an Aroclor 1242-isobutanol mixture. System monitoring demonstrated fast aerobic degradation of biphenyl combined with significant methane production. The onset of PCB feeding (12 mg/L in the influent) did not impair the methanogenesis. The effluent was found to contain not more than 0.06 mg/L of PCBs which are much less chlorinated homologs than the original compound. PCB was dechlorinated at a rate of 1.4 mg PCB/g volatile suspended solid (VSS).day. At the end of the reactor operation, 16S rDNA-sequencing identified the presence of *Pseudomonas*, *Xanthomonas* and *Rhodococcus* spp. other than *Rhodococcus* sp. strain M5.

INTRODUCTION

Polychlorinated biphenyls (PCBs) were used intensively in many industrial applications until mid-seventies when the use of these compounds was banned due to their potential carcinogenicity. While PCBs are not appreciably degraded under conventional aerobic conditions (Zitomer and Speece 1993), they can be biologically dechlorinated under reduced, anaerobic conditions (Abramowicz 1990, Mohn and Tiedje 1992). This results in the formation of less chlorinated congeners that are more amenable to further aerobic mineralization. Aerobic microorganisms have dioxygenase enzyme systems that can cleave lightly chlorinated or dechlorinated biphenyls (Abramowicz 1990). Combined anaerobic-aerobic treatment is thus required to achieve complete mineralization of PCBs.

While traditional approach of sequential biotreatment involves a train of reactors, coupled anaerobic-aerobic conditions can be easily established in a single upflow anaerobic sludge bed (UASB) reactor with limited aeration (Shen and Guiot 1996). The performance of the coupled system is expected to be superior to that of a sequential set-up due to the close proximity of the source and the sink of intermediates and the elimination of toxic intermediate accumulation.

Objective. The objective of this study is to compare the PCB dechlorination efficiency of a *Rhodococcus* M5 strain-augmented coupled anaerobic-aerobic reactor system (CANOXIS) to a strict anaerobic system.

MATERIALS AND METHODS

Experiments were performed in a CANOXIS bioreactor that consisted of a 1 L UASB-type column and a 0.75 L aeration column. Bulk liquid was recycled from the top of the UASB column through the aeration column and back to the reactor base, for the purpose of oxygen transfer to the entire system. Design details are given in Shen and Guiot (1996) and Guiot (1997). The oxygenation rate was 0.5 $L/L_{reactor}$.day. The reactor was inoculated with anaerobic granules enriched with a pure culture of *Rhodococcus* sp. strain M5 (Wang et al. 1995). The biomass amount was 5.5 g volatile suspended solids (VSS) in the system at steady state. The temperature was 30°C, the pH ranged from 7.0 to 7.5, and hydraulic residence time was 2.5 days. The influent contained either 24 mg/L of biphenyl and 2.4 g/L of isobutanol (first 34 days) or 12 mg/L of Aroclor 1242 and 2.4 g/L of isobutanol (days 35-63). The nutrients composition is given in Michotte (1997). Reactor monitoring included measurements of CH_4, CO_2 in the off-gas and isobutanol, biphenyl, PCBs in the effluent. Performance of the CANOXIS reactor was compared with that of a strict anaerobic UASB reactor. The latter contained 9.5 g $VSS/L_{reactor}$ of similar anaerobic granular sludge and was operated at the conditions given above, but aeration.

Microorganisms. Anaerobic sludge was obtained from a UASB reactor treating wastewater from a food industry (Champlain Industries, Cornwall, Ontario). A strain of *Rhodococcus* sp. M5 was grown on a minimal salt medium with biphenyl as the sole carbon source (Kimbara et al. 1989). Initial attachment of *Rhodococcus* M5 to the surface of anaerobic granules was achieved by coating the granules with a layer of chitosan-lignosulfonate polymer containing *Rhodococcus* sp. M5 according to the procedure described by Tartakovsky et al. (1996). The ratio of anaerobic sludge to *Rhodococcus* M5 was 10:1 by weight. Along with the reactor operation, the population of the biphenyl-degraders was characterized. Representative samples were taken from the CANOXIS biomass bed, and cultivated in the presence of biphenyl. The enrichments were then plated; 16S rDNA-sequencing of the colonies was performed as described by Wang et al. (1995) and Michotte (1997).

Chemicals and analytical methods. Aroclor 1242 was obtained from Monsanto (St.Louis, MO). Defined PCB congeners were purchased from AccuStandard (New Haven, CT). Chitosan (high molecular weight) was purchased from Fluca Chemical Corp. (Ronkonkoma, NY). Lignosulfonate (lignosite 458) was provided by Georgia-Pacific Corp. (Bellingham, WA). All other chemicals were of analytical grade. Off-gas composition (CH_4 and CO_2) was determined by gas chromatography (GC) (Sigma 2000, Perkin-Elmer, Norwalk, CT). Details are given in Shen and Guiot (1996). Isobutanol and biphenyl concentrations in the reactor effluent were determined using gas chromatography and high performance liquid chromatography (HPLC) (Thermo Separation Products, San Jose, CA), respectively. Chemical oxygen demand (COD) was measured by colorimetry using a DR/3000 spectrophotometer (Hach, Loveland, CO) at 650 nm. The inorganic chloride content was determined using the mercuric thiocyanate method (Standard Methods 1989). PCBs from the samples were extracted with hexane and analyzed using a GC according to the method described by Hawari et al. (1992).

RESULTS AND DISCUSSION

In the CANOXIS reactor, the methane yield ranged between 0.04 and 0.15 L CH_4/g COD over the experimental period as opposed to 0.20-0.30 L CH_4/g COD in the anaerobic UASB reactor. The lower methane yield that was observed in the CANOXIS reactor indicated that part of the carbon source was utilized by aerobes, yet a portion was available to anaerobes. This also demonstrated that the peripheral aerobic and facultative bacteria limited the O_2 penetration into the CANOXIS granules which allowed for the innermost methanogens to remain notably active.

To establish steady population of PCB-degraders, for the first 34 days the reactor was fed with the biphenyl-isobutanol mixture. The pathway of PCB mineralization is similar to that of biphenyl and proceeds via cleavage of the biphenyl ring (Furukawa 1994). At the influent concentration of biphenyl of 24 mg/L the effluent concentration was less than 1 mg/L. Also, biphenyl did not accumulate in the sludge. This suggests complete mineralization of biphenyl by *Rhodococcus* M5 and by other aerobic biphenyl degraders which were expected to be present in the sludge and could proliferate in the presence of oxygen.

The onset of PCB feeding did not affect the rates of methane and carbon dioxide production (data not shown). At an influent concentration of Aroclor 1242 of 12 mg/L, the PCB effluent concentration was not more than 0.06 mg/L. This corresponds to a removal efficiency of 99.5%. Due to the low solubility of PCBs in water, their accumulation in the sludge was anticipated. However, only 4% of the total amount of PCBs fed in the reactor accumulated in the sludge by the end of the experiment. The PCBs in the effluent mostly consisted of trichlorobiphenyls (0.03-0.07 mg/L) as shown in Fig. 1b. This observation can be explained by combined anaerobic-aerobic degradation of PCBs. Reductive dehalogenation reduced the number of highly chlorinated congeners while aerobes mineralized the resulting lightly chlorinated congeners.

The PCB dechlorination rate estimated using a chloride balance over the reactor was 0.6±0.2 mg Cl -/g VSS.day. This would be equivalent to a complete dechlorination rate of 1.4±0.5 mg PCB/g VSS.day. The dechlorination efficiency was 74.3 % while the biphenyl concentration in the effluent was below 0.1 mg/L.

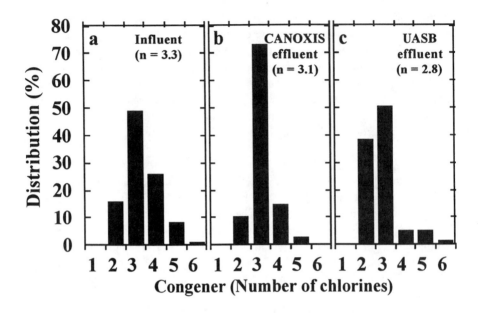

FIGURE 1. Distribution of homologs of PCB, in the feed (Aroclor 1242) (a), in the CANOXIS effluent (b) and in the anaerobic UASB effluent (c), at end of experiment. n denotes the average number of chlorines per biphenyl.

In comparison, the effluent of the anaerobic reactor, as expected, contained a large number of lightly chlorinated congeners with an average number of chlorines per biphenyl of 2.8 (Fig. 1c). The influent contained a Aroclor 1242 concentration of 20 mg/L, solubilized by the addition of a non-toxic surfactant (Tween 80, 1 g/L). Chloride balance over the anaerobic reactor showed a dechlorination rate of 0.28±0.1 mg Cl-/g VSS.day, i.e. twofold lower than that in the coupled CANOXIS system. This dechlorination rate corresponded to the maximum dechlorination specific capacity of the anaerobic biomass, since further increase in the PCB loading did not result in an higher specific dechlorination rate, while PCB concentration in the effluent increased significantly. In fact, at the highest PCB specific load (1.3 mg PCB/g VSS.day) the anaerobic reactor started to fail, releasing an effluent concentration of PCB close to the influent one. In comparison, the above performance of the CANOXIS was obtained at a PCB specific load of 1.7 mg PCB/g VSS.day.

Analysis of bacterial populations at the end of the reactor operation, using 16S rDNA-sequencing showed the presence of *Pseudomonas*, *Xanthomonas* and *Rhodococcus* spp. other than *Rhodococcus* ap. strain M5. It appeared that

Rhodococcus M5 strain was not the dominant strain but outgrown by other aerobic bacteria. Thus optimization of the start-up procedure and control of the oxygenation rate may allow for natural development of anaerobic-aerobic consortia for mineralization of PCBs.

CONCLUSION

The coexistence of anaerobic and aerobic populations within a single reactor system optimized the coupling of reductive and oxidative metabolisms of PCB degradation, thus leading to complete mineralization. Anaerobic granules - or more generally anaerobic biofilms - can be used as precursors for coupling anaerobic-aerobic populations within a single system. This may be easily feasible at large scale and with large volumes of pollution.

Acknowledgment. Authors thank J. Breton, J.J. Cadieux and A. Michotte, for reactors' control; D. Labbé, for molecular biology contribution; A. Corriveau, C. Rohfir, and E. Zhou for analytical chemistry support.

REFERENCES

Abramowicz, D. A. 1990. "Aerobic and anaerobic biodegradation of PCBs: a review." *Biotechnology 10*:241-251.

APHA, AWWA, and WPCF. 1989. *Standard Methods for the Examination of Water and Wastewater.* American Public Health Association: Washington, DC.

Furukawa, K. 1994. "Molecular genetics and evolutionary relationship of PCB-degrading bacteria." *Biodegradation 5*:289-300.

Guiot, S. R. 1997. "Anaerobic and aerobic integrated system for biotreatment of toxic wastes (CANOXIS)." United States Patent No. 5,599,451.

Hawari, J.A., A. Demeter, and R. Samson. 1992. "Sensitized photolysis of polychlorobiphenyls in alkaline 2-propanol: dechlorination of Aroclor 1254 in soil samples by solar radiation." *Environ. Sci. Technol. 26*:2022-2027.

Kimbara, K., T. Hashimoto, M. Fukuda, T. Koana, M. Takagi, M. Oishi, and Y. Yano. 1989. "Cloning ans sequencing of two tandem genes involved in the degradation of 2,3-dihydroxybiphenyl to benzoic acid in the polychlorinated biphenyl-degrading soil bacterium *Pseudomonas* sp. strain *KKS102*." *J. Bacteriol. 171*:2740-2747.

Michotte, A. 1997. "Dégradation de biphényls polychlorés par un consortium aérobie/microaérobie (*Rhodococcus* sp. M5) dans un réacteur à alimentation continue." Mémoire, Faculté des Sciences Agronomiques, Gembloux, Belgique.

Mohn, W. W., and J. M. Tiedje. 1992. "Microbial reductive dehalogenation." *Microbiol. Rev. 56*:482-507.

Tartakovsky, B., L. Petti, and S. R. Guiot. 1996. "Removal of chlorinated compounds using mixed bacterial cultures immobilizeed in chitosan gel." In H. Dautzenberg and D. Poncelet (Eds.), Proc. of the Int. Workshop on Bioencapsulation V : from fundamentals to industrial applications. Sept. 23-15, 1996, Potsdam, Germany, pp. T5:1-5.

Shen, C. F., and S. R. Guiot. 1996. "Long-term impact of dissolved oxygen on the activity of anaerobic granular sludge." *Biotech. Bioeng. 49*:611-620.

Wang, Y., J. Garnon, D. Labbe, H. Bergeron, and P. C. K. Lau. 1995. "Sequence and expression of the *bpdC1C2BADE* genes involved in the initial steps of biphenyl/chlorobiphenyl degradation by *Rhodococcus* sp. M5." *Gene 164*:117-122.

Zitomer, D. H., and R. E. Speece. 1993. "Sequential environments for enhanced biotransformation of aqueous contaminants." *Environ. Sci. Technol. 27*:227-244.

PHOTOCHEMICAL DEGRADATION OF DIOXINS IN SOIL

Pirjo Isosaari (National Public Health Institute, Kuopio, Finland), Tuula Tuhkanen and Zhanghua Yan (University of Kuopio, Finland), Terttu Vartiainen (National Public Health Institute and University of Kuopio, Finland)

ABSTRACT: Experiments on direct and indirect photochemical degradation of PCDD/Fs in soils were conducted in two summertime studies, in 1996 and 1997. The reagents used in the experiments were ethanol, titanium dioxide and hydrogen peroxide, either alone or together with iron, as Fenton's reagent. In further experiments, the depth of photolysis was investigated and a surfactant solution was used to immobilize soil-bound contaminants. As a preliminary test for the latter experiments, soil washing capacities of five surfactants and alkaline water solution were evaluated.

The best reduction in the PCDD/F concentration of the soil, 34% of I-TEqs, was obtained with titanium dioxide powder which was spread on the soil surface. Furans were more degradable than dioxins, and the reductions in the total PCDD/F concentrations were higher than those in the concentrations of the toxic 2,3,7,8-chlorinated congeners, measured in I-TEqs. Sunlight exposure alone was unable to degrade PCDD/Fs, which was also observed from the results of a soil column test. Some degradation of PCDD/Fs seems to occur in the surfactant-amended soil samples exposed to sunlight. The soil washing experiments with various surfactants resulted in 1-10% removal of PCDD/Fs from the soil.

INTRODUCTION

Polychlorinated dibenzo-*p*-dioxins and dibenzofurans (PCDD/Fs) are known to be so persistent toward chemical and microbial degradation that incineration tends to be the only generally acceptable means of treating contaminated soil. However, after realizing the extent of the contamination problem, it has turned out to be impossible to remediate all the contaminated areas by incineration. In Finland, the main source of PCDD/Fs at industrial sites is the production and use of a chlorophenolic fungicide.

In order to be able to remediate contaminated sites, cost-effective treatment methods have to be developed. The use of photochemical methods could provide a solution. The occurrence of PCDD/F photodegradation in the atmosphere has been demonstrated, and evidence on their photodegradation in solutions has been obtained in laboratory experiments, using either direct or indirect (sensitized) photochemical methods. Even though PCDD/Fs absorb light at those wavelengths reaching the soil surface, the low intensity of the radiation limits the rate of degradation. The problem can be overcome by adding such reactants which absorb light at higher wavelengths and act as hydrogen donors in the reaction system, e.g., the semiconductor titanium dioxide and ethanol. Ethanol may also act as a solubilizing agent that makes the particle bound PCDD/Fs less shielded from sunlight and possibly transfers them to the soil surface.

Other reagents commonly used for the immobilization of contaminants include various surfactants. Although PCDD/Fs are considered practically insoluble, it has been shown that in the presence of surfactants their leaching rates in soils increase (Schramm et al. 1995). Solutions of various surfactants have been used to enhance the photodegradation of hydrophobic contaminants (Chu and Jafvert, 1994).

The additional energy provided by sunlight has been shown to increase the rates at which Fenton's reagent degrades organic contaminants. The yields of hydroxyl radicals are higher in the photochemically enhanced Fenton's reaction than in the conventional method (Venkatadri and Peters, 1993).

METHODS

Soil material. The soil used was obtained from a compost pile which had been constructed some years ago. The purpose of the composting was to remediate contaminated sawmill soil, since the chlorophenols were biodegradable (Laine and Jørgensen, 1996). It was not known, however, that the soil contains nondegradable impurities, PCDD/Fs, at concentrations as high as 50 ng/g (given in international toxicity equivalents, I-TEqs). The main congeners were hepta- and octachlorinated furans. The soil was used in the photochemical experiments after sieving it to the particle size smaller than 2 mm. Ignition loss, representing the organic matter content of the soil, was 10% of dry weight.

First summer's experiments. The first summer's screening of methods included the use of the following reagents: titanium dioxide, ethanol and Fenton's reagent consisting of ferrous iron solution or iron powder with hydrogen peroxide additions. Dark controls and untreated soil were used as reference samples. The total exposure time was three months, from June to August, and sampling was done every second week. Five-gram portions of soil were spread on small petri dishes so that the soil layer was about 5 mm thick, and after this the reagents and water were added. The dishes were placed outdoors, onto a roof of a building located at 63 °N latitude. During the exposure more ethanol, hydrogen peroxide, ferrous iron solution and water (pH 5) were added, and data of temperature and the intensity of UV-irradiation on the roof were recorded. Borosilicate glass was used to protect the samples from wind and rain.

Soil washing experiments. Five surfactants and an alkaline water solution at pH 10 were tested for their capability to immobilize PCDD/Fs present in weathered soil. The surfactants used were sodium dodecylsulfate, Tween 80 (polyoxyethylene(20) sorbitan monooleate), soft tall oil soap and two detergents sold for household use: one of them was a dish-washing concentrate and the other was a washing powder without bleaching agents. Soil-surfactant slurries were stirred twice, and the foam, liquid and soil phases were analyzed for PCDD/Fs.

Second summer: effects of exposure time and soil depth. This summer the sampling was done more frequently to find out how fast the degradation occurs. The samples were rapidly extracted after the sampling, so that no degradation would occur after removing the samples from the roof. To find out whether the degradation is restricted to the surface of the soil or present in deeper layers, too, we packed columns with about 8 cm of soil, exposed them to sunlight, cut the soil in eight slices and analyzed the slices for PCDD/Fs. These experiments were done with untreated soil and soil watered with a washing powder solution.

Sample analysis. All of the samples from first summer's photochemical experiments and the soil washing samples were extracted by sonication and purified using three columns packed with silica gel-sodium sulfate, activated carbon and alumina (Vartiainen et al., 1997). For the extraction and purification of the other samples a more rapid method, Supercritical Fluid Extraction (SFE), was introduced. Method development was made to optimize extraction conditions and to separate PCBs from PCDD/Fs. In the analyses, a mixture of 16 C-13 labeled PCDD/Fs was used as an internal standard; a recovery standard was also used. The measurements were carried out with a high resolution mass spectrometer (HRGC/HRMS).

RESULTS AND DISCUSSION
Temperature and UV-intensity. The average daily temperatures measured during the 12-week exposure periods were 14.5 °C and 17.2 °C in summer 1996 and 1997, respectively. The temperatures under the borosilicate glass were 6-7 degrees higher, the highest temperatures being 48.8 °C and 58.2 ° C in 1996 and 1997. These temperatures are too low to volatilize PCDD/Fs. The highest UV-light intensities measured were 30-35 W/m^2, and during the 12 weeks' exposure the total energy inputs were 178 MJ/g of soil (yr. 1996 petri dishes), 147 MJ/g (yr. 1997 petri dishes) and 75 MJ/g (yr. 1997 soil columns). However, these values do not account for the shielding effect of the borosilicate glass. Based on the data given by the manufacturer, transmissions are 20-80% at wavelengths 290-320 nm. 2,3,7,8-TCDD has an absorption maximum at the wavelength 300-315 nm, and PCDFs have absorption maxima at regions 290-305 nm and 315-335 nm. Although the energy received by the columns was lower than that received by the petri dishes, it should be noted that in practice the light only reaches the surface and affects there.

Soil washing. The soil washing experiments led to the distribution of 4-22% of the total PCDD/F amount into the liquid and foam fractions, thus reducing the concentrations in the soil fraction. Reductions measured in I-TEqs were 1-10%. Liquid and foam fractions contained fine particles, so that it is likely that the removal of PCDD/Fs was both due to the enhanced solubility of PCDD/Fs and the transfer of particle-bound PCDD/Fs into these fractions. The best removal efficiencies were obtained with the soft tall oil soap (from Havi Co., Finland) and the detergent "Fairy" (from Procter & Gamble Ltd., UK). These two surfactants were effective in immobilizing furans, especially, whereas the washing powder "Omo" (from Lever, UK) seemed to immobilize dioxins, too.

Photochemical degradation. Reductions in PCDD/F toxicity (given in international toxicity units, I-TEqs) obtained after a 12-week exposure to sunlight are shown in Table 1. Measured in absolute (total) concentrations, the reductions were higher than those measured in I-TEqs. The concentrations in the photochemically treated soil samples were compared with untreated reference soil that was only stored in a cool place. No reduction in PCDD/F toxicity took place in the soil sample which was only exposed to sunlight without further additives - only the additions of water at pH 5 were made three times a week, as for the other samples, too. In those experiments where degradation was observed, furans were more degradable than dioxins. The best reduction, 34%, was obtained with the higher dosage of titanium dioxide (Degussa P25) spread on the surface of the soil. However, after 8 weeks of exposure this method was still not effective, and some problems such as the shielding effect and the poor contacts of titanium particles with soil particles may retard degradation. In cases of ethanol and Fenton's reagent treatments, increasing the reagent dose did not seem to improve the reductions. The difference between the two doses of Fenton's reagent was small, since high concentrations of hydrogen peroxide would have caused a violent reaction.

TABLE 1. Reductions in the toxicity (I-TEq, pg/g) of PCDD/Fs after the 12 -week photochemical experiments.

Treatment	Reduction %
Light only	-4
Ethanol 1%	25
Ethanol 10%	0
TiO_2, surface, 20 mg	21
TiO_2, surface, 100 mg	34
TiO_2, mixed, 100 mg	24
Fe powder, 500 mg	28
Fe powder, 500 mg + 1 ml of 3% H_2O_2	13
0,25 mmol Fe(II) + 0.5 ml of 3% H_2O_2	24
0.25 mmol Fe(II) + 1 ml of 3% H_2O_2	24
1 ml of 3% H_2O_2	25

Effects of exposure time and soil depth. In Table 2, results of one series of petri dish experiments and two different column experiments are presented. The variations in the PCDD/F concentrations are quite large, but there are no clear trends in the degradation. There are two reasons for this: first, referring to the results obtained in the earlier experiments (Table 1), the sunlight only -treatment is not effective enough to degrade PCDD/Fs. Secondly, the SFE method has not given sufficient repeatability, most probably because of the small size of the soil sample (100 mg).

TABLE 2. Reductions in the toxicity (I-TEq) of PCDD/Fs after the 12-week exposure to sunlight, with and without the surfactant "Omo".

Treatment (petri dish)	Reduction %	Treatment (column)	Reduction %
Light only, 3 days	-2	Light only, 0-9 mm	15
Light only, 1 week	17	Light only, 9-18 mm	25
Light only, 4 weeks	32	Light only, 18-27 mm	13
Light only, 12 weeks	8	Light only, 27-36 mm	32
		Light only, 36-45 mm	16
Surfactant, 3 days	42	Light only, 45-54 mm	16
Surfactant, 1 week	30	Light only, 54-63 mm	20
Surfactant, 2 weeks	23	Light only, 63-72 mm	-12
Surfactant, 3 weeks	33		
Surfactant, 4 weeks	61		
Surfactant, 8 weeks	18		
Surfactant, 12 weeks	-2		
Surf., 12 weeks, dark	3		

CONCLUSION

The results of the photochemical experiments show that it is not very easy to remediate PCDD/F-contaminated soil. Within three months, 20-30% reductions in PCDD/F concentrations could be achieved. The reason why the reductions did not exceed this level could be the fact that the rest of the PCDD/Fs are so tightly bound to the soil particles that they are not available to photochemical reactions. Similarly, the negative reductions obtained in some experiments might be explained by the immobilizing effect of the reagents, or by the degradation which leads to the formation of more toxic congeners. Movement of PCDD/Fs may also have occurred within the layers of the soil columns.

REFERENCES

Chu, W. and Jafvert, C.T. 1994. "Photodechlorination of Polychlorobenzene Congeners in Surfactant Micelle Solutions." *Environmental Science and Technology* 28(13): 2415-2422.

Laine, M.M. and Jørgensen, K. S. 1996. "Straw Compost and Bioremediated Soil as Inocula for the Bioremediation of Chlorophenol-Contaminated Soil". *Applied and Environmental Microbiology 62(5)*: 1507-1513.

Schramm, K.-W., Wu, W.Z., Henkelmann, B., Merk, M., Xu, Y., Zhang, Y.Y. and Kettrup, A. 1995. "Influence of Linear Alkylbenzene Sulfonate (LAS) as Organic Cosolvent on Leaching Behaviour of PCDD/Fs from Fly Ash and Soil." *Chemosphere 31(6)*: 3445-3453.

Vartiainen, T., Mannio, J., Korhonen, K., Kinnunen K., and Strandman, T. 1997. "Levels of PCDDs, PCDFs and PCBs in Dated Lake Sediments in Subarctic Finland." *Chemosphere 34*: 1341-1350.

Venkatadri, R. and Peters, R. 1993 ."Chemical Oxidation Technologies: Ultraviolet Light/Hydrogen Peroxide, Fenton's Reagent, and Titanium Dioxide-Assisted Photocatalysis." *Hazardous Waste & Hazardous Materials 10(2)*: 107-149.

ANAEROBIC / AEROBIC BIOREMEDIATION OF PCB-CONTAMINATED SOIL AND SEDIMENT

Timothy M. Vogel [1] and Thomas Schlegel [2]
[1] Rhodia Eco Services [2]Antipollution Techniques Entreprise
Meyzieu, France

ABSTRACT

PCB-contaminated soil can be remediated by sequential anaerobic/aerobic biodegradation. This sequential process was applied to both soils and sediments in order to determine the efficiency and treatment duration. The anaerobic phase was initiated by the addition of nutrients and an external carbon source. The aerobic phase was initiated by either hydrogen peroxide or air plus other nutrients. In soils, the concentration dropped from about 50 ppm to 10 ppm in five months. In sediments, the process was considerably longer. Analytical and regulatory aspects play a major role in the determination of the applicability of this technique.

INTRODUCTION

Polychlorinated biphenyls (PCBs) are a family of potentially 209 related chemical compounds synthesized by the direct chlorination of biphenyl. These compounds were manufactured and sold as complex mixtures (under trade names Aroclor, Phenoclor, Clophen, and Kanechlor) with formulations, produced for different applications, that differed in their average chlorine content. The individual PCB isomers, or PCB congeners, in these mixtures are referred to according to the position of the chlorine substituent (e.g., Figure 1). Obviously up to ten chlorine substituents could theoretically exist, although averages run between 5 and 7 generally. All 209 possible configurations are not available even in mixtures.

The desirable physical and chemical properties of PCBs (dieletric, flame resistence, thermal and chemical stability) led to their extensive industrial use as heat transfer fluids, hydraulic fluids, solvent extenders, plasticizers, flame retardants and dielectric fluids (Hutzinger, Safe, and Zitko, 1974). Extensive application of these chemically and thermally stable compounds has resulted in their widespread contamination of soils and sediments (Buckley, 1982). Several hundred million kilograms of PCBs have been estimated to have been released into the environment (Huzinger and Veerkamp, 1981) (Figure 2). Although, in general, PCBs are considerably hydrophobic, their octanol/water partition coefficients, Kow, range over 5 orders of magnitude from for example log Kow of 5.22 for a tricholorobiphenyl to 10.44 for a nonachlorobiphenyl. Hence, they tend to concentrate in sediments and soils. Their solubilities range from about 0.1 ug/L for decachlorobiphenyl to about 7500 ug/L for chlorobiphenyl. Yet, as the world's oceans represent the largest sink, although relatively dilute, they contain about 80 % of the PCBs found in the mobile environmental reservoir (Figure 3).

ANALYTICAL DIFFICULTIES

Critical to predicting the existance and fate of PCBs in the environment is an understanding of the range of chemical and physico-chemical characteristics. PCBs are not monolithic and different mixtures result in different behavoirs as a group and also as individual molecules (or congeners). For example, the vapor pressure of individual congeners is a function of the number of chlorine substituents but also the location of these chlorine substituents. PCBs with fewer chlorines and more ortho-chlorines tend to have higher vapor pressures. Higher vapor pressures are often correlated with chormatographic retention times (Figure 4). Due to this variablility in vapor pressure, several trends can be considered. The first is the effect this has on the volatilization of different congeners with time from soils causing a final chromatographic imprint inconsistent with commercial mixtures. The second is the chromatographic analyses themselves. The separation of all PCB congeners in a mixture is not observed. With the use of capillary columns, over 40 peaks can be separated. For example, a mixture of 67 molecules can result in 44 individual chormatographic peaks. The double effect of the diversity of PCB physico-chemical characteristics on the evolution of congener importance with time and the difficulties of analytical measurements can lead to misunderstanding.

In addition, the accepted methods for PCB analyses is not standardized from one country to another. For example, in the USA, the total area of all the PCB peaks (in the USA even monochlorobiphenyl is considered a PCB) are quantified relative to an internal standard (octachloronaphthalene) even though the response factors are obviously not at all similar between different PCB congeners. Concentrations are reported as mass PCBs per mass sediment or soil and not as molar concentrations. In France, the PCBs total chromatographic surface area is compared to the surface area of a known mass of a commercial mixture. Of course, if the mixtures are different or the environmental sample has evolved, then the results can be misleading. Concentrations are reported as ppm on a mass basis as the referenced commercial mixture (e.g., Aroclor 1254). In Germany, 6 specific congener standards are used to quantify accurately the respective six chromatogrphic peaks (2,4,4'-trichloro-, 2,5,2',5'-tetrachloro-, 2,4,5,2',5'-pentachloro-, 2,3,4,2',4',5'-hexachloro-, 2,4,5,2',4',5'-hexachloro-, and 2,3,4,5,2',4',5'-heptachloro-biphenyl). The smallest is a trichlorobiphenyl. These six peaks are used for subsequent regulatory decisions. Obviously, changes in the relative importance of these peaks would drastically change the results. Concentrations are also reported on a mass per mass basis. Unfortunately, these methods are also applied to environmental samples, where changes in congener importance can occur as the result of physico-chemical and biological processes. Biological processes are especially troublesome as "new" congeners (those not observed in the original mixture) can be produced.

BIODEGRADATION

PCB have been shown to undergo biodegradation under a variety of conditions in the laboratory and in the environment similar to other chlorinated organics (Abramowicz, 1990; Montgomery, Assaf-Anid, Nies, Anid and Vogel, 1991). Two distinct biological systems capable of degrading PCBs have been identified: anaerobic reductive dechlorination and aerobic oxidation.

Naturally occurring anaerobic bacteria can attack more highly chlorinated PCB congeners. In general, this reductive dechlorination process preferentially removes meta and para chlorine substituents (Figure 5 from Nies, 1993) resulting in lower chlorinated ortho-substituted PCBs. This process has been shown to occur naturally and slowly in the laboratory (Quensen, Tiedje, and Boyd 1988) and in river sediments (Brown and Wagner, 1990) and by phototrophic anaerobes (Montgomery and Vogel, 1992). Studies have demonstrated that the PCB reductive dechlorination activity under anaerobic conditions does not require special adapted microorganisms. Common bacterial co-factors (such as vitamin B12) have been shown to be capable of catalyzing PCB dechlorination (Assaf-Anid, Nies, and Vogel, 1992). The source of hydrogen for replacing the chlorine is derived biologically from water (Nies and Vogel, 1991). This anaerobic process produces some congeners that are not normally found in commercial mixtures. In addition, as the congeners are partly dechlorinated they produce less response analytically and, therefore, less chromatographic surface area and, thus, apparently "reduce" the PCB concentration on a mass per mass basis. In fact, even if the molar concentration of all PCBs did not change, the dechlorination would lead to a smaller mass as the chlorine substituent is replaced by a hydrogen. Thus, the same quantity of PCB molecules after partial dechlorination are "reduced" in concentration when expressed as ppm (mg/kg).

Aerobic bacterial degradation of PCBs has been extensively studied. Several microorganisms have been isolated from PCB-contaminated soil demonstrating a preference for degrading lightly chlorinated PCBs (Anid, Ravest-Webster, and Vogel, 1993). Most of these microorganisms attack PCB congeners via the 2,3-dioxygenase pathway to the corresponding chlorobenzoate, which is subsequently degraded by other indigenous bacteria to carbon dioxide, water, and chloride. Aerobic biodegradation has been observed for PCB congeners with up to four chlorine substituents. Naturally occurring aerobic biodegradation has been observed in several cases (Flanagan and May, 1993). Contrary to the anaerobic process, the aerobic process tends to degrade the less chlorinated compounds, specifically the mono- and dichlorobiphenyls (these are not even analyzed by the German method). The result of this degradation is a relative increase of the more chlorinated congeners (since the lightly chlorinated congeners are no longer observed) and possibly some confusion regarding the type of original commercial mixture (i.e., the mixture might appear more heavily chlorinated).

In any case, from a complete treatment viewpoint, a first anaerobic step reducing the chlorine substituents followed by an aerobic step to break up the molecule is required. Natural biodegradation of PCBs does not always occur due to limits in appropriate nutrients or the lack of the proper sequential anaerobic/aerobic processes. Even in the documented cases where natural degradation does occur, the process is very slow (years).

Recently, techniques have been developed to actively bioremediate soils and sediments contaminated by chlorinated compounds (reviewed in Adriaens and Vogel, 1995) including PCBs. The technology for highly chlorinated compounds involves the sequential anaerobic/aerobic processes. The two processes are separated either spatially or temporally. For PCBs, which are not very mobile, the processes are temporally separated (Anid, Nies, and Vogel, 1991). The engineering design can be separated into two parts: anaerobic and aerobic, although consideration of the effect of the anaerobic phase on the subsequent aerobic phase is required.

The first phase requires the induction of active anaerobic conditions if they are not already in evidence as is the case in many soils. This can be acheived by limiting oxygen diffusion from air and by the addition of a harmless suitable simple organic molecule. This compound when degraded by the aerobes present consumes all of the dissolved oxygen, provides energy and carbon for the anaerobes, and has an effect on the subsequent microbial ecology. Significant process performance variations have been linked to different carbon sources (Nies and Vogel, 1990). The correct choice is important, in addition to possible other nutrient requirements. In different in-situ pilot tests, this phase has varied from 5 to 18 months.

The second phase requires the induction of aerobic conditions. This can be acheived by various techniques of aeration (Anid, Ravest-Webster, and Vogel, 1993). These techniques need to be in conformity with the rescitation and maintenance of the proper aerobic microorganisms. This process has taken from 2 to 4 months in pilot tests. As in the anaerobic process, only natural indigenous microorganisms are used.

This technology has been applied to both soils and sediments with PCB contamination ranging from 10 to about 500 ppm with no obvious upper limit yet reached. An example of the type of results observed is shown in Figure 6, where the PCB congeners are lumped into isomeric categories in order to observed the degradation trends (mono-, di, trichloro, etc.). The anaerobic phase significantly reduced the PCB concentration and shifted the general pattern to the less chlorinated biphenyls. The subsequent aerobic phase further reduced the PCB total attacking in preference the mono-, di- and trichlorobiphenyls. Often the target degradation levels are set either by limits for putting soil into landfills (below about 100 ppm depending on the local regulations) or leaving the soil in place (less than 10 ppm also depending on the local regulations

REFERENCES

Abramowicz, D.A. 1990. Aerobic and anaerobic biodegradation of PCBs: a review. In: CRC Critical Reviews in Biotechnology, G.G Steward and I. Russell, eds. CRC Press, Boca Raton, FL, USA.

Adriaens, P. and Vogel, T.M. 1995. Biological treatment of chlorinated organics. In: Microbial Transformation and Degradation of Toxic Organic Chemicals, eds. L.Y. Young and C.E. Cerniglia. Wiley-Liss, New York.

Anid, P.J., Nies, L. and Vogel, T.M. 1991. Sequential anaerobic/aerobic biodegradation of polychlorinated biphenyls (PCBs) in the laboratory model. In: In Situ and On-Site Bioreclamation, Eds. R.E. Hinchee and R.F.Olfenbuttel. Butterworth-Heinemann Publishers.

Anid, P.J., Ravest-Webster, B.P., and Vogel, T.M. 1993. Effect of hydrogen peroxide on the biodegradation of PCBs in anaerobically dechlorinated river sediments. Biodegradation, 4:241-248.

Assaf-Anid, N., Nies, L., Vogel, T.M. 1991. Reductive dechlorination of a PCB congener and hexachlorobenzene by vitamin B_{12}. Applied and Environmental Microbiology. 58:1057-1060.

Brown, J.F., Jr. and Wagner, R.E. 1990. PCB movement, dechlorination, and detoxication in the Acushnet Estuary. Environ. Toxicol. Chem. 9:1215-1233.

Flanagan, W.P. and May, R.J. 1993. Metabolite formation as evidence for in situ aerobic biodegradation of polychlorinated biphenyls. Environ. Sci. Technol. 27:2207-2212.

Hutzinger, O., Safe, S.. and Zitko, V. 1974. The chemistry of PCBs. CRC Press, Cleveland, OH USA.

Montgomery, L. and Vogel, T.M. 1992. Dechlorination of 2,3,5,6-tetrachlorobiphenyl by a phototrophic enrichment culture. FEMS Microbiology Letters, 94:247-250.

Montgomery, L., Assaf-Anid, N., Nies, L., Anid, P.J. and Vogel, T.M. 1991. Anaerobic biodegradation of chlorinated organic compounds. In: Biodegradation and Bioreclamation Technologies, K.G. Rasul Chaudry (ed.). Dioscordes Publ.

Nies, L. and Vogel, T.M. 1990. Effects of organic substrates on dechlorination of Aroclor 1242 in anaerobic sediments. Applied and Environmental Microbiology, 56:2612-2617

Nies, L. and Vogel, T.M. 1991. Identification of the proton source for the microbial reductive dechlorination of a chlorinated aromatic compound (2,3,4,5,6 penta chlorobiphenyl PCB). Applied and Environmental Microbiology, 57:2771-2774.

Quensen, J.F., III, Tiedje, J.M., and Boyd, S.A. 1988. Reductive dechlorination of PCBs by anaerobic microorganisms from sediments. Science, 242:752-754.

AEROBIC BIODEGRADATION OF PCBS IN PHOTOLYZED AND NON-PHOTOLYZED SURFACTANT SOLUTIONS

Keith A. La Torre, Zhou Shi, Mriganka M. Ghosh, University of Tennessee , and Alice C. Layton, Center for Environmental Biotechnology, University of Tennessee, Knoxville, TN, USA

Abstract: The enhancement of aerobic biodegradation of polychlorinated biphenyls (PCBs) by UV irradiation was investigated in batch reactors. The PCBs were dissolved in a micellar solution of a nonionic surfactant, polyoxyethylene 10 lauryl ether [POL (10)]. UV irradiation was used to reductively dechlorinate highly chlorinated PCB congeners to make them more vulnerable to aerobic biochemical attack. In an integrated bioremediation scheme, (1) surfactants solubilize PCBs; (2) photolysis is used to make highly chlorinated congeners more amenable to biodegradation; and finally, (3) genetically engineered aerobic microorganisms (GEMs), capable of using surfactants for growth while cometabolizing PCBs, are used for biodegradation. At a POL (10) dosage of 10 g/L, percent degradation ranged from 52-64% in PCB solutions containing 22-269 mg/L of total PCB, the maximum removal occurring in a solution containing 46 mg/L of PCB. For 50 mg/L PCB solutions in 0.5 - 10 g/L POL (10), a maximum biodegradation of 74% was obtained at a POL (10) concentration of 2 g/L. With UV irradiation for 40 min, 63% of the PCBs in a 210 mg/L solution could be degraded, and an additional degradation of 60% of the remainder was realized in a subsequent biodegradation step. By comparison, only 52% of PCBs could be aerobically biodegraded without photolysis.

INTRODUCTION

Anaerobic microorganisms are known to biodegrade PCBs by reductive dechlorination (Brown *et al.*,1987), albeit slowly. Aerobic microorganisms can degrade PCBs more rapidly, but most often they are unable to degrade highly chlorinated PCB congeners with more than 4 chlorine substitutions (Hickey *et al.*, 1993), particularly if the substitutions occur at *ortho* positions of the biphenyl ring (2,2',6, and 6') (Furukawa *et al.*, 1978).

At concentrations above the critical micelle concentration (CMC), surfactants greatly increase the solubility of PCBs (Abdul *et. al.*, 1992). Thus, they may significantly contribute to biodegradation of PCBs mainly by increasing solubilities in water (Lajoie *et. al.*, 1994). Highly chlorinated PCB congeners in aqueous solutions, especially those with *ortho* -substituted chlorines, are most resistant to aerobic biodegradation (Sawhney, 1986). Photolysis has been shown to degrade PCBs in surfactant solutions (Epling *et. al.*, 1988; Shi *et. al.*, 1994). While *ortho*-chlorines are most resistant to aerobic biodegradation, fortuitously they are most vulnerable to photochemical attack. Thus, surfactant washing followed by photolysis and aerobic biodegradation holds promise for satisfactory remediation of PCB-contaminated soils. Genetically engineered aerobic microorganisms (GEMs) which can grow on POL (10) and concurrently cometabolize PCBs can be used for biodegradation (Lajoie *et. al*, 1994). This research investigates the feasibility of such a treatment scheme.

MATERIALS AND METHODS

Aroclor 1242, a commercial mixture of approximately 59 PCB congeners (Bedard *et. al.*, 1987), was dissolved in various concentrations (0.5-4.0 g/L) of POL (10) and used for all experiments. The molecular weight and CMC of POL (10) were 626 g mol^{-1} and 8.39 x 10^{-5} M, respectively. Based on a weight-averaged molecular weight of 260.3 for Aroclor 1242, the molar solubility ratio (MSR) and the water-micelle partitioning coefficient (K_m) for the POL (10)/Aroclor 1242 system were, respectively, 0.6165 and 2.75x 10^7. The test solutions contained total PCBs in the range of 22 to 269 mg/L. The PCB solutions were photolyzed for various time periods in a Rayonet photoreactor using an irradiance of 9.59 x 10^{-5} einstein m^{-2} sec^{-1} at 254 nm.

All biodegradation experiments were conducted in 1-L batch reactors. PCB-degrading microbial strains, *Psuedomonas putida* IPL5::TnPCB and *Ralstonia eutropha* B30P4::TnPCB, were constructed at the Center for Environmental Biotechnology (CEB), University of Tennessee, by inserting a transposon that codes for PCB degrading enzymes into two POL (10) degrading bacterial strains, *Psuedomonas putida* IPL5 and *Ralstonia eutropha* B30P4 (Lajoie *et. al.*, 1994, 1997). A 1:1 volumetric mixture (total volume = 30 mL) of the GEMS resulting in an initial bacterial concentration of approximately 4.6 x 10^8 cells/ml in all biodegradation experiments. These two organisms were found to degrade POL (10) better in tandem than alone. An identical inoculum made with the surfactant-degrading organisms lacking the *TnPCB* gene was used for control (Layton *et. al.*, 1998). The progress of biodegradation was monitored by measuring POL (10) and total PCB in solution as a function of time. Aqueous PCB was measured by gas chromatography (GC) (Bedard *et. al.*, 1987; Shi *et. al.*, 1997a) and POL (10) was measured by the CTAS method (Standard Methods, 1989). To monitor cell growth, optical density of the cell suspension was measured at 600 nm over time along with total cell using the Pierce BCA Protein Assay Kit (Pierce Instructions, product #23225). Concurrently, the activity of 2,3 dihydroxybiphenyl dioxygenase, an enzyme encoded by the *bphC* gene, was also monitored. Whole cell samples (100 µL) of the growing population were added to buffered 2,3 dihydroxybiphenyl solutions (0.11 mM) and the absorbance (A) at 434 nm was spectrophotometrically measured over time. The slope of the resulting line ($\Delta A/\Delta t$) divided by the molar extinction coefficent (ε = 221 $mM^{-1}cm^{-1}$) yielded the rate of change in concentration of *bphC* with time (Kuhm *et. al.*, 1991).

RESULTS AND DISCUSSION

Effect of PCB on POL (10) Biodegradation. The activity of the surfactant-degrading organisms (*Psuedomonas putida* IPL5 and *Ralstonia eutropha* B30P4) was not affected by the presence of Aroclor 1242. As shown in Table 1, the specific substrate utilization rate (q) of POL (10) at two concentrations of the surfactant remained unaffected by the presence of 50 mg/L of Aroclor 1242. In fact, q was linear between POL (10): Aroclor molar ratio of 16.7 and 80 indicating that the bacteria were only substrate-limited.

TABLE 1. POL (10) utilization in the presence of Aroclor 1242

POL (10) (g/L)	Aroclor 1242 (mg/L)	Specific substrate utilization rate (hr^{-1})*
2	0	0.098
2	50	0.101
10	0	0.239
10	50	0.222

*Initial POL (10) concentration was 4 g/L

To determine the optimum concentration of PCB that can be cometabolized by the surfactant-degrading GEMs in solutions containing 10 g/L of POL (10), 7-day biodegradation studies were conducted. For each PCB concentration, total PCB degraded as a function of time was obtained from the difference of PCB remaining in solution in parallel bioreactors, the experimental reactor containing the GEMs and the control. Table 2 indicates that best biodegradation was obtained at a starting PCB concentration of 46 mg/L. A sudden decrease in the activity of 2,3 dihydroxybiphenyl dioxygenase was detected after 5 days at a PCB concentration of 269 mg/L with the concurrent appearance of black metabolites in the growth culture, presumed to be chlorocatechols which were inhibitory to the GEMs. PCB degradation was the poorest in this bioreactor.

TABLE 2. Biodegradation (7 days) of Aroclor 1242 dissolved in 10 g/L of POL (10)

Initial PCB concentration (mg/L)	PCB remaining (mg/L)	Percent degraded[1]
22	4.5	57.1
46	11.9	63.7
110	40.2	58.3
269	152.1	52.4

[1]As compared with the PCB concentration in the control reactor

To determine the optimum concentration of POL (10) necessary to sustain biodegradation of Aroclor, degradation of 50 mg/L of Aroclor in the presence of various amounts of POL (10) (0.5-10 g/L) was monitored over a 7-day period. Best biodegradation was obtained at a starting POL (10) concentration of 2 g/L (Table 3). In reactors where 0.5-1.0 g/L of the surfactant was used initially, metabolism of POL (10) was sufficiently large after 2 days to cause release and subsequent precipitation of PCBs. Therefore, PCB biodegradation after seven days was not calculated in these reactors. However, as seen in Table 3, fastest degradation of PCB was realized at a starting POL (10) concentration of 1 g/L during the first 2 days.

TABLE 3. Biodegradation of PCBs in 7 days with varying initial POL (10) concentrations

Initial POL (10) concentration (g/L)	PCB remaining[1] (mg/L)	Percent degraded (%)
10	20.0	60.0
7	15.0	70.0
4	15.1	69.9
2	12.8	74.4
1	11.3	57.8
0.5	7.9	37.2

[1]Initial concentration of PCB was 50 mg/L in all bioreactors

Photolysis of Aroclor 1242. The GC of a 210 mg/L solution of Aroclor 1242 in 4 g/L of POL (10), before and after photolysis, is shown in Figure1. The initial rate of photolysis (t < 10 min) was found to be 2.076 h^{-1} with a resultant quantum yield (Φ) of 4.740x 10^{-5}, calculated by the method of Shi et al., (1997). A large fraction of the highly chlorinated congeners, with peak numbers of 31 and higher, were destroyed after only 10 minutes of photolysis.

PCB Destruction by Photolysis and Biodegradation. To study the effect of photolysis on the overall removal of PCBs by biodegradation, a 210 mg/L solution of Aroclor in 4g/L of POL (10), was subjected to UV-irradiation for 40 min at 254 nm followed by aerobic biodegradation. After adding nutrients and bacterial inoculum, the initial concentrations of PCB at the onset of the biodegradation experiment were 154 mg/L and 56 mg/L, respectively, in the unphotolyzed and photolyzed solutions. The POL (10) concentration in both solutions was 3.2 g/L. As can be seen in Table 4, photolysis alone removed 63% of the PCBs. Of the remainder, 6 days of biodegradation removed 60.1%, the total removal by the combined treatment being nearly 85.2% compared to only 51.6% by biodegradation alone. The specific substrate utilization rate (q) for the photolyzed and unphotolyzed solutions were 0.140 hr^{-1} and 0.132 hr^{-1}, respectively. Further, biodegradation of POL (10), and the activity of *bphC* remained unchanged after photolysis. Seemingly, photolysis merely dechlorinated PCB congeners, especially the highly chlorinated congeners, to different degrees depending on the number and location of the substituted chlorines (Shi *et. al.*, 1998). Thus, the congeners remaining after photolysis were more amenable to biodegradation. Also, photolysis did not appear to produce products that were toxic to biodegradation.

Table 4. Biodegradation of PCBs* with and without photolysis

Treatment	Percent of PCBs Removed
Photolysis only (40 min)	63.0
Biodegradation only (6 days)	51.6
Biodegradation only (6 days) of photolyzed (40 min) Aroclor 1242 solution	60.1
Total removal by biodegradation (6 days) and photolysis (40 min)	85.2

*210 mg/L Aroclor 1242 dissolved in 4 g/L POL (10)

ACKNOWLEDGEMENTS

The research reported here was supported by the U.S Department of Energy, Contract No. DE-FG02-97ER62350. Additional financial support was provided by the Waste
Management Research and Education Institute, University of Tennessee.

REFERENCES

Abdul, A.S., Gibson, T.L., Ang, C.C., Smith, J.C., and Sobcynski, R.E. 1992. "*In Situ* Surfactant Washing of Polychlorinated Biphenyls and Oils From a Contaminated Site." *Ground Water*. 30: 219-231.

Bedard, D.L., Wagner, R.E., Brennan, M.J., Haberl, M.L., Brown, Jr., J.F. 1987. "Extensive Degradation of Aroclors and Environmentally Transformed Polychlorinated Biphenyls by *Alcaligenes eutrophus* H850." *Appl. Envr. Micro*. 53(5): 19094-1102.

Brown, J.F., Bedard, D.L., Brennan, M.J., Carnahan, J.C., Feng, H., and Wagner, RE. 1987. "Polychlorinated Biphenyl Dechlorination in Aquatic Sediments." *Science*. 236:709-712.

Epling, GA, Florio, EM; Bourque, AJ, Quian, X; Stuart, JD. 1988. "Borohydride, Micellar, and Exiplex-Enhanced Dechlorination of Chlorobiphenyls." *Environ. Sci. Technol*. 22(8): 952-956.

Furukawa, K., Tonomura, K., Kamibayashi, A. 1978. "Effect of Chlorine Substitution on the Biodegradability of Polychlorinated Biphenyls." *Appl. Envr. Micro.* 35(2): 223-227.

Hickey, W.J., Searles, D.B., Focht, D.D. 1993. "Enhanced Mineralization of Polychlorinated Biphenyls in Soil Inoculated with Chlorobenzoate-Degrading Bacteria." *Appl. Envr. Micro.* 59(4): 1194-1200.

Kuhm, A.E., Stolz, A., Knackmuss, H-J. 1991. "Metabolism of Napthalene by the Biphenyl-degrading Bacterium *Psuedomonas paucimobilis* Q1." *Biodeg.* 2(2): 115-120.

Lajoie, CA, Layton, AC, Sayler, GS. 1994. "Cometabolic Oxidation of Polychlorinated Biphenyls in Soil with a Surfactant-Based Field Aplication Vector." *Appl. Envr. Micro.* 60(8): 2826-2833.

Lajoie, C.A., Layton, A.C., Easter, J.P., Menn, F-M., Sayler, G.S. 1997. "Degradation of Nonionic Surfactants and Polychlorinated Biphenyls by Recombitant Field Application Vectors." *J. Indust. Micro. Biotech.* 19(X): 252-262.

Sawhney, B.L. 1986. "Chemistry and Properties of PCBs in Relation to Environmental Effects." In PCBs and the Environment. Waid, J.S., and Biol, F.I. (ed.). CRC Press, Inc., Boca Raton, FL 33431, pp. 47-64.

Standard methods for the examination of water and wastewater. 1985, 16th ed. American Public Health Association, Washington, D.C.

Shi, Z., Ghosh, M.M., Cox, C.D., Robinson, K.G. 1994. "Photodegradation of Polychlorinated Biphenyls (PCBs) dissolved in Micellar Pseudophase at the Irradiated TiO$_2$ Surface." *Special Symposium on Emerging Technologies in Hazardous Waste Management VI.* Abstracts. Vol. II, pp938-944. American Chemical Society. Atlanta, GA. September 19-21, 1994.

Shi, Z., Sigman, M.E., Ghosh, M.M., Dabestani, R. 1997. "Photolysis of 2-Chlorophenol Dissolved in Surfactant Solutions." *Environ. Sci. Technol.*, 31(12):3581-3587.

Shi, Z, Ghosh, MM, Sigman, ME. 1998. "Surfactant-Enhanced Photolysis of Polychlorobiphenyl Congeners." *Wat. Res.* In Review.

A NATIONAL TSCA OPERATING PERMIT APPLICATION OF ISV

B.E. Campbell (Geosafe Corporation, Richland, WA)
C.L.Timmerman (Geosafe Corporation, Richland, WA)

ABSTRACT: Geosafe Corporation of Richland, WA has recently completed the testing and evaluation process to allow for a modification to its National Toxic Substance Control Act (TSCA) Operating Permit for the In Situ Vitrification (ISV) technology. Testing was performed at a site located in Spokane WA, which was contaminated with PCBs and PAHs from activities associated with the maintenance and disposal of used transformers. This paper presents the process through which Geosafe was granted the original National TSCA Operating Permit and its subsequent modification including: 1) project planning, 2) site preparation activities, 3) treatment of the contaminated soil and debris, 4) extensive sampling and analysis, and 5) evaluation of the data and granting of the permit.

INTRODUCTION

In support of an initial application for a National TSCA Operating Permit for the treatment of PCBs using the ISV process, demonstration testing was performed at a private superfund site in EPA Region 10. An aerial photo of the site is shown in Figure 1. Submission of the application occurred in 1989 and testing was performed between July and October 1994 and additionally in October and November of 1996. The project was designed to demonstrate the capabilities of the ISV process for treating PCB-contaminated soils and associated site debris (e.g., concrete, asphalt, and ruptured drums). The demonstration testing involved the treatment of 3321 tons of contaminated soil and debris during 1994 followed by an additional 2459 tons treated late in 1996.

The vitrified soils and off-gases were subsequently sampled and analyzed to determine the resulting destruction and removal efficiency (DRE) for PCBs. Air emissions were analyzed to confirm compliance with Washington State standards. Analyses of the surrounding soils were performed to verify the expected lack of movement of PCBs into the soils next to the treatment zones.

The results of this testing enabled Geosafe to receive a national permit to treat PCBs using the ISV process. Included in this permit is the ability to treat PCB contaminated materials up to a concentration of 1.78 weight percent PCBs, contaminated debris, and ruptured 55-gal drums. As a result of this permit, application of the ISV process can be performed to a wide range of configurations and waste profiles. Additionally, if particular configurations or waste concentrations fall outside of the established permit, procedures are in-place to gather additional data to support expansion of permit conditions.

Geosafe currently offers the ISV technology in four different application configurations, collectively called GeoMelt vitrification technologies. The four GeoMelt treatments include: 1) GeoMelt-ISV for in situ treatment, 2) GeoMelt-Staged ISV for treating materials that have been staged for processing, 3) GeoMelt-Stationary Batch for repetitive melt cycling at a single location, and 4) GeoMelt-Continuous vitrification for material feeding and melt withdrawal at a stationary facility. The GeoMelt technology is a demonstrated, commercially proven process that involves the electric melting of contaminated soil and/or other earthen materials to permanently destroy, remove, and/or immobilize hazardous and radioactive contaminants. ISV was invented by Battelle, Pacific Northwest National Laboratories in 1980 for the U.S. Department of Energy. More than 300

FIGURE 1. Aerial View of Geosafe's Large-Scale ISV Equipment at a PCB Contaminated Soil Site.

developmental tests and demonstrations of the technology have been performed since that time at various scales ranging from bench- (to 3 kg) to large-scale (to 1400-tonne).

The process involves forming a melt, which then serves as the heating element for the process. Electrical energy is directly converted to heat as it passes through the molten soil between electrodes. Continued application of energy results in the melt growing deeper and wider until the desired treatment volume has been encompassed. When electrical power is shut off, the molten mass solidifies into a vitrified glass and crystalline monolith with unequalled physical, chemical, and weathering properties compared to alternative solidification/stabilization technologies. The GeoMelt processes and the vitrified product also have advantages relative to other vitrification processes which are derived from its versatile application modes, tolerance for debris, higher melt temperatures, and monolithic waste forms.

The GeoMelt applications typically involve molten soil temperatures in the range of 1600 to 2000°C. The high processing temperatures results in the removal of organics from the treatment volume by vaporization followed by pyrolysis within the dry zone immediately adjacent to the melt. No organics remain in the melted volume due to the inability of organics to exist at the temperatures involved. A broad range of organic contaminant types has been successfully treated using the GeoMelt applications.

The predominant disposition of heavy metals and radionuclides during ISV involves physical and chemical incorporation into the vitrified product. Most heavy metals are actually encapsulated within the glass and crystalline matrix. This incorporation produces a permanent immobilization result. It is important to note that the GeoMelt family of applications simultaneously processes both organic and heavy metals (including radioactive) contaminants, which is a capability largely limited to vitrification technologies.

Initial Demonstration Project

The project was initiated by applying for a National TSCA permit to the U.S. EPA Office of Toxic Substances. Following submission of the application, Geosafe provided several documents to support the application and to govern demonstration testing. These documents included a TSCA Test Plan, a sampling and monitoring plan and procedures, a quality assurance project plan, health and safety plans, site preparation plans, and other additional guidance documents. As a result of data generated during testing prior to the TSCA demonstration, a modification to the initial demonstration testing approach was determined to be necessary. Modifications of the above mentioned documents were performed and resubmitted as addendums to the originals. As this particular site was a state lead remediation, the Washington State Department of Ecology (WDOE) also participated in the approvals of the project documentation and planning.

The site, which was not originally on EPA's National Priority List (NPL) when the project was initially planned, was classified as an NPL site midway during preparation of the project. The site selected for the demonstration testing was a private, formerly used transformer service facility. Business operations at the site resulted in PCB contamination of site soils, drainage sumps, a drywell, concrete pads, and asphalted areas. Due to the concentration levels of PCBs at the site and the various types of debris, the site was selected as a good location for performance of a demonstration test. Due to the timing of the demonstration testing, the WDOE agreed to allow the demonstration to proceed in advance of the CERCLA remediation of the site.

The initial demonstration testing was performed by collecting contaminated soil from the site and passing the material through a screening plant to remove all of the course material, since the rock fraction had contamination below the required PCB treatment levels. The soil passing through the screening plant was then placed in five 26-ft square by 16-ft deep treatment cells. The average PCB concentration of this soil was approximately 140 ppm. One of the treatment cells was configured to include three layers of soils that were spiked with PCBs to levels up to 17,860 ppm. The four other cells contained 20-drum arrays of sealed drums containing investigation-derived soil and water. Of these four cells, one contained 8 weight percent asphalt and another contained 11 weight percent concrete. In both cases, the asphalt and concrete were distributed randomly throughout the treatment cells. Prior to processing the treatment cells containing the sealed drums, a vibratory beam technique was used to rupture the drums below grade and in place so that water vapor could readily escape to the surface during processing. A clean soil corridor was constructed adjacent to the cell with the spiked PCB concentrations. This corridor was used to evaluate the presence of migration of PCBs out of the treatment volume and into the surrounding soils. A clean corridor was needed as the concentration of PCBs in the soil used to backfill around the treatment cells ranged from 3 to 44 ppm.

Samples of the vitrified product and off gases were collected periodically during the vitrification of the five cells. In addition, soil samples were collected from within the clean corridor a few months after completing treatment of all five cells. These samples were collected to allow confirmation that the GeoMelt process is technically applicable to remediating soils and debris contaminated with PCBs. Off-gas sample results and vitrified product testing indicated full capability to achieve TSCA performance standards. However, clean zone soil samples were inconclusive due to pre-test contamination of the clean zone. Therefore, EPA granted a National TSCA Operating Permit with the provision that the next application of the process must involve repeat clean zone testing.

Follow-on Demonstration Test

The repeat clean soil zone test was performed in 1996 during remedial activities at the same site to further evaluate the lack/presence of contaminant migration from the treatment zone into the surrounding clean soil. The remedial phase of work which began in July 1996 involved the treatment of 1900 yd³ of soil found to be the highest levels of contaminated soils on the site. The origin of this soil was an on-site drywell that extended from grade down to a depth of 45 ft. Prior to placement in an efficient treatment configuration, the soil was excavated and processed through a screening plant to remove all cobble with a diameter greater than 1 in. The soil fines were then placed into four excavated treatment cells. Three of the cells were built with the dimensions 30-ft square by 15-ft deep and the fourth cell was 26-ft square by 8.5-ft deep. A total of 2459 tons of contaminated soils were placed into the treatment cells. The average PCB concentration of the contaminated soil was 500 ppm with the exception of a small amount of soil that was contaminated to a level of 5000 ppm, which was placed in the first treatment cell.

A clean soil zone was constructed adjacent to the edge of the first cell to assist in the further evaluation of possible contaminant migration away from the treatment area. This zone, which was 30 ft wide by 20 ft long and 15 ft deep, was filled with certified clean soil, as all pre-existing site soils contained various low levels of PCBs. Analytical sampling relative to TSCA requirements during the remedial phase of the project consisted of pre- and post-test samples of the clean soil zone only.

Processing data gathered during operations and analytical results were used to evaluate the technology for technical and commercial viability. The results of these tests are presented in the following section.

Demonstration Testing Results

The five TSCA demonstration test cells were completed successfully and without difficulty. No significant downtime or process upsets were experienced during the demonstration testing. Samples collected of the off gas during the testing and analyzed for PCBs and dioxin/furans were all found to be below the analytical method detection limit. These results, when used to determine the DRE for the process, resulted in levels equivalent to >99.9999%. Samples of the vitrified product taken during treatment of cells 1 and 3 were also reported as less than the detectable level. This is expected as organic compounds cannot exist within a melt due to typical melt temperatures of 1600 - 2000°C. Surrounding soil samples taken after the demonstration testing indicated that the process effectively removed pre-existing low levels of contamination present in the surrounding soils up to a distance of four feet from the melts. No change in background concentrations were detected in more distant adjacent soil. Most samples taken from the clean zone proved a lack of contaminant migration into the clean soil during treatment. Other samples due to questionable experimental procedures were inconclusive due to the inability to prove whether low contamination levels were derived from the adjacent contaminated site soils or from the treatment zone.

Based on the demonstration test results, EPA granted Geosafe a National TSCA Operating Permit for the nationwide treatment of PCBs on October 31, 1995. The permit includes numerous provisions that were established as a result of the testing approach and the equipment used during the test. These provisions define the scope of acceptable applications, equipment operating requirements, and other operational and administrative requirements as necessary to ensure the appropriate application of the GeoMelt processes. Examples of limitations established in the permit are the maximum average PCB concentration of 14,700 ppm and the maximum allowable hot spot concentration up to 17,860 ppm. The permit also includes as treatable materials various debris materials included in the test (i.e., asphalt, concrete, and metal drums) and system and operating conditions. The permit does require that all sealed containers be ruptured, although provisions are made to allow changes to the existing permit.

As a conditional requirement of the Operating Permit, Geosafe was required to perform a clean soil zone test during its first commercial PCB project. The purpose of this test was to confirm that PCBs do not migrate away from the treatment zone. As a result, a clean soil zone test was performed during the first melt of the remedial portion of the project, which was performed in October of 1996. This test was designed to resolve the experimental uncertainties associated with clean soil placement position, size, and sampling techniques encountered during the demonstration test. Measures taken to ensure that accurate data was obtained included: accurate survey of the boundaries of the clean zone, thorough and representative sampling of the clean soil zone both pre- and post-test, and widening of the clean soil zone to ensure all of the samples were taken from within the clean zone and further ensure that no surrounding site soils became intermixed with the certified clean soil.

After treatment of the cell adjacent to the clean soil zone was completed, vertical core samples were taken within the boundaries of the clean zone. All of the composite samples collected from the cores, which were split and sent to two independent laboratories, were determined to be below the TSCA Operating limit of <2 ppm. Additionally, samples taken and analyzed by representatives of the site owner resulted in all samples being below the detection limit for the method used. The results of these samples conclusively proved the lack of contaminant migration in surrounding soils either directly adjacent to the contaminated soils or at further distances away.

Based upon the results of this test that indicated the lack of PCBs at concentrations ≥2 ppm in the clean soil area, the U.S. EPA has since removed the clean soil zone test requirement through a modification to Geosafe's National TSCA Operating Permit.

While performing the remedial phase of the project in 1996, Geosafe was able to evaluate and further demonstrate its ability to adhere to the provisions outlined in its TSCA permit which was granted a year earlier. All conditions specified in the permit which included such measurements as thermal oxidizer operating temperature and destruction efficiency, minimum oxygen levels, and off-gas hood containment vacuum levels, where met continuously without exception. No reportable events as defined in the TSCA operating permit occurred and melting operations were continued without unscheduled downtime. Significant experience of operating under the provisions of the TSCA operating permit was gained during the remedial phase of work during this project. As a result of this operating experience, Geosafe is confident that adherence to the TSCA Operating Permit can and will be continually achieved regardless of the site or application.

Applicability to Other Sites or Configurations

Geosafe's National Operating Permit covers all operations within the U.S. while operating under the provisions outlined in the permit. Some particular applications may require modifications to the Operating Permit. Major modifications are described as changes to capacity, design, efficiency, waste type, or any other changes that may affect overall PCB destruction performance or environmental impact. These changes are made by notifying and receiving written approval of EPA's Office of Toxic Substances. In some cases additional testing and sample confirmation may be required depending upon the nature of the modification. For instance, when changes are made that may impact the stack emission products, additional monitoring of the stack emission discharges may be required.

As a result of the broad basis of Geosafe's existing Operating Permit and the relative ease in which modifications can be obtained, Geosafe is capable of applying the GeoMelt processes at PCB contaminated sites within the U.S.

ANAEROBIC DECHLORINATION OF PCB IN MARINE SEDIMENTS - POSSIBILITIES FOR ENHANCEMENT

T. Briseid[1], R. Sørheim[2], O. Bergersen[1], G. Eidså[1], A. Kringstad[1] and T. Røneid[1]

[1] SINTEF, Applied Chemistry, Oslo, Norway
[2] Centre for Soil and Environmental Research, Ås, Norway

Abstract: Marine sediments in several Norwegian harbours and fjords are in many cases contaminated with PCB. The aim of this work has been to study the influence of sulphate on anaerobic dechlorination of PCB. The application objective is to use this knowledge to evaluate the possibilities for bioremediation of PCB-contaminated marine sediments.

Inocula from PCB contaminated and non-contaminated marine sediments, aquatic sediments and contaminated river sediments have been used. The cultures were added a mixture of defined congeners of PCB. Different carbon sources like methanol, acetate, propionate, butyrate, and malate were added to enhance dechlorination. After 5 months GC-analysis showed dechlorination and formation of new lower chlorinated congeners in cultures added freshwater and artificial seawater without sulphate. Reduction varied from 9 to 88% on mol basis. The highest reductions were observed in the lowest chlorinated congeners. No dechlorination was observed when the artificial seawater with sulphate was used. As far as we know, this is the first time where dechlorination of PCB have been demonstrated in laboratory experiments, using inocula from Norwegian sediments.

The results of this work show that dechlorination of PCB in marine sediments may be enhanced by initial removal of surplus sulphate, and by adding sediments containing PCB dechlorinating microorganisms.

INTRODUCTION

Marine sediments in several Norwegian harbours and fjords near shipyards are in many cases contaminated with PCB (Winther-Larsen. and Iversen, 1997). PCB is also found in bottom fauna and liver of fish (Knutzen, 1995). The most contaminated harbours have to be dredged. In some cases the sediments are deposited in special areas, in other cases the sediments are cleaned and the small particle fractions have to be treated or deposited. This treatment involve handling of the sediments that might be used to enhance the biodegradation of PCB during subsequent deposition. Partly inhibition of PCB dechlorination by sulphate has been showed earlier (Rhee et al., 1993) but other kinds of reductive dechlorination seem to happen at both sulphate reducing and nitrate reducing environments (Mohn and Tiedje, 1992).

The aim of this work has been to study the influence of sulphate on anaerobic dechlorination of PCB. The application objective is to use this information to evaluate possibilities for bioremediation of PCB-contaminated marine sediments. Significant rates of dechlorination in marine sediments may involve a necessary initial removal of surplus sulphate, and may be enhanced by adding sediments containing PCB dechlorinating microorganisms to the PCB contaminated sediments.

MATERIALS AND METHODS

Growth media. Freshwater medium was prepared partly modified as described by Shelton and Tiedje (1984) by adding 0.27g KH_2PO_4; 0.35g K_2HPO_4; 0.53g NH_4Cl; 0.075g $CaCl_2 \cdot 2H_2O$; 0.10g $MgCl_2 \cdot 6H_2O$ and 0.02g $FeCl_3 \cdot 4H_2O$ and 1 mL micro-nutrient solution per L destilled water. pH was adjusted to 7.0. The mineral growth medium was autoclaved for 15 min at 121°C and put into an anaerobic chamber (Don Whitley MK III) with an anaerobic atmosphere (80% N_2, 10% CO_2 and 10 % H_2) for cooling overnight. Thereafter 1.2 g $NaHCO_3$ and 0.5 g $Na_2S \cdot 9H_2O$ was added per l medium.

Artificial seawater medium was prepared by adding 0.75g $NaH_2PO_4 \cdot 2H_2O$; 0.5g NH_4Cl; 23.4g NaCl; 12.3g $MgSO_4 \cdot 7\ H_2O$; 0.8g KCl; 0.14g $CaCl_2 \cdot 2H_2O$ and 4mL resazurin (0.1% solution) and 1 mL micro-nutrient solution per L destilled water. pH was adjusted to 7.0. The seawater medium was autoclaved for 15 min at 121°C and put into an anaerobic chamber (Don Whitley MK III) with an anaerobic atmosphere (80% N_2, 10% CO_2 and 10 % H_2) for cooling overnight. Thereafter 3.0 g $NaHCO_3$ and 0.5 g $Na_2S \cdot 9H_2O$ was added per L medium, each of them autoclaved separately in 50 mL water. Per L seawater media was added 10 mL of the following filter sterilized vitamin solution: 5.0 mg p-aminobenzoic acid, 2.0 mg folic acid, 2.0 mg biotin, 5.0 mg nicotinic acid, 5.0 mg calcium pantothenate, 5.0 mg riboflavine, 5.0 mg thiamin HCl, 10.0 mg pyridoxine HCl (vitamin B_6), 0.1 mg Cyanocobalamin (vitamin B_{12}), 5.0 mg thioctic acid.

Artificial seawater without sulphate was prepared as above but 12.3g $MgSO_4 \cdot 7\ H_2O$ was replaced by 11.0g $MgCl_2 \cdot 6H_2O$. Micro-nutrient solution was prepared by adding 0.5g $MnCl_2 \cdot 4H_2O$; 0.05g H_3BO_3; 0.05g $ZnCl_2$; 0.03g $CuCl_2$; 0.01g $NaMo_4 \cdot 2H_2O$; 0.5g $CoCl_2 \cdot 6H_2O$; 0.05g $NiCl_2 \cdot 6H_2O$ and 0.05g Na_2SeO_3 per L distilled water.

Organic substrates. Adding organic substrates to anaerobic sediments has earlier been shown to enhance PCB dechlorination (Nies and Vogel, 1990). The media were added 1.0g per L medium of each of the following organic substrates: methanol, Na-propionate, Na-acetate, Na-butyrate and Na-malate.

Inocula. Inocula from PCB contaminated and non-contaminated marine sediments (Bygdøy and Bergen havn), aquatic sediments (Sognsvann), PCB contaminated river sediments (Alna), PCB contaminated marine harbour sediments (Herdla) were used as inocula. 50g sediment of each sediment were added 50 ml sterile physiological salt water (0.9% NaCl). The physiological salt water was made anaerobic by cooling after the autoclave treatment in the anaerobic chamber to avoid oxygen uptake during cooling to room temperature. The sediments and the water was added 0.1g Na-dithionitt and mixed together by shaking the bottles for 5 minutes. The mixture standed quit for 10 minutes and the water phase was transfered to another bottle. 2.5mL of this water phase was used as inoculum.

PCB specific congeners. Stock solutions of 234-tri-CB (K21); 2345-tetra-CB (K61); 23456-penta-CB (K116); 23456-2-heksa-CB (K142); 23456-26-hepta-CB (K186) and 23456-3-heksa-CB (K160) were prepared either separately or as a mixture. The concentrations in the stock solutions were 20 000ppm in acetone.

Degradation tests. Sediments from a nonpolluted lake (Sognsvann) was dried at room temperature for some days and sifted through a sieve (1 mm). 7.5 g of this sludge (growth matrix) was added 40 mL glass bottles (Supelco Cat. No. 2-780, Clear Vial, Screw Top 29x81, Hole Cap with PTFE/Silicone Septa) equipped with teflon sealed screw caps. 20 mL selected growth medium and small amounts of sediments from a nonpolluted lake were added each bottle. The handling were performed within an anaerobic chamber. The bottles were incubated at room temperature for some weeks. After methane production was detected, the bottles were autoclaved submerged in water to inhibit oxygen diffusion through the rubber caps during cooling. These bottles were added a mixture of specific PCB congeners, inoculated and incubated at room temperature for PCB dehalogenation studies.

Chemical analysis of PCB. The whole samples were quantitatively transferred from test bottles to centrifuge tubes and rinsed several times by isopropanol. The isopropanol was used in 1. extraction. The samples were added internal standards and extracted twice with a mixture of cyclohexan/ispropanol using ultrasonic bath and shaking table. The cyclohexan-extracts were isolated and consentrated to 1ml. Clean-up: Consentrated sulfuric acid, tetrabutylammoniumsulphite (TBA) and etanolic KOH. The analysis were performed using gas chromatography with Electron Capture detector (GC/ECD). The identification was based upon comparison of chromatographic pattern and retention times for the samples and commercial PCB-oils and standard-solutions of the single PCB-congeners. For quantification both internal and external standards were used.

RESULTS AND DISCUSSION

After 5 months incubation no significant dechlorination was observed in cultures with artificial seawater (Figure 1a and Table 1). In the cultures where artificial seawater without sulphate added or with freshwater, dechlorination was detected (Figure 1b and 1c). All the added congeners showed some degree of dechlorination and new congeners were formed.

In cultures added seawater without sulphate 73 % reduction of the lowest tri-chlorinated congener K21 was observed. The tetra- and penta-chlorinated congeners K61 and K142 were reduced with 30% and 24% respectively. The hexa-chlorinated compounds K142 and K160 were reduced with 15% and 9% respectively while the hepta-chlorinated K186 was reduced with 9%.

TABLE 1. Reduction (% mol) of different added congeners after 5 months incubation in different media: (A) Artificial seawater, (B) Artificial seawater without sulphate added, (C) Artificial freshwater.

		mol %		
		A	B	C
K21	234´	3,8	-73	-88
K61	2345´	0,4	-30	-51
K116	23456´	2,5	-24	-51
K142	23456´2	-1,0	-15	-
K160	23456´3	1,2	-9	-30
K186	23456´26	7,6	-9	-36

The highest dechlorination rate was observed in the cultures with freshwater, where 88 % reduction of the lowest tri-chlorinated congener K21 was observed (Table 1). The tetra- and penta-chlorinated congeners K61 and K142 were both reduced with 51% while the highest chlorinated compounds K160 (hexa-chlorinated) and K186 (hepta-chlorinated) was reduced with 30% and 36% respectively. The hexa-chlorinated congener K142 was not added these cultures.

It has earlier been shown that anaerobic dechlorination of highly chlorinated PCB congeners results in increased amounts of lower chlorinated PCB (Abramowicz, 1990). It has been several documentations that dechlorination of PCB most commonly are observed in microbial consortia well adapted for methane production. The results of these experiments show that the presence of sulphate inhibits the dechlorination of PCB. Partly inhibition of PCB dechlorination has been showed earlier (Rhee et al., 1993) but other kinds of reductive dechlorination seem to happen at both sulphate reducing and nitrate reducing environments (Mohn and Tiedje, 1992).

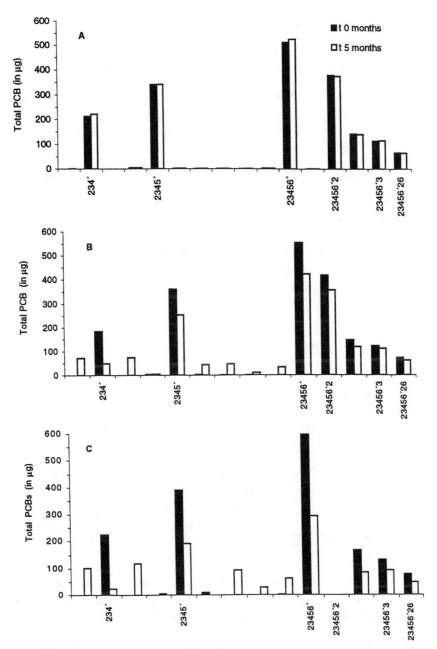

FIGURE 1. Amounts of different added congeners and not identified dechlorination products at the beginning of the experiment (t = 0 months, diagonal lines columns) and after 5 months incubation (t = 5 months, white columns) in different media: (A) Artificial seawater, (B) Artificial seawater without sulphate added, (C) Artificial freshwater.

CONCLUSION

The results shows that dechlorination of PCB is inhibited by the presence of sulphate at concentrations normally found in marine environments. This may explain why PCB is accumulated in marine sediments. However, it might be possible to dechlorinate PCB in marine sediments provided that handling of the sediments involve removal of surplus of sulphate, and provided that microorganisms able to dechlorinate PCB are present or might be added during a washing process.

As far as we know, this is the first time where dechlorination of PCB have been demonstrated in laboratory experiments, using inocula from Norwegian sediments.

AKNOWLEDGMENTS

This project was funded by the State Pollution Control Authorities (SFT) and the Norwegian Defence Construction Service (NODCS). We will also thank Jens Laugesen and Vidar Ellefsen in Environmental Consultants A.S for theoretical contribution and interesting discussions.

REFERENCES

Abramowicz, D. A. 1990. "Aerobic and anaerobic biodegradation of PCBs: A review." *Crit. Rev. Biotechnol.*, 10(3): 241-252.

Knutzen, J. 1995. "Summary report on levels of polychlorinated dibenzofurans/dibenzo-p-dioxins and non-ortho polychlorinated biphenyls in marine organisms and sediments in Norway." *NIVA Report*, No. O-95161, Serial No. 3317, ISBN82-577-2845-4

Mohn, W. W. and Tiedje, M. T. 1992. "Microbial reductive dehalogenation." *Microbiological Reviews*, 56(3): 482-502.

Nies, L. and Vogel, T. M. 1990. "Effects of organic substrate on dechlorination of Arochlor 1242 in anaerobic sediments." *Appl. Environ. Microbiol*, 56(9):2612-2617

Rhee, G-Y., Bush, B. Bethoney, C. M., DeNucci, A., Oh., H. M. and Sokol, R. C. 1993. "Anaerobic dechlorination of Arochlor 1242 as affected by some environmental conditions." *Appl. Environ. Microbiol.* 12: 1033-1039.

Shelton, D. R. and Tiedje, J. M. 1984. "General method for determining anaerobic biodegradation potential." *Appl. Environ. Micerobiol.* 47(4): 850-857.

Winther-Larsen,T. and Iversen, P. E. 1997. "The Norwegian action plan on contaminated sediments." Poster presentation at the International Conference on Contaminated Sediments, Rotterdam, 1997.

PRELIMINARY EXPERIMENT FOR REMEDIATION OF PCBs CONTAMINATED SOIL BY ELECTROOSMOSIS

Soon-Oh Kim, Seung-Hyeon Moon and Kyoung-Woong Kim
(Kwangju Institute of Science and Technology, Kwangju, Korea)

ABSTRACT : The electroosmotic remediation is one of the promising soil decontamination processes due to its high removal efficiency and time-effectiveness with particular reference to low-permeability soil such as clay. PCBs is a toxic and poorly-biodegradable contaminant, and its removal process by electrochemical technique has been studied. In order to investigate the feasibility and the applicability of electroosmosis for purging PCBs from saturated soil and to find out optimum conditions for removal process, the processes of electroosmotically purging phenol and PCP from the soil has been examined. It is suggested that the removal efficiency is significantly influenced by applied voltage & current, type of purging solutions, soil pH, permeability and zeta potentials of soil. The removal efficiency for phenol and PCP was higher than 85% in the duration of 4 days, and this remediation technique may be applied to PCBs contaminated soils.

INTRODUCTION

Soil contaminations are derived from various sources. Contaminants migrating from those sources may threaten human health in the local area and contaminate the ground-water supply. However, technologies for decontaminating those sites have not been well developed yet, although it has been recently reported that soil contaminations are increasing in various sites-residential areas near industrial complexes and reservoirs of drinking water.

Remediation of hazardous waste sites is one of the most important technological challenges, and can be generally classified into two groups. The first one is the biological remediation that has been mainly used to detoxify organic contaminants, and the other is the physico-chemical decontamination that has been usually applied to remove organic and inorganic contaminants-excavation, soil washing and flushing, solidification and stabilization, electroosmotic remediation and so on (Charmbers, 1991).

Electroosmotic remediation is an effective technology to remove contaminants in low-permeability soils ranged from clay to clayey sand. The significance of the technology is its low operation cost and its potential applicability to a wide range of contaminant types (Pamukcu and Wittle, 1994). Electroosmotic remediation using low-level direct currents (DC) envisioned for the removal separation of organic and inorganic contaminants and radionuclides.

Since PCBs is good conductors of heat and bad conductors of electricity, the insulating oil contained PCBs can be applied to transformers and capacities (Eduljee, 1988). About 205 ton of insulating oil containing PCBs are still in use and stored in Korea (Chosunilbo, 1995). Although PCBs is chemically inert and stable, there have been a number of reports on the microbial degradation of PCBs. However, the fact that PCBs are not single compound but a mixture of different isomers is main barrier to study about the interaction between PCBs and microorganisms (Madsen, 1991). In this study, electroosmotic remediation, one of the phyico-chemical decontamination techniques, was applied to the remediation of the soils contaminated with PCBs. In order to investigate the feasibility and the applicability of electroosmosis for purging PCBs from saturated soil and to find out optimum conditions for removal process, the experiments on the removal of phenol and PCP were conducted preliminarily.

With these experiments, the feasibiltiy of electroosmtic remediation on the removal of PCBs from soils, and the optimum condition for the most efficient removal have also been invstigated.

MATERIALS AND METHOD

The soils used in this experiment were the kaolinite soils obtained commercially. The kaolinite soil samples were artificially contaminated by phenol and PCP. Two types of test were conducted to investigate the feasibility of electroosmotic remediation. One was the removal of phenol (C_6H_5OH) from kaolinite soil to find out the efficient purging solution, and the other was designed to remove PCP (Pentachlorophenol;

Cl$_5$C$_6$OH) from kaolinite soils.

For the removal of phenol from kaolinite soil specimens, two types of anode purging solutions were used to compare the efficiency of anode purging solutions between the neutral and acid solution. The first was 0.1M NaCl solution and the other was 0.1M NaOH solution. Another test for the removal of PCP was conducted using 0.1M NaOH anode purging solution which was confirmed to be an efficient solution by previous tests of phenol. The measured parameters for three tests are summarized in Table 1. A schematic diagram of experimental apparatus used in this study is shown in Fig. 1.

TABLE 1. Summary of testing program and measured parameters for removal experiments of organics

Parameter	Test 1	Test2	Test 3
Soil Specimen	kaolinite	kaolinite	kaolinite
Contaminants	phenol (C$_6$H$_5$OH)	phenol (C$_6$H$_5$OH)	PCP (Cl$_5$C$_6$OH)
Initial Con. of Contaminants(μg/g)	178.2	333.8	9.91
Applied Current (A)	0.1	0.1	0.1
Area of Soil Cell (cm^2)	81	81	81
Length of Soil Cell (cm)	15	15	15
Duration (hr)	96	96	96
Anode Purging Solution	0.1M NaCl Solution 2L	0.1M NaOH Solution 4L	0.1M NaOH Solution 4L
Cathode Electrolyte Solution	0.5N H$_2$SO$_4$ Solution 1L	0.5N H$_2$SO$_4$ Solution 1L	0.5N H$_2$SO$_4$ Solution 1L

During the experiments, the overall voltage drops of soil cell soil, pH variation, and transported porewater volume by electroosmosis were measured every 4 hours. Phenol and PCP were extracted with Supercritical Fluid Extractor (SFE, ISCO -SFX^{TM}3560) from wet soil samples obtained in the soil cell. The extracted solutions (organics and modifier solutions) wereanalyzed by HPLC (column ; Supelcosil LC-8,).

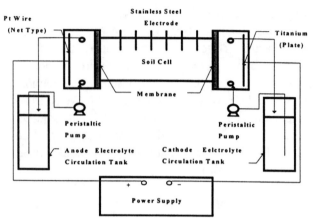

FIGURE 1. Schematic diagram of experimental apparatus

RESULTS AND DISCUSSION

Variation of pH in soil cell. Variation of pH in soil cell are shown in Fig. 2. The removal efficiency for organics from soils was significantly dependent on the electroosmotic purging, and electroosmotic flow was predominantly influenced by soil pH. It means that control of soil pH was an important factor in removing organic contaminants from soils. In order to control soil pH, two kinds of purging solutions were used ; one was neutral solution (NaCl), and the other was alkaline solution (NaOH).

The soil pH was regarded as the major factor to make   maximum electroosmosis.

Variation of amount of contaminants in soil cell and removal efficiency. The variations of amount of contaminants in soil cell for each test are shown in Fig. 3. Overall amount of contaminants decreased during the treatments, and the migration of species in soil bed was observed in Fig. 3.

The removal efficiency was calculated for each test (Table 2). Compared with the removal efficiency of test 1 of NaCl purging solution, that in test 2 of NaOH purging solution was significantly higher inremoving phenol from kaolinite soil by electrokinetic soil processing. It suggests that control of soil pH influencing on the electroosmotic flow was an important factor in removing organic contaminants from soils. In order to maintain soil pH suitable for the development of the maximum electroosmotic flow, it was necessary to find out the suitable anode purging solution.

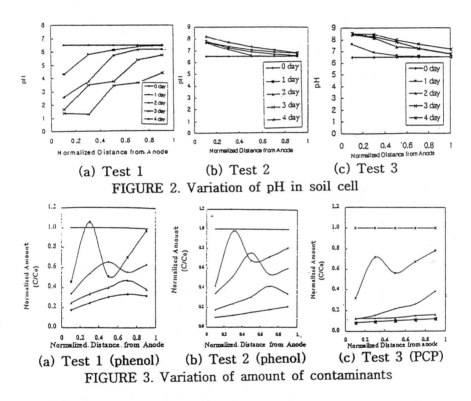

(a) Test 1	(b) Test 2	(c) Test 3

FIGURE 2. Variation of pH in soil cell

(a) Test 1 (phenol)	(b) Test 2 (phenol)	(c) Test 3 (PCP)

FIGURE 3. Variation of amount of contaminants

TABLE 2. Removal efficiency for each test (unit ; %)

Time (day)	Test 1	Test 2	Test 3
1	25.8	27.8	38.8
2	48.7	54.8	77.1
3	61.2	70.2	86.2
4	75.6	85.6	89.8

CONCLUSION

The following conclusions are obtained from the study on the removal of phenol and PCP from soils by electroosmotic remediation ;

1. Since the rate of electroosmotic flow was influenced by the soil pH, the soil pH was the important parameter on removing non-polar organic contaminants from soils by electroosmotic remediation.

2. The removal efficiency for phenol using NaOH purging solution was higher than that using NaCl purging solution. It was explained by the cease of the electroosmotic flow from the decrease of soil pH.

With the result of this study, it suggests that the electroosmotic remediation will be the efficient technique for the removal of PCBs from saturated soils.

REFERENCES

Chambers, D. C. 1991. In Situ Treatment of Hazardous Waste Contaminated Soils. 2nd edition, ndc, New Jersey, p. 98.

Probstein, R. F., and P. C. Renaud. 1986. *Proc. Workshop on Electrokinetic Treatment and its Application in Environmental Geotechnical Engineering for Hazardous Waste Site Remediation.* Univ. of Washington, Seattle, Aug. 4-5. Hazardous Waste Engineering Research Laboratory, EPA, Cincinnati, Ohio.

Shapiro, A. P. 1990. Ph. D. Dissertation: Electroosmotic Purging of Contaminants from Saturated Soils. Massachusetts Institute of Technology, Cambridge, MA.

Pamukcu, S., and J. K. Wittle. 1994. Electrokinetically enhanced in situ soil decontamination. Remediation of Hazardous Waste Contaminated Soils, Marcel Dekker, Inc., New York.

Putnam, G. 1988. Thesis for M.S Degree : Development of pH gradients in electrochemical processing of kaolinite, Louisiana State University, Baton Rouge, LA.

Khan, L. I., S. Pamukcu, and I. Kugelman. 1989. Electro-osmosis in fine-grained soil. *2nd International Symposium on Environmental Geotechnology*, Shanghi, China, Envo Publishing, Bethlehem, Pa., 1, 39-47.

Mitchell, J. K., and T. C. Yeung. 1991. Electro-kinetic flow barriers in compacted clay. *Transpotation Research Records, No. 1289*, National Research Council, Washington DC.

Eduljee, G. H. 1988. PCBs in the environment. *Chemistry in Britain*, March, p.241-244.

Chosun-ilbo : daily newspaper in Korea, 1995. p.35. Oct. 23. Seoul, Korea.

Madsen, E. L. 1991. Determining insitu biodegradation. Environ. Sci. Technol., vol. 25, No. 10, p. 1663-1673.

REMEDIATION OF 2,4-DINITROTOLUENE USING PSEUDOMONAS, SPIZIZEN MEDIUM AND LOAM

Steven Kornguth, Glen Chambliss, Kristine Gehring (Univ. Wisconsin-Madison)
George Shalabi and Louis Unverzagt (Olin Corp., Baraboo, WI)
Lyman Wible and Jack Anderson (RMT Corp., Madison, WI)
Bruce Brodman (US Army, Picatiny, NJ)

ABSTRACT: The degradation of 2,4 dinitrotoluene (DNT), in sandy soils and clay soils, is shown to be affected by salts in Spizizen nutrient medium, by addition of *Pseudomonas* organisms recovered from soil contaminated with DNT, and by addition of prairie silt loam to contaminated soils (80:20 ratio). 110 kilogram batches of soil were used per treatment. Addition of Spizizen medium to the aged clay soil, containing plasticized propellant, initially resulted in a marked increase in the DNT that was extracted by methylene chloride. DNT in contaminated sandy soils was rapidly degraded when *Pseudomonas* and Spizizen medium were added (85% degraded in 20 days). *Pseudomonas* isolated from DNT contaminated soils have particular utility for the *in situ* degradation of 2,4 DNT in clay and sandy soils because they metabolize Spizizen medium, thrive in diverse climates, and have been selected for their ability to grow in soils contaminated with DNT. The loam retained water throughout the mixture and precluded clumping of the clay soil. We propose that citrate in Spizizen medium chelates metals in aggregates of humin in aged clay soils, thereby releasing propellant components. The estimated cost of remediation, using this process is $95 per ton of contaminated soil when remediating 10,000 tons or more.

INTRODUCTION

Soils, sediment and ground water at facilities used for the production of munitions are contaminated by 2,4 and 2,6 dinitrotoluene (DNT), which are components of certain propellants. The 2,4 and 2,6 DNT are toxic to mammals. Mineralization of energetic compounds, including nitrotoluenes, by bacteria and fungi has been demonstrated earlier by several research teams. *Pseudomonas* is one of the most studied organisms for bioremediation of DNT (Haigler et al., 1994) and was used in this study. The degradation of the energetic compounds by bacteria requires the addition of salts and a carbon source; to avoid generating reducing conditions that yield aromatic amines, a non-fermentable carbon source such as succinate is recommended. Because amine derivatives are more water soluble than aromatic nitrocompounds, the former may enter the ground water more readily and thereby increase environmental pollution concerns. The clay soils in this study were contaminated with plasticized propellant containing DNT, DPA (diphenylamine), DBP (di-n-butyl phthalate) and NC (nitrocellulose) for over 20 years. The sandy soils contained DNT, DPA and DBP.

MATERIALS AND METHODS

Sandy soil contaminated with uncomplexed DNT, DBP, DPA, or clay soil contaminated with plasticized propellant containing DNT, DPA, DBP and NC were mixed with prairie silt loam and placed in bins; the total amount of mixed soils in each bin approximated 110 kilograms (kgs). Soils were aerated and kept moist by pumping high moisture air through the pile. *Pseudomonas* consortia from BAAP clay soils contaminated with propellant were added to the soil samples at specific intervals for each experimental protocol listed below. Spizizen nutrient medium was added to the bins containing the added Pseudomonas and to the control bins. To determine the amount of DNT, DPA and DBP after treatments, the soils were extracted with methylene chloride and the extract analyzed by GC-MS.

RESULTS

Degradation of DNT by *Pseudomonas* in Sandy Soils. The first experiment describes the effect of added *Pseudomonas* organisms and prairie silt loam on the degradation of DNT in sandy soils. The sandy soil, which contained 2,4 DNT, 2,6 DNT, DBP, DPA but no NC, was mixed with prairie silt loam at a ratio of one part sand to four parts loam. Addition of *Pseudomonas* organisms and Spizizen medium resulted in extensive degradation of DNT (Figure 1) at low (95 mg DNT/kg) or high concentrations of DNT (465 mg DNT/kg). At high concentrations of 2,4 DNT, 85% of the DNT in the sand soil was degraded by day 32 following addition of *Pseudomonas*. Almost 80% of the DNT was degraded at day 20. In the absence of added Pseudomonas, 55% of the DNT was degraded at day 20 and 66% was degraded at day 32. A portion of the material reported here was published (Kornguth et al., 1997).

Only 1% of the 2,4 DNT was converted to aminonitrotoluene (ANT) which is significant because ANT is more water soluble than DNT and may present increased risk of ground water contamination.

Degradation of Plasticized DNT in Clay Soils. The second experiment describes the effect of added *Pseudomonas* organisms, Spizizen medium and prairie silt loam on the degradation of plasticized propellant (mixture contained DNT, DBP, DPA and NC) in clay soils. The soil mixture consisted of 8 parts prairie silt loam and 2 parts of clay soil. The average concentration of the 2,4 DNT, was 74 mg/ kg soil. Figure 2 reveals that the addition of Spizizen medium caused a marked increase in the amount of 2,4 DNT that was recovered after incubation for ten days. Spizizen medium facilitated the mobilization and extraction of 2,4 DNT present in aged clay soils. Addition of *Pseudomonas* organisms caused an increase in the rate of DNT degradation.

Cost Benefits of the Demonstrated Technology. The costs of remediation per ton of soil contaminated with DNT was estimated from the experience obtained in the current study. The basis of the calculation is shown in Table 1.

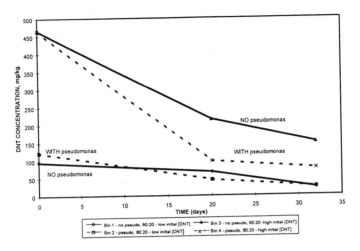

FIGURE 1. Bioremediation of sandy soil contaminated with energetic materials. (From Kornguth et al., 1997. "Bioremediation of Dinitrotoluene in Aged Clay and Sandy Soils Using *Pseudomonas* Organisms, Spizizen Nutrient Medium and Prairie Silt Loam." *Remediation.* 8: 7-18. Reprinted with permission of John Wiley & Sons, Inc.)

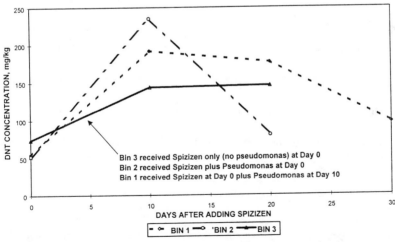

FIGURE 2. Bioremedication of clay soil contaminated with energentic materials. (From Kornguth et al., 1997. "Bioremediation of Dinitrotoluene in Aged Clay and Sandy Soils Using *Pseudomonas* Organisms, Spizizen Nutrient Medium and Prairie Silt Loam." *Remediation.* 8: 7-18. Reprinted with permission of John Wiley & Sons, Inc.)

Table 1. Cost of the Demonstrated Technology

Clean-up Process	Treatment with Spizizen Salt Medium and *Pseudomonas* Cost per ton of Contaminated Soil		
	2 Tons	1000 Tons	10,000 Tons
Excavate contaminated soil	$500	$6	$6
Transport prairie silt loam	$1,000	$48	$48
Mix and treat soil	$300	$8	$8
Engineering and analytical cost	$500	$8	$3
Reagent costs	$50	$50	$30
Total Cost/Ton	$2350	$120	$95

CONCLUSIONS

These studies demonstrated that:

1) Addition of *Pseudomonas* organisms and Spizizen medium to sand soils contaminated with DNT resulted in 85% degradation of DNT in 20-32 days;

2) Addition of Spizizen medium to clay soil increases the amount of 2,4 DNT and of 2,6 DNT initially extracted from the aged soil and propellant grains;

3) the rate at which DNT is removed from sandy soils is faster than removal from clay soils and more rapid when higher initial concentrations of DNT are present than at lower concentrations;

4) less than 1% of DNT was converted to ANT during the degradation of DNT in sandy or clay soils;

5) the addition of *Pseudomonas* increased the rate of degradation of DNT in sandy and clay soils;

6) the addition of uncontaminated prairie silt loam (20:80 ratio) to contaminated clay or sandy soil facilitated the management of the treatment process.

REFERENCES

Haigler, B. E., W. H. Wallace, and J. C. Spain. 1994. "Biodegradation of 2,4 Dinitrotoluene by *Pseudomonas* sp strain JS 42." *J. Bacteriology* 176: 3466-3469.

Kornguth, S. E., G. Chambliss, K. Gehring, L. Wible, J. Anderson, G. Shalabi, L. Unverzagt, and   B. Brodman. 1997. "Bioremediation of Dinitrotoluene in Aged Clay and Sand Soils Using *Pseudomonas* Organisms, Spizizen Nutrient Medium and Prairie Silt Loam." *Remediation: The Journal of Environmental Cleanup Costs, Technologies and Techniques.* 8: 7-18.

TOXICITY REDUCTION OF EXPLOSIVES-CONTAMINATED SOILS BY COMPOSTING

Roy Wade (USACE Waterways Experiment Station, Vicksburg, MS)
Dr. Kyoung S. Ro (Louisiana State University, Baton Rouge, LA)
Steven Seiden (North Carolina State University, Raleigh, NC)

ABSTRACT: The treatment of explosive-contaminated soil was evaluated using the compost technology. Two laboratory-scale compost reactors were used to provide enough composted material to conduct mutagenic and toxicological experiments and chemical analyses. Two explosive-contaminated soil mixtures containing cow manure and alfalfa were composted for 30 days. The Mutatox[TM] bacterial mutagenicity assay and the earthworm acute toxicity tests were used to evaluate the effectiveness of the non-composted and composted soils. The non-composted explosive-contaminated soil was lethal to all exposed earthworms, as were both composted soils prior to composting. Extracts of the initial materials were toxic to bacteria in the Mutatox[TM] assay. Dilutions of the extracts to sublethal concentrations showed a low level of mutagenicity. Extracts of the finished composts showed reduced bacterial toxicity. Composting also reduced the lethality to earthworms. Serial dilutions of the finished composts with artificial soil had earthworm 14-day LC_{50}'s of 100% and 35.3%. The reduction in lethality paralleled the evaluation of explosives caused by composting, as indicated by chemical analyses. However, the increased mutagenicity was a result that might not have been indicated by chemical analysis alone.

INTRODUCTION

The production and handling of conventional munitions have resulted in the generation of explosives-contaminated soils at various military installations. The principle explosive contaminants are 2,4,6-trinitrotoluene (TNT), hexahydro-1,3,5-trinitro-1,3,5-triazine (RDX), and octahydro-1,3,5,7-tetranitro-1,3,5,7-tetrazocine (HMX). One of the more promising remediation technologies being developed is the composting of soils for microbial degradation.

Premiliniary studies (Preston et.al, 1997) of composting explosive contaminated soil have shown that composting significantly reduces the TNT, RDX, HMX, and the TNT metabolites. Although composting decreased the extractable explosives, it does not mineralize a significant portion of the removed explosives. Pennington (1995) reported that more than half of the initial radiolabeled TNT was found in the cellulose and humic fractions after 20 days of composting. Several other researchers reported the similar results (Kaplan and Kaplan, 1982). The crucial question is whether the humified explosives and biotransformed products (i.e., those chemicals that have been incorporated into humic material) are environmentally safe. Understanding of the complete biogeochemical fates of many different explosives and their harmful effects on ecosystem surrounding the finished compost may not be possible, nor practical.

However, toxicity may be used as a simple gross parameter to quantify the quality of the finished compost. This information can be used as the initial framework for determining the risk of the final composting material.

Two assays were employed to evaluate the efficacy of composting in reducing the toxicity of explosive-contaminated soils, the Mutatox™ assay and earthworm acute toxicity. The Mutatox™ assay determines the mutagenicity of organic solvent extracts. The earthworm acute toxicity test measures the lethality of solid materials to *Eisenia sp.* in a 14-day static bioassay (Greene et.al., 1989).

MATERIAL AND METHODS

Composting. An explosives-contaminated soil was collected from a military installation site. A portion of the soil was reserved for toxicity and mutagenicity testing (Raw soil). The remaining soil was amended for composting with 40% cow manure, 40% alfalfa, and 20% soil mixture and tested for 30 days using two 40-liter adiabatic reactors (Compost Mixes A and B). Compost mixture samples were collected in triplicate on days 0, 15, and 30 and analyzed for explosives. On day 30, samples of the finished compost were collected for toxicity and mutagenicity testing. Five samples were compared.

Mutatox™. The Mutatox™ assay is a proprietary assay that determines the mutagenic potential of sample extracts. It utilizes a dark mutant of the bacterial strain, *Photobacterium (P.) phosphoreum*, which will normally bioluminesce (similar to fireflies). These dark mutants of *P. phosphoreum* revert to the wild type in the presence of mutagens. The mutation causes *P. phosphoreum* to bioluminesce, and the light produced were easily measured with a luminometer.

Five soil samples (raw soil, Mix A Initial, Mix B Initial, Mix A Final, and Mix B Final) were evaluated. Aliquots of the samples were oven dried overnight at 130°C for determination of wet/dry weights. Four replicate one gram samples of each soil sample were combined with 3 ml of pesticide grade acetonitrile in 25 ml glass centrifuge tubes and sonicated for 18 hours in a sonicating water bath at 9°C. The soil extracts were centrifuged for 5 minutes at 2,000 rpm and 14°C for sedimentation of the soils. An aliquot of the extract was pipetted into an amber vial and solvent exchanged under a stream of ultrapure nitrogen into an equal volume of spectrophotometric grade dimethyl sulfoxide (DMSO).

As per the Mutatox™ testing protocol, 10μl of each soil extract was added to 250μl of Mutatox™ medium and serially diluted over a wide range with extra media into cuvettes. A ten μl portion of bacterial reagent was pipetted into cuvettes, mixed, and incubated at 27°C for 21 hours. Phenol was used as a positive control in addition to DMSO and acetonitrile solvent controls. Each extract was measured and recorded after the incubation period using the Microbics M500 Toxicity Analyzer. A soil extract was considered mutagenic when light levels produced were greater than two times the DMSO control for two consecutive dilutions, e.g., modified twofold rule (Chu *et al.*, 1981).

Earthworm Acute Toxicity. The earthworm acute toxicity test estimates the acute toxicity of solid material to earthworms (*Eisenia sp.*). The responses measured include the synergistic, antagonistic, and additive effects of all the chemical, physical, and biological components that adversely affect the biochemical and physiological functions of the test animal. The test uses soil as the exposure medium because the exposure conditions closely mimic natural conditions. The test soil is serially diluted with a non-contaminated "artificial soil" in which mature earthworms are placed for 14-days. At the end of the exposure, the number of live earthworms is counted and the concentration that is expected to kill 50% of a test population (LC_{50}) is calculated.

Adult earthworms were purchased and held in moistened sphagnum peat for at least 3 weeks prior to testing on a diet of Magic Worm Food. Test soil samples were diluted with artificial soil (70% industrial sand, 20% kaolin clay, and 10% sphagnum peat on a dry weight basis). The artificial soil pH was adjusted to 6.0 ± 0.5 with calcium carbonate. The dilutions were 100, 50, 25, 12.5, 6.3, and 3.1 percent sample soil to artificial soil on a dry weight/dry weight basis. Aged tap water was added, if needed, to increase the percent moisture of the soil mixtures to 25%. Each soil sample was split into three replicates of 100 g. The replicates were placed in 250-ml polypropylene beakers, covered with polyethylene wrap, sealed with a tight-fitting rubber band, and allowed to sit overnight.

On day zero of the 14-day exposures, mature earthworms were removed from the peat bedding, rinsed with aged tap water, and four earthworms were placed in each beaker. The beakers were re-covered with polyethylene wrap and held under continuous illumination at room temperature (23 ± 2 °C) for 14 days. The earthworms were not fed during exposure. At the end of the exposure period, the earthworms were removed and the number alive per beaker was recorded. Earthworms were considered dead if they did not respond to gentle probing with a blunt probe. A reference toxicant control was run simultaneously with the test soils by spiking artificial soil with 2-chloroacetamide. The calculated LC_{50} for 2-chloroacetamide was compared with published data for performance validation of the tests (Edwards 1984). The test data were found to be acceptable.

RESULTS AND DISCUSSION

Concentration Profile. TNT, RDX, and HMX were measured in Compost Mix A and B at Day 0. Table 1 shows the explosive concentration profiles of raw soil and each mixture during composting. TNT concentration was reduced 99.9 and 99.8% for Mixes A and B, respectively. RDX concentration was reduced 82 and 88% for Mixes A and B, respectively. HMX concentration was reduced 25 and 28% for Mixes A and B, respectively. The TNT biotransformation products (2A- and 4A-DNT) concentrations increased initially, but later decreased to less than 100 mg/kg, except for 4A-DNT in Mix A.

TABLE 1. Average explosive analysis for Mix A and Mix B

Mix A explosive concentrations, mg/kg					
Day No.	TNT	RDX	HMX	4A-DNT	2A-DNT
0	21,203	1,633	7,412	360	382
15	246	1,176	5,486	866	342
30	20	291	5,580	129	22
raw soil	18,867	1,290	26,567	<10.0	<10.0
Mix B explosive concentrations, mg/kg					
0	9,599	906	6,950	644	304
15	22	839	5,066	420	56
30	17	107	5,003	55	13
raw soil	13,050	905	23,933	<10.0	<10.0

Mutatox[TM]. Mutatox[TM] results are presented in Table 2. Each number represents the average degree of light output from four replicate samples. In some instances, the dilutions were acutely toxic to the bacteria. The values indicate the range of mutagenicity of the extract dilution series as determined by the modified two-fold rule. A higher number indicates a higher degree of mutagenicity. Table 2 shows that the raw soil was acutely toxic at 1:1 to 1:64 dilutions and the three lower dilutions (1:80, 1:128, and 1:160) were mildly mutagenic. The Mix A initial and Mix B initial samples caused bacterial mortality, but to a lesser degree than the Raw soil, probably due to the dilution of the Raw soil with the compost amendments. The degree of mutagenicity in the raw soil and two initial Mixes are relatively comparable. Composting reduced the bacterial acute toxicity of the extracts as shown by the lack of toxicity in the two final mixes. However, the degree of mutagenicity increased dramatically at 1:2 to 1:8 for Mix A. This can probably be attributed to TNT degradation products formed during composting.

TABLE 2. Mutatox[TM] results

		Mixture A		Mixture B	
Dilution	Raw Soil	Initial	Final	Initial	Final
1:1	Toxic	Toxic	972	Toxic	2,176
1:2	Toxic	Toxic	5,077	Toxic	2,961
1:4	Toxic	Toxic	2,578	Toxic	1,282
1:8	Toxic	183	3,644	Toxic	199
1:16	Toxic	124	489	Toxic	51
1:32	Toxic	56	55	Toxic	13
1:40	Toxic	38	13	147	7
1:64	Toxic	37	8	25	7
1:80	54	--	7	36	7
1:128	19	--	6	6	4
1:160	12	--	--	--	--

Note: Value indicates range of mutagenicity

Earthworm Acute Toxicity. Table 3 contains earthworm acute toxicity results. Each number represents the mean percent mortality of three replicate samples. The Raw soil was extremely toxic, with 100% mortality at all dilutions. The Mix A initial and Mix B initial samples yielded LC_{50}'s of 4.4% as determined by linear interpolation (Stephen 1977). Therefore, a mixture of 4.4% of either Mix A initial or Mix B initial in non-contaminated soil would be expected to kill 50% of the earthworms exposed to the mixture. The decrease in toxicity can be contributed to the dilutional effect of adding the composting amendments to the raw soil. Mix A final and Mix B final samples yielded LC_{50}'s of 35.3% and 100%, respectively. Apparently, composting reduced the acute toxicity of the raw soil to the earthworms (the higher the LC_{50}'s value, the less toxic).

TABLE 3. Percent mortality from the earthworm acute toxicity results

Dilution	Raw Soil	Mixture A Initial	Mixture A Final	Mixture B Initial	Mixture B Final
1:1	100	100	100	100	50
1:2	100	100	100	100	0
1:4	100	100	0	100	0
1:8	100	100	0	100	0
1:16	100	100	0	100	0
1:32	100	0	0	0	0

CONCLUSIONS

The efficacy of the compost mixture was evaluated in reducing the toxicity of explosive contaminated soils by the Mutatox[TM] assay and earthworm acute toxicity. Composting reduced the HMX concentration by 25 and 28% for Mixes A and B, respectively. The RDX concentration was reduced 82 and 88% for Mixes A and B, respectively. TNT concentration was reduced 99.9 and 99.8% for Mixes A and B, respectively. The transformation products, 2A- and 4A-DNT, concentrations increased initially, but later decreased to less than 100 mg/kg, except for 4A-DNT in Mix A.

The toxicity study identified the mutagenic potential and acute toxicity to the earthworms of the raw soil, and the effectiveness of the two-composted material in reducing toxicity. The Raw soil was extremely toxic with 100% mortality even after dilutions. The composting process apparently reduced the acute toxicity of the raw soil, however the finished compost product was still acutely toxic. The increased mutagenic potency of the compost extracts points to the production of reactive degradation products during the composting process. The Mutatox assay and earthworm toxicity tests provide a sensitive means of monitoring the effectiveness of various composting techniques for remediating explosive-contaminated soils.

REFERENCES

Chu, K.C., K.M. Patel, A.H. Lin, R.E. Tarone, M.S. Linhart, and V.C. Dunkel. 1981. "Evaluating Statistical Analyses and Reproducibility Of Microbial Mutagenicity Assays" *Mutatation Research*. 85:119-132.

Edwards, C.A. 1984. *Report Of The Second Stage In Development Of A Standardized Laboratory Method For Assessing The Toxicity Of Chemical Substances To Earthworms*. European Economic Communities EUR 93600 EN 99. Brussels, Belgium.

Greene, J.C., C.L. Bartels, W.J. Warren-Hicks, B.R. Parkhurst, G.L. Linder, S. A. Peterson, and W.E. Miller. 1989. *Protocols For Short Term Toxicity Screening of Hazardous Waste Sites*. EPA 600/3-88/029. USEPA, Corvallis, Oregon.

Kaplan, D.L. and A.M. Kaplan. 1982. "Thermophilic Biotransformations of 2,4,6-Trinitrotoluene Under Simulated Composting Conditions." *Applied & Environmental Microbiology*, 44(3).

Pennington, J.C., et al. 1995. "Fate of 2,4,6-Trinitrotoluene In A Simulated Compost System." *Chemosphere*. 30 (3), 429.

Preston, K.T., R. Wade, S. Seiden, and K.S. Ro 1997. *Bench-Scale Compost Investigation of Explosive-Contaminated Soil at Naval Surface Warfare Center Crane*. U.S. Engineer Waterways Experiment Station, Vicksburg, MS.

Stephen, C.E. 1977. "Methods For Calculating LC_{50}." *Aquatic Toxicology and Hazard Evaluation ASTM STP 634*, F.L. Mayer and J.L. Hamelink, eds., American Society for Testing and Materials.

FIELD DEMONSTRATION OF MULTIPLE BIOSLURRY TREATMENT TECHNOLOGIES FOR EXPLOSIVES-CONTAMINATED SOILS

Mark L. Hampton (U.S. Army Environmental Center, A.P.G., MD)
Dr. John F. Manning (Argonne National Laboratory, Argonne, IL)

ABSTRACT: The U.S. Army Environmental Center field tested both aerobic/anoxic and anaerobic soil slurry biotreatment processes at Iowa Army Ammunition Plant to evaluate their performance and cost in remediation explosives contamination in soils. The trials were conducted in a RCRA quality lagoon and a concrete trench reactor, employing a variety of commercial impeller and hydraulic mixing systems. Test results described both degradation kinetics and metabolic fate, as well as the capabilities and limitations of different reactor configurations and mixing strategies. The demonstration was conducted in conjunction with a CERCLA removal action going on at the site, so excavated soils were treated at the test site, and the demonstration results will be used by environmental managers in their remedy selection.

INTRODUCTION

The past production and handling of conventional munitions has resulted in explosives contamination (TNT, RDX, and HMX) of the soils at various military facilities. Depending upon the concentrations present, these explosives contaminated soils can pose both a reactivity and toxicity hazard, and a potential for groundwater contamination.

In 1995, the U.S. Army Environmental Center (USAEC) demonstrated the feasibility of soil slurry biotreatment as an alternative to incineration, using soils contaminated with explosives from the Joliet Army Ammunition Plant. The process involved suspending the contaminated soil in an aqueous slurry, and treating it in a reactor in which a co-substrate (molasses) was added to boost the metabolic activity of the microbial consortium native to the soil. Laboratory and the field pilot tests showed that operating the reactors in an aerobic/anoxic mode resulted in more complete degradation of TNT. The demonstration showed highly promising results, with the removal of TNT from about 1300 mg/kg to < 10 mg/kg . In addition, RDX, HMX, and dinitro-toluene in the Joliet soil were also removed in the process to < 20 mg/kg. Metabolic fate studies using radiolabeled TNT showed about 23% of the TNT was mineralized to carbon dioxide, with another 55% degraded to organic acids and carbon fragments in the biomass, indicating ring cleavage. Additionally, the microbial consortium functioned effectively with a wide range of soil concentrations, and in ambient temperatures down to 15-20°C. This provides a significant degree of flexibility in designing and operating systems to achieve treatment goals.

Objective. To accelerate the fielding of bioslurry technology, the Department of Defense's Environmental Security Technology Certification Program sponsored the current study, which tested the aerobic/anoxic bioslurry process simultaneously with J.R. Simplot Company's Simplot Anaerobic Bioremediation Ex-situ (SABRE$_{tm}$) process. The objective of the demonstration was to evaluate the performance and cost of soil slurry biotreatment as an alternative to incineration. The demonstration was conducted in conjunction with a Comprehensive Environmental Response Compensation and Liability Act removal action at Iowa Army Ammunition Plant (IAAAP), which hosted the demonstration.

MATERIALS AND METHODS

At Iowa, two types of bioslurry systems were evaluated: 1) the aerobic/anoxic process in a 55,000 gallon lagoon and 2) the SABRE$_{tm}$ process in a 10 to 12,000 gallon concrete trench. Clean up standards were 196 mg/kg for TNT and 53 mg/kg for RDX.

Laboratory studies conducted in support of the field demonstration at Iowa showed that TNT and RDX could be removed from the soil to concentrations of less than 20 mg/kg in approximately 30-40 days. The design of the laboratory systems indicated that batch type processes were the most effective systems for use at Iowa.

The aerobic/anoxic lagoon system was designed to meet all Resource Conservation and Recovery Act standards, including a clay layer, dual high density polyethylene (HDPE) liners, and a leachate collection system. The lagoon was approximately 82' by 47' and included an appropriate two to one side slope for anchoring the liner system.. The working depth ranged from five to six feet. No leachate was detected in the collection system. Figure 1 shows the reactor was designed to use off-the-shelf fan type mixers and a diffuser mounted on a traversing bridge. Although the bridge was designed primarily to support sampling throughout the reactor, its traveling mechanism also allowed for complete mixing and distribution of air. Approximately 55 to 60 cubic yards (yd^3)of soil were added to the lagoon at the beginning of the demonstration. Water was added to the system to provide a soil slurry of approximately 40 percent by weight. Additionally, about 150 gallons of molasses was added to the system three times during the study.

The SABRE$_{tm}$ system was tested in a concrete trench to utilize available mixing equipment without modifications for gantry width. The concrete trench was 50' long and 8'4" wide. The trench had an external HDPE liner and leachate collection system. It contained approximately 40 yd^3 of soil and water to generate a 40% soil slurry. The system was operated by Simplot personnel and all additions were conducted by Simplot personnel. The proprietary amendments were added at the beginning of the process, based on the results of the laboratory treatability studies.

In both processes, explosives-contaminated soils and water are biologically treated in a reactor. Contaminated soils are excavated and screened to remove large rocks and debris that might interfere with mixers and pumps. In the reactor, mixers maximize mass transfer of the contaminant, and process controls for amendments,, temperature, and aeration are designed to optimize the growth and activity of microbes which are capable of degrading explosives. The soils are mixed with water to produce a slurry that is typically 25 to 40% solids by weight. In both cases the additives, either molasses or the Simplot proprietary mixture, provide co-substrate and nutrients that are necessary for microbial growth and degradation of the explosives. This offers the potential for greater rates of degradation that can be seen with solid phase biotreatment processes. Reactor processes are also inherently flexible; it is theoretically possible to change the conditions of the reactor so that a treatment train alternative can be designed.

The biotreatment processes in both reactors were tracked intensively by sampling at ten locations in each reactor every week and analyzing for explosives and their degradation products using EPA method 8330. Other process variables, temperature, dissolved oxygen (DO), and redox potential, were monitored two or three times per week.

RESULTS AND DISCUSSION

Aerobic/anoxic process. As can be seen in Figure 2, TNT was reduced from 1500 mg/ kg to below treatment levels in approximately eight weeks while limiting the accumulation of metabolic intermediates to less than 40 mg/kg. The temperature remained above 22°C, the DO fluctuated between 1.0 and 0 mg/L, and the redox potential of this system varied in the range of -50 to +50 mV, indicating the aerobic/anoxic nature of process. The variability in successive readings can be attributed to the characteristic heterogeneity of explosives in soil, and indicates the need for thorough mixing. The mixer pattern in the lagoon was altered by week seven; and following that, degradation of TNT proceeds in a timely fashion. Figure 2 also shows the degradation of HMX and RDX follows that of TNT, with much of the RDX removal generally occurring after that of TNT.

SABRE$_{tm}$ process. Figure 3 shows results from Simplot's SABRE$_{tm}$ process. It took approximately eight weeks for TNT to be removed in this operation. The temperature remained above 22°C, and the SABRE$_{tm}$ system maintained a redox potential generally less than -100 mV throughout the process. RDX was removed from the system after approximately 10 weeks, much of this occurring after TNT had been degraded. Process kinetics and degradation byproducts appear to be similar between the two processes.

Further testing. Currently, the treated slurry from both reactors are undergoing a variety of tests: radiolabeled TNT metabolic fate studies, leaching stability tests, plant growth analysis, and toxicity testing, including Ames assay and earthworm tests. Test results will be combined with the slurry sample analyses to determine the disposition of the treated slurry. Results from previous field trials indicate the slurry is suitable for direct land application.

Slurry disposition. The demonstration also evaluated options for disposing of the treated soil and process water, if land application is not an option. Slurry was gravity dewatered to 40% soil moisture in 2-3 weeks. Options for disposing of the process water depend on discharge standards at the site. Residual explosives in the treated water were under the local National Pollutant Discharge Elimination System standards of 2 μg/L TNT, but the biotreatment processes in both reactors drone the biological oxygen demand (BOD) in the water above 10000 mg/L. The results of analytical and toxicity tests can support land application of the process water. Otherwise, high BOD levels can require additional aerobic treatment or processing through activated carbon to bring it down to discharge levels.

Design and cost. Additionally, experience gained in operating these biotreatment processes has been applied to developing a conceptual engineering design for full scale treatment of Iowa's 10000 yd^3 of explosives-contaminated soil. Both aerobic/anoxic and SABRE$_{tm}$ processes were incorporated as biotreatment alternatives, and the design included both aboveground tank and sunken lagoon reactor options. Cost projections from this design put the unit cost of bioslurry treatment around $300-350 per yd^3, in the same range as other biotreatment options such as windrow composting. This information was provided to the environmental managers at IAAAP for consideration in selection of a treatment process for their contaminated soil.

LESSONS LEARNED
(1) Laboratory scale treatability studies are an important first step in determining whether site conditions will support biotreatment. Treatability studies can give the project planners the ability to estimate treatment periods before the project commences. The treatment times obtained from laboratory studies should be considered as estimates.

(2) Mixing in rectangular lagoons requires additional operator time to handle dead spots. The mixing system used in this demonstration was theoretically oversized for the soil slurry. It still required extra attention to thoroughly mix material that contained 25-40% clay.

(3) Removal of the slurry from the lagoon needs to be considered when locating the treatment system. Simply pumping the slurry from the reactor is effective for the majority of the contents, but a fraction remained that required mechanical removal.

(4) The environment that this equipment must function is very extreme. Generally, acidic environments, low DO levels, and constant particle friction wear heavily on mixing and pumping equipment. Long term operations in these conditions require particular attention to be paid to equipment selection.

CONCLUSIONS

Soil slurry systems have been successfully demonstrated in lagoon and trench configurations. The results indicate that TNT and RDX can be removed to standards that are necessary for the direct land application of soil. The bioslurry process is robust and may be adapted to operate in a relatively wide range of slurry concentrations and temperatures.

Process times and cost projections indicate that it is also a cost effective remediation option. Unit costs estimated for bioslurry treatment are at least half the cost of incineration, particularly for sites with less than 25000-30000 yd^3 of soil to treat. Based on experience from this field demonstration, the concept design and cost estimates for full scale application of soil slurry biotreatment will be updated and published by USAEC later this year

ACKNOWLEDGMENTS

The authors appreciate that this demonstration would not have been possible without the close cooperation of our hosts at IAAAP, and the participation of the J.R. Simplot Company.

REFERENCES

Manning, J.F., Boopathy, R., and Kulpa, C.F., 1995, *A Laboratory Study in Support of the Pilot Demonstration of a Biological Soil Slurry Reactor*, U.S. Army Environmental Center report number SFIM-AEC-TS-CR-94038.

Manning, J.F., Boopathy, R., and Breyfogle, E.R., 1996, *Field Demonstration of Slurry Reactor Biotreatment of Explosives-Contaminated Soils,* U.S. Army Environmental Center report number SFIM-AEC-ET-CR-96178.

FIGURE 1. Lagoon reactor with mixers on sampling gantry

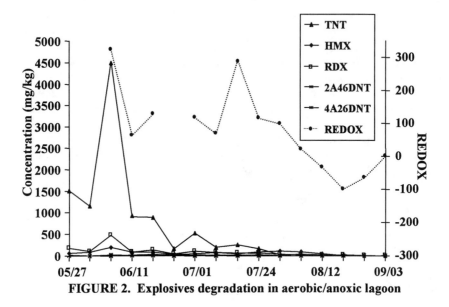

FIGURE 2. Explosives degradation in aerobic/anoxic lagoon

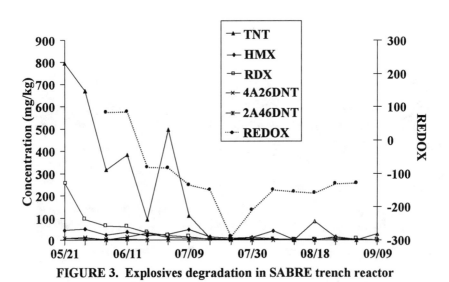

FIGURE 3. Explosives degradation in SABRE trench reactor

HYDROLYSIS/BIODEGRADATION OF CHEMICAL WARFARE VESICANT AGENT HT: TWO APPROACHES

Steven P. Harvey, Linda L. Szafraniec and William T. Beaudry
U.S. Army ERDEC, Edgewood, MD 21010-5423

ABSTRACT: HT is a powerful vesicant produced as a chemical warfare agent. It is comprised of 60 wt % 2,2'-dichlorodiethyl sulfide (H) and 40 wt % bis-(2-(2-chloroethylthio)ethyl) ether (T). Since HT reacts with water to form primarily the alcoholic compounds 2,2'-thiodiethanol (thiodiglycol or TDG) and bis-(2-(2-hydroxyethylthio)ethyl) ether (T-OH), disposal might be accomplished by hydrolysis followed by biodegradation. The half-lives of H and T in a 3.8% HT/water reaction at 90° C were 1.4 and 1.6 minutes, respectively. Concentrations of both were reduced below 1 ppm within 30 minutes. TDG is readily biodegradable. However, T-OH biodegradability has not previously been reported. HBr treatment converted HT ether-alcohol products to TDG. A comparative study of two hydrolysis/biodegradation approaches is reported here. HT was hydrolyzed (1) in water, and (2) in water then with HBr. Products were used as feed for separate aerobic Sequencing Batch Reactors. Bioreactor performances were compared for carbon removal efficiency, effluent toxicity, chemical composition and nitrogen levels, mixed liquor and effluent suspended solids, loading, and oxygen uptake rate. Although both feeds were detoxified in the SBRs, water hydrolysis yielded better overall bioreactor operation.

INTRODUCTION

HT ("sulfur mustard") is a major component of the U.S. chemical stockpile. A total of 3,715.96 tons of HT are stored in 3,591 ton containers at Pine Bluff, AK, 183,552 mortars at Anniston, AL and 20,384 mortars at Pueblo, CO (OSD News Release, 1996).

MATERIALS AND METHODS

HT purity, bioreactor feedstock preparation, sampling and analysis of HT/water reaction, low level analysis for H and T, nuclear magnetic resonance spectroscopy and bioreactor equipment and operation, calculation of mass balance and oxygen uptake rate, and determination of nitrate, nitrite and ammonia concentration measurements were as described (Harvey et al. 1997).

RESULTS

4.1 HT Hydrolysis. H and T hydrolysis rates were measured in a well-agitated reactor. HT (3.8 weight %) was added batchwise to 200 ml water at 90° C with vigorous agitation. The hydrolysis of both compounds proceeded to levels below 1 ppm in about 20 minutes. Half-lives were calculated at 1.4 minutes for H and 1.6 minutes for T in this system.

The essential overall reaction of T hydrolysis is shown in Equation 1:

$$(ClCH_2CH_2SCH_2CH_2)_2O + \underline{2}\,H_2O \rightarrow (HOCH_2CH_2SCH_2CH_2)_2O + \underline{2}\,HCl \quad (1)$$

Ethers undergo the acid hydrolysis reaction (Morrison and Boyd, 1976):

$$R\text{-}O\text{-}R' + H\text{-}X \rightarrow R\text{-}X + R'\text{-}OH \qquad\qquad (2)$$

The order of reactivity of the acid halides is: HI > HBr > HCl.

In order to maximize T-OH biodegradability one approach is to cleave the ether bond prior to biodegradation (Equation 3). Since HCl is a product of the initial T hydrolysis the acidic hydrolysis is shown here using HCl as the acid.

$$(HOCH_2CH_2SCH_2CH_2)_2O + \underline{2}\,HCl \rightarrow (ClCH_2CH_2)_2S + (HOCH_2CH_2)_2S + HCl \ (3)$$

$(ClCH_2CH_2)_2S$ is known to further hydrolyze to TDG and HCl (Yang et al., 1988) giving an overall equation for the reaction as shown in Equation 4:

$$(ClCH_2CH_2SCH_2CH_2)_2O + \underline{2}\,H_2O \rightarrow \underline{2}\,HOCH_2CH_2SCH_2CH_2OH + \underline{2}\,HCl \quad (4)$$

HBr and HI effects on T-OH to TDG conversion were investigated (Harvey et al. 1997) and showed an enhancement of conversion rate with both acids. The initial rates with HBr and HI were similar although the HBr reaction progressed further toward completion (a reduction in T-OH from 80.7% to 23.8% in three hours as compared to a reduction from 80.7% to only 37.6% with HI).

4.2. HT Biodegradation. The water-hydrolysis feed was generated by reacting 526 g HT in 90° water at 3.8 wt % for 2 hr. The HBr-hydrolysis feed was generated by subjecting half the initial batch to the acidic hydrolysis step in 0.29 M HBr. Both feeds were neutralized with NaOH and provided as sole carbon source to the bioreactors. Products were analyzed by gas chromatography and ion trap detection for residual H or T. No H or T were detected with a method detection limit of 1.8 μg/ml. ^{13}C NMR analysis (Harvey et al, 1997) showed that only a relatively minor portion of the T-OH was converted to TDG in the acidic hydrolysis (HBr) step (a reduction of 28 to 26 area %). However, the majority of the "other" portion apparently was converted to TDG, (a reduction from 16.7 to 3.04 area %) while the TDG portion was increased from 56 to 73 area %.

The biodegradation study was conducted with two bioreactors operated under essentially identical conditions. The objective was to determine which, if either, hydrolysis process would permit stable bioreactor operation, efficient carbon removal and acceptable effluent quality (organic composition, solids content, and aquatic toxicity).

SBRs were seeded with activated sludge; biomass concentrations were measured as mixed liquor suspended solids (MLSS). Starting MLSS was adjusted to 4.5 g/L. Effluent TOC concentration, ammonia, nitrate, nitrite, MLSS, and effluent suspended solids (ESS) levels were monitored several times a week. Details of the operational periods were described elsewhere (Harvey et al. 1997).

Initially, during the start-up period, MLSS decreased, as expected, due to biomass digestion. The decrease continued somewhat during the 20 day HRT, generally leveled off during the 15 day HRT and increased during the 10 day HRT, consistent with the increased levels of daily feed during each subsequent operating period. Excess biomass was wasted each day at the end of the React period with an 80 day SRT during the 20, 15 and 10 day HRT periods. No biomass was wasted during the start-up period.

With the exception of an initial ESS excursion to over 500 mg/L in the HBr-hydrolyzed feed reactor, levels in both reactors remained below about 200 mg/L and decreased to below about 100 mg/L during the 15 and 10 d HRT operating periods. Results generally showed a regular pattern of effective biomass settling, consistent with stable bioreactor operation.

Overall efficiency of TOC removal was 86.1 % for the water-hydrolyzed feed and 79.3 % for the HBr-hydrolyzed feed. Effluent TOC concentrations averaged 517.6 ppm for the water-hydrolyzed feed and 819 ppm for the HBr-hydrolyzed feed. Efficiencies generally decreased with the decrease in HRT although it is not possible from these data to ascertain to what extent that change might have been due to equilibration as opposed to actual differences in the efficiency of degradation.

Loading started at about 0.04 -0.05 mg TOC/mg MLSS in both reactors, then climbed to 0.9 - 1.0 as the HRT was decreased. The MLSS initially dropped during start-up then remained fairly steady during the 20 and 15 day HRT periods.

NH_4Cl feed levels were continually adjusted to maintain the lowest possible effluent concentrations. The overall average total nitrogen levels (nitrate-N + nitrite-N + ammonia-N) for the 10 d HRT period were 3.2 ppm for the water-hydrolyed feed and 3.7 ppm for the HBr-hydrolyzed feed.

Effluents were analyzed for their toxicity toward luminescent test bacteria (Microtox test). Results of Microtox tests on the HT feeds and the effluents at the end of each test period are shown in Figure 1. Toxicity was measured as the effective concentration of test material causing a 50% decreased light output from the test organisms (EC_{50}). The biodegradation process achieved a 28-44 fold reduction in toxicity of the water-hydrolyzed feed and a 19-21 fold reduction in toxicity of the HBr-hydrolyzed feed. Controls used were phenol (EC_{50} of 27.3 mg/L), beer, included as a control because of a salt content similar to the bioreactor effluents and organic components comprised largely of alcohols, and a 1% aqueous dilution of TDG which was relevant because TDG comprises a major component in both feeds and has a known LD_{50} value of 3000-4000 mg/kg for neat TDG (Registry of Toxic Effects of Chemical Substances, 1981-82). The toxicity of the TDG solution and the beer were intermediate to those of the bioreactor feed and effluent solutions. In summary, the bioreactor effluents were consistently less toxic than either beer or 1% TDG towards Microtox test bacteria. The phenol control was far more toxic than any of the test materials. HT is far more toxic than phenol. Although human LD_{50} values are not established for T,

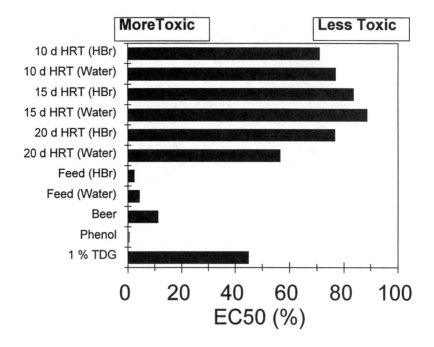

FIGURE 1. Aquatic toxicity of bioreactor feed, effluent and controls.

the H component has an oral LD_{50} of 0.7 mg/kg which is 4,200 - 5,700 times more toxic than TDG (U.S. Army, 1996). Thus, the hydrolysis process reduces the human toxicity of the major component by a factor greater than 4,000 and the biodegradation process further reduces the aquatic toxicity of the hydrolysate by a factor of 19-44, depending on the particular feed and the HRT of the reactor.

Effluents from both reactors were also analyzed for semi-volatile priority pollutant compounds by U.S. Environmental Protection Agency (EPA) toxicity characteristic leachate protocol (TCLP) procedures. No semi-volatile priority pollutant compounds were detected at or above the EPA-specified detection limits.

DISCUSSION

The hydrolysis of 3.8 wt % HT in water proceeds to completion (H and T levels below 1 ppm) within about 30 minutes in a well-agitated system at 90° C. The half-lives of H and T are about 1.4 and 1.6 min, respectively in this system. This reaction yields a product containing 56% TDG, 28 % T-OH and 17 % "other" (linear sulfonium ions, hydrolysis products of impurities, etc.). No chlorinated products were detected by NMR analysis. The primary component of the hydrolysate (TDG), has an oral LD_{50} value more than 4,000 times less than the primary starting material (HD). Therefore, hydrolysis provides a simple, effective means of dechlorinating and detoxifying the starting material.

Hydrolysis of 3.8 wt % HT in water followed by acid hydrolysis with 0.38 M aqueous HBr yielded 73% TDG, 26 % T-OH and 1% "other". Thus, the HBr substantially increased the yield of TDG and decreased the yield of compounds in the "other" category. With respect to inorganic components, the primary difference between the two feeds is the presence of NaBr in the HBr-hydrolyzed feed. The water-hydrolyzed HT contains approximately 1% NaCl whereas the HBr-hydrolyzed HT contains approximately 1% NaCl plus approximately 1% NaBr.

Both bioreactors showed generally stable operation throughout the different test cycles (start-up, 20, 15 and 10 day HRT periods). Higher HRTs were generally correlated with higher efficiencies of TOC removal. However, since the higher HRTs were completed earlier in the test, it is not certain to what extent the difference was due to equilibration and to what extent it was due to actual differences in degradation of the feed compounds. At any rate, the relative efficiency of TOC removal did not necessarily correspond to differences in aquatic toxicity; for example, Microtox data actually showed slightly lower toxicity for the 15 d HRT effluent than for that from either the 10 or 20 d HRTs, although the differences were relatively minor. In general though, toxicity from both reactors at all HRTs was consistently low relative to the feed and the controls (a 19 to 44-fold reduction in aquatic toxicity of the feed).

Detailed comparisons between the two bioreactors with respect to mixed liquor and effluent solids concentrations, efficiency of TOC removal, loading, and toxicity indicated that the water-hydrolyzed feed had a generally more favorable outcome. The difference may have been due at least in part to the differences in salt concentration. The actual salt output from the reactor fed water-hydrolyzed feed was 1.61 mass units (0.82 units Na_2SO_4 and 0.79 units NaCl) per mass unit HT. For the HBr-hydrolyzed feed, there was 1.61 units Na_2SO_4, 0.78 units NaCl plus 1.09 units NaBr for a total of 2.63; a 63% increase over that from the other reactor (Harvey et al., 1997 for detailed calculations). In addition to the improved performance, the use of the water-hydrolyzed feed offers the significant advantages of a simpler hydrolysis process (no requirement for the HBr-mediated acid hydrolysis step) and a more favorable mass balance due to the elimination of HBr from the process input and NaBr from the output. In summary, hydrolysis in hot water followed by neutralization with NaOH and aerobic biodegradation in a Sequencing Batch Reactor seeded with activated sludge offers the more promising of the two approaches tested for the combined chemical/biological treatment of the chemical warfare agent HT.

REFERENCES
Office of Assistant Secretary of Defense News Release, January 22, 1996: *U.S. Chemical Weapons Stockpile Information Declassified.*

Harvey, S.P., Szafraniec, L.L., Beaudry, W.T., Rohrbaugh, D.K., Haley, M.V., Kurnas, C.W. and Rosso, T.E. "Sequencing Batch Reactor Biodegradation of HT: A Detailed Comparison of the Results of Two Different Approaches. U.S. Army Technical Report, ERDEC-TR-437, Edgewood Research, Development and Engineering Center, APG, MD.

Morrison, R.T. and R.N. Boyd. 1976. *Organic Chemistry, Third Edition*, Allyn and Bacon, Inc., New York University.

Yang, Y-C., L.L. Szafraniec, W.T. Beaudry, and J.R. Ward. 1988. "Kinetics and Mechanism of the Hydrolysis of 2-Chloroethyl Sulfides" *J. Org. Chem.* 53:3293-3297.

Registry of Toxic Effects of Chemical Substances (1981-2). Azur Environmental, MICROTOX manual, model 500 toxicity test system, 2232 Rutherford Rd., Carlsbad CA 92008, May, 1991

U.S. Army, 1996. HT Material Safety Data Sheet.

SEQUENTIAL ANAEROBIC-AEROBIC TREATMENT OF TNT-CONTAMINATED SOIL

Roger B. Green[1], Paul E. Flathman[2], Gary R. Hater[1], Douglas E. Jerger[2], and Patrick M. Woodhull[2]

[1] Waste Management, Inc., Cincinnati, Ohio
[2] OHM Remediation Services Corp., Findlay, Ohio

ABSTRACT: The microbial degradation of TNT is a reductive process and occurs most rapidly under anaerobic conditions. The humification of the primary amino transformation products of TNT likely requires aerobic conditions. The effectiveness of the sequential anaerobic-aerobic treatment of TNT-contaminated soil was investigated in a laboratory study. TNT contaminated soil, with a starting concentration of approximately 3,000 mg/kg, was supplemented with one of three carbon sources: starch, starch and glucose, or molasses and combined with an inoculum of anaerobic digester sludge and cow manure. The resulting mixture was incubated under anaerobic conditions. The high solids anaerobic phase of treatment resulted in between 97.0 and 99.3% removal of TNT in 29 days. Over 95% of the TNT removal was achieved within 7 days. The first order rate constant and half-life for TNT in the anaerobic phase were 0.17 day^{-1} and 4 days respectively.

Aerobic composting of a select number of the anaerobically treated samples was performed using three different compost mixtures. Aerobic treatment is important for the irreversible incorporation of reduced amino transformation products into the humic matrix of the soil. Additional removal of TNT and substantial removal of amino intermediates was observed after 28 days of composting. The combination of treatments resulted in between 98.6 and 99.8% removal of TNT in 57 days. The final concentrations of total mono- and diamino-transformation products were less than 5 mg/kg and 35 mg/kg respectively.

In conclusion the sequential anaerobic-aerobic treatment of TNT-contaminated soil was very effective in rapidly reducing concentrations of TNT and the mono- and diaminotoluene transformation products.

INTRODUCTION

While the mineralization of TNT in laboratory-scale experiments has been reported (Fernando et al., 1990), mineralization is not held to be the primary fate of TNT in complex natural environments (Major et al., 1996). However, substantial biotransformation of TNT has been demonstrated. For example, TNT is known to be susceptible to reduction of the nitro groups to amino groups via a multistep process involving nitroso and hydroxylamine intermediates (Walsh, 1990). It is also known that the predominant metabolite produced under aerobic conditions is 4-amino-2,6-dinitrotoluene (4A26DNT), while under mildly acidic (pH 6) anaerobic conditions the 2,4-diamino-6-nitrotoluene (24DA6NT) reduction product is most prevalent (Funk et al., 1993). The complete conversion of TNT to 2,4,6-triaminotoluene (TAT) has been reported in samples maintained under strictly anaerobic conditions and low redox potentials (E_h < −200mV) in the presence of a carbon source (Funk et al., 1993, Lenke et al., 1994).

The reduction of TNT to TAT is of interest because of the ability of aromatic amines to be incorporated into the organic fraction present in soil or compost by forming covalent bonds with a variety of functional groups (Major et al., 1996). The irreversible incorporation of TAT into the humic matrix of soils is believed to be facilitated by aerobic processes (Rieger and Knackmuss, 1995). Aerobic treatment following anaerobic treatment has also been suggested as means accelerate the degradation of intermediates produced during anaerobic conditions (Funk et al., 1993) and to scavenge excess carbon which, due to its high oxygen demand, has been reported to cause toxicity to *Daphnia magna* (Roberts et al., 1996).

The objective of the treatability study reported here was to evaluate a novel two stage treatment process for TNT contaminated soil. The process combines an initial anaerobic treatment stage operated at a high solids level followed by an aerated composting stage. TNT contaminated soil was combined with anaerobic digester sludge, cow manure, and one of three different carbon substrates, and water. After anaerobic incubation, a select number of samples were combined with three different compost mixtures. The rationale for sequencing the treatment is to facilitate the aerobic degradation of any residual substrate and intermediates remaining after anaerobic treatment and to promote binding of amine intermediates to the humus-rich yard waste compost matrix.

METHODS AND MATERIALS

TNT-contaminated soil and uncontaminated background soil samples were collected at the Pueblo Chemical Depot, Pueblo, CO, in clean five-gallon plastic containers and transported to the OHM Treatability Laboratory in Findlay, OH. TNT and amino metabolites were analyzed according to USEPA Method 8330. Liquid chromatography analysis was performed with a Hewlett-Packard model HP1090M HPLC equipped with a PV-5 solvent delivery system, auto-injector system, a 5 μm C_{18} reverse-phase column, and diode array detector. A minimum of 12 g of sample was extracted in acetonitrile using sonication. HPLC results were confirmed by GC/MS. Standards for 2,4-diamino-6-nitrotoluene, and 2,6-diamino-4-nitrotoluene were obtained from AccuStandard Inc., New Haven, CT.

Anaerobic Treatment

The experimental design for the initial anaerobic phase of the study is presented in Table 1. For the purpose of the study, an initial TNT concentration of 3,000 mg/kg was targeted. To achieve the desired starting concentration, 0.70 kg of the TNT-contaminated soil was mixed with 3.50 kg of the uncontaminated background soil. The study was performed using 8.0 L closed glass reactors for each of the six treatments. Final volume of the soil mixture in each reactor was approximately 4.0 liters. The reactors were initially monitored on a daily basis for pH and redox potential.

TABLE 1. Anaerobic Treatment Experimental Design

Parameter	Reactor Number					
	1	2	3	4	5	6
Soil, kg (L)	4.2 (2.8)	4.2 (2.8)	4.2 (2.8)	4.2 (2.8)	4.2 (2.8)	4.2 (2.8)
50 mM PO_4 Buffer, kg	1.2	---	---	---	---	1.2
50 mM PO_4-Buffered Inoculum, kg	---	1.2	1.2	1.2	1.2	---
Starch, g (% w/v)	10 (0.25)	20 (0.50)	40 (1.0)	40 (1.0)	---	---
Glucose, g (% w/v)	---	---	40 (1.0)	80 (2.0)	---	---
Molasses, g (% w/v)	---	---	---	---	80 (2.0)	---
Vitamins, mg (mg/kg)	---	---	---	54 (10)	---	---

Aerobic Treatment.

On Day 29, the soil was removed from Reactors 4, 5, and 6 and subdivided (by volume) into three equal portions in preparation for aerobic biological treatment of any remaining TNT and/or its transient transformation products. Aerobic composting techniques, using three compost mixtures, were used for the evaluation. The Colorado and Louisville Composts were prepared in a 1:1 ratio (by volume) of finished yard waste compost:active yard waste compost. The green waste compost was prepared in a 1:1:½:½ ratio (by volume) of horse manure:straw:bailed alfalfa:fresh alfalfa.

RESULTS

Anaerobic Phase Results

During the anaerobic phase of treatment no removal of TNT or production of reduced intermediates was observed in Reactor 6, the uninoculated and unamended control sample. The redox potential for this sample was initially measured at -50 mV but rapidly increased to approximately 100 mV and held at that level for the remainder of the test. In Reactor 1, the other control sample, supplemented with 10 g starch and no inoculum, a slight reduction in TNT and low concentrations of reduced intermediates were observed after 41 days of incubation. The measured redox potential for this sample remained around 200 mV throughout the 57 days of monitoring. The remaining reactors, inoculated and supplemented with an external carbon source of starch, starch and glucose, or molasses, attained redox values of approximately -200 mV after six days and in general maintained those values over the course of incubation. Reactor 5, supplemented with 80 g of molasses attained the most rapid reduction in redox potential.

Results, in descending order of TNT removal efficiency, were: Reactor 3 at 99.3%, Reactor 5 at 98.4%, Reactor 4 at 97.0%, and Reactor 2, at 91.4%. These results are generally consistent with the redox monitoring results which showed that Reactor 3 maintained the lowest redox potential during the anaerobic incubation. Reactor 3, which was held under anaerobic conditions for a total of 57 days, ultimately achieved >99.6 % removal of TNT and contained a combined amino metabolite concentration of 35 mg/kg. For each of the carbon supplemented reactors the majority of TNT removal was observed within the first 7 days of incubation. Figure 2 illustrates the sequential reduction of TNT through ADNT and diaminonitrotoluene intermediates. The formation of 24DA6NT in preference to the 26DA4NT isomer was observed as previously reported (Funk et al., 1993). Diaminonitrotoluene concentrations peaked by 14 days of incubation with a subsequent decline.

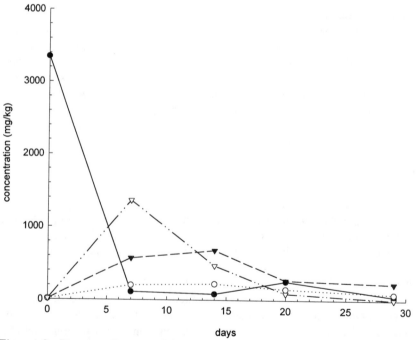

Figure 2. Reactor 5 anaerobic phase results for (●)TNT, (▽)ADNT, (▼)24DA6NT, and (○)26DA4NT.

Aerobic Composting Results

Aerobic composting resulted in additional removal of residual TNT. Overall reduction in TNT for soil from Reactor 4 and Reactor 5 combined with the three compost mixtures ranged from 98.6% to 99.8%. Substantial removal of TNT from Reactor 6, the anaerobic control, ranged from 79.1% to 92.3% over the three compost mixtures. Of the three mixtures evaluated, the green waste containing horse manure, straw, and alfalfa yielded the greatest TNT removal. In the sample containing unamended, uninoculated soil mixed with the green waste mixture, ADNT was detected within 2 hours after mixing. After 17 days, the presence of 26DA4NT was also detected in this sample. Samples containing the unamended, uninoculated soil combined with the two yard waste composts also produced ADNT, but at an order of magnitude lower concentration and after a longer incubation (17 days) than the sample containing the green waste mix.

Samples containing soil from Reactor 4 and Reactor 5 combined with the yard waste composts from Louisville and Colorado demonstrated greater removal of ADNT and DANT than samples with soil combined with the green waste mixture. This result may be explained by the fact that the yard waste compost contains a high humus content and can reasonably be expected to contain organisms that participate in the humification process.

DISCUSSION

The coupling of anaerobic and aerobic treatment for the TNT-contaminated soil was shown to be highly effective, removing up to 99.8% of the starting TNT concentration. Monitoring of TNT and reaction intermediates over the course the anaerobic treatment indicates TNT was sequentially reduced to monoamino- and then diamino- intermediates, with 24DA6NT being the predominant diammino isomer as previously reported (Funk et al., 1993). These results, coupled with the low redox potential maintained in the reactors, suggest that a large percentage of TNT was converted to TAT (Lenke et al., 1994).

Aerobic composting results demonstrated that each of the compost mixtures evaluated provided further reduction TNT and amino metabolites. The green waste mixture provided better TNT removal while the yard waste composts demonstrated better removal of the amino intermediates. This may be attributable to the higher humus content of the yard waste composts providing a better medium for binding of the TNT reduction products. Several studies have provided evidence that the products of TNT biotreatment bind to soil organic matter (Held et al., 1997; Pennington et al., 1992; Kaplan and Kaplan, 1982). While the mechanism of removal for the amino intermediates during the aerobic composting phase cannot be assessed from the methods used in this study, a combination of binding and biotransformation is a reasonable explanation.

REFERENCES

Fernando, T., J.A. Bumpus, and S.D. Aust. 1990. "Biodegradation of TNT (2,4,6-Trinitrotoluene) by *Phanerochaete chrysosporium.*" *Appl. Environ. Microbiol.* 56:1666-1671.

Funk, S.B., D.J. Roberts, D.J. Crawford, and R.L. Crawford. 1993. "Initial-Phase Optimization for the Bioremediation of Munition-Contaminated Soils." *Appl. Environ. Microbiol.* 59:2171-2177.

Held, T., G. Draude, F.R.J. Schmidt, A. Brokamp, and K.H. Reis. 1997. "Enhanced Humification as an In-Situ Bioremediation Technique for 2,4,6-Trinitrotoluene (TNT) Contaminated Soils." *Environmental Technology.* 18:479-487.

Kaplan, D.L., and A.M. Kaplan. 1982. "Thermophilic Biotransformations of 2,4,6-Trinitrotoluene Under Simulated Composting Conditions." *Appl. Environ. Microbiol.* 44:757-760.

Lenke, H., B. Wagener, G. Daun, and H.-J. Knackmuss. 1994. "TNT-contaminated soil: A Sequential Anaerobic/Aerobic Process for Bioremediation." P.456. Abstracts of the 94th Annual Meeting of the American Society of Microbiology.

Major, M.A., W.H. Griest, J.C. Amos and W.G. Palmer. 1996. *Evidence for the Chemical Reduction and Binding of TNT during the Composting of Contaminated Soils.* U.S. Army Center for Health Promotion and Preventive Medicine Technical Report No. A324278.

Pennington, J.C., C.B. Price, E.F. McCormick, and C.A. Hayes. 1992. *Effects of Wet and Dry Cycles on TNT Losses from Soil.* U.S. Army Corps of Engineers Waterways Experiment Station Technical Report #EL-92 37.

Rieger, P.-G. and H.-J. Knackmuss. 1995. "Basic Knowledge and Perspectives on Biodegradation of 2,4,6-Trinitrotoluene and Related Nitroaromatic Compounds in Contaminated Soil." In: J.C. Spain (ed.), *Biodegradation of Nitroaromatic Compounds..* Plenum Press, New York. pp.1-17.

Roberts, D.J., F. Ahmad and S. Pendharkar. 1996. "Optimization of an Aerobic Polishing Stage to Complete the Anaerobic Treatment of Munitions-Contaminated Soils." *Environ. Sci. Technol.* 30:2021-2026.

Walsh, M.E. 1990. *Environmental Transformation Products of Nitroaromatics and Nitramines.* U.S. Army Corps of Engineers Cold Regions Research and Engineering Laboratory, Special Report 90-2.

SIMULTANEOUS ABIOTIC-BIOTIC MINERALIZATION OF
PERCHLOROETHYLENE (PCE)

Fatih Büyüksönmez, Thomas F. Hess, Ronald L. Crawford, Andrzej Paszczynski
(University of Idaho, Moscow, Idaho) Richard J. Watts (Washington State
University, Pullman, Washington)

ABSTRACT: Simultaneous abiotic and biotic reactions were investigated for the mineralization of perchloroethylene, PCE. Modified, phosphate-buffered, Fenton reactions were combined with biological treatment by a hydrogen peroxide-resistant strain of *Xanthobacter flavus*, which was able to mineralize dichloroacetic acid, DCAA, a Fenton-degradation product of PCE. The extent of mineralization of PCE was increased by approximately 15% with the addition of microorganisms over that of a noninnoculated control. The results of this study showed that both abiotic and biotic reactions leading to PCE mineralization can occur simultaneously and may be a viable alternative for the treatment of waters and wastewaters containing PCE.

INTRODUCTION

Combined abiotic-biotic reactions can be used to mineralize recalcitrant xenobiotic compounds. Organic chemicals with electron-withdrawing substitutions, such as halogens and nitro groups, are often recalcitrant to stand-alone biological degradation. Abiotic reactions, however, can be used to initiate transformation of such biorecalcitrant molecules and provide intermediary products susceptible to further biodegradation and mineralization. Sequential abiotic-biotic transformation processes have been used for the destruction of recalcitrant xenobiotic compounds in industrial effluents and contaminated sites (Lee and Carberry, 1992), and a literature review of sequential treatment processes is available (Scott and Ollis, 1995). We are aware of no attempts to combine abiotic and biotic reactions to simultaneously degrade the more resistant compounds. Leung et al., (1992) showed that PCE was degraded via Fenton reactions to dichloroacetic acid (DCAA), a biodegradable product (Heinze and Rehm, 1998). We investigated simultaneous use of modified Fenton's reaction, an abiotic advanced oxidation process, along with biological degradation, using a hydrogen peroxide-resistant strain of *Xanthobacter flavus* FB71, to mineralize PCE.

Fenton reaction, a widely used abiotic transformation process (Watts and Dilly, 1996), is, formally, iron (II)-catalyzed decomposition of dilute hydrogen peroxide to yield hydroxyl radical, a strong, nonspecific oxidant. However, most environmental applications of Fenton chemistry involve modifications of the original process including the use of high concentrations of peroxide (low percent range), phosphate buffered media, chelated metal ion solutions, iron (III) or heterogeneous catalysts (Watts et al., 1997). Fenton reactions have been investigated for use in sequential, abiotic-biotic degradation processes (Carberry and Benzing, 1991).

It is likely that Fenton reactions occurred simultaneously with biological reactions during *in situ* bioremediation applications using hydrogen peroxide to increase dissolved oxygen concentrations. Brown and Norris (1994) noted that iron and copper salts present in the soil matrix catalyzed hydrogen peroxide decomposition. Additionally, the data provided by Tyre et al. (1991) suggested that naturally occurring iron oxyhydroxides served as an effective Fenton catalyst.

A limitation to combining abiotic and biotic systems was the potential toxicity of reactive oxygen species of the abiotic reactions, such as hydrogen peroxide and hydroxyl radicals, to the biotic system. We have previously shown that the toxic effects of reactive oxygen species can be reduced through pre-acclimation of the microorganisms to high concentrations of hydrogen peroxide, so that during the treatment a significant number of microorganisms survive for biodegradation (Büyüksönmez et al., 1998).

Objective. The objective of this study was to investigate simultaneously combined Fenton reactions and microbiological treatment to promote mineralization of PCE.

MATERIALS AND METHODS

Chemicals. Compounds used in the experimentation were the following: perchloroethylene (PCE), dichloroacetic acid -sodium salt (DCAA), and D6-benzene (all from Aldrich Chemical Co., Milwaukee, WI); ^{14}C [1,2]-PCE (2.7 µCi/mmol) (Sigma Chemical Co., St. Louis, MO); hydrogen peroxide (30%) (Fisher Scientific, Fair Lawn, NJ); Ecolite scintillation cocktail (ICN Pharmaceuticals, Costa Mesa, CA), Carbo-Sorb (Packard Instruments, Meriden, CT); and Ready Organic (Beckman Instruments, Fullerton, CA).

Organism and culture conditions. A hydrogen peroxide-resistant strain of *Xanthobacter flavus* FB71 (Büyüksönmez et al., 1998) was used for biotransformation of DCAA, the major product of PCE degradation by modified Fenton reactions. Cells were grown in a continuous-flow fermentor (New-Brunswick Scientific BioFlo III) in 1.5 liter volume, at steady state, with 1.0 l/d growth medium at 30°C, stirred at 200 rpm and with 3.0 l/d sterile air flow. The medium, buffered with half-strength M9 solution, contained 300 mg/l DCAA, 200 mg/l H_2O_2, and a vitamin and mineral solutions (Büyüksönmez et al., 1998).

Analytical techniques. Analyses of PCE and DCAA were performed using a HP series II, 5890 gas chromatograph-HP 5989A mass spectrometer (GC-MS) equipped with a 30 m x 0.25mm WCOT fused silica column. PCE concentrations were determined using a HP 7695 purge and trap concentrator equipped with a tenax silica gel/charcoal column. Samples of 5 ml were purged for 10 minutes with nitrogen. The column was desorbed for 2 minutes at 150°C and baked for 4 minutes at 220°C between samples. Results of PCE analyses were normalized based on known concentrations of D6-benzene added as an internal standard. A micro liquid-liquid extraction, derivatization and gas chromatographic techniques were used for DCAA analyses as outlined in Standard Methods (Greenberg et al., 1992) but with the following modifications: 10 ml sample volume, 1.5 ml of concentrated H_2SO_4, 2 g. $CuSO_4 \cdot 7H_2O$ and 8 g. $NaSO_4$ were added prior to extraction. Extraction was achieved by shaking the samples with methyl *tert*-butyl ether on a rotary platform shaker at 200 rpm for 30 minutes.

Experimental procedure. All PCE degradation experiments, including radiotracer experiments, were conducted in 55 ml serum bottles containing half-strength M9 buffer solution. To minimize autooxidation, and loss of PCE by volatilization, the following experimental order was followed: (1) hydrogen peroxide was added, (2) microorganisms were added (if required), (3) bottles were crimped, (4) PCE was added, and (5) reactions were initiated by the addition of Fe(II)-NTA (nitrilotriacetic acid) prepared under anaerobic conditions. The bottles were then shaken on a rotary platform shaker at 200 rpm at ambient temperature. Identical amounts of hydrogen peroxide and Fe(II)-NTA were added every hour for the first 5 hours of the experiments with a syringe and needle through the bottle septum.

Radiotracer experiments. Cultures containing approximately 95,000 dpm of ^{14}C-PCE at a concentration of 5 mg/l were incubated at ambient temperature for 108 hours with shaking at 200 rpm. Samples were acidified with 1 ml of 1.5 N HCl and purged with nitrogen (60 ml/min) for 30 minutes. Volatile organic compounds were captured by three organic traps (in series) containing 15 ml of Ready Organic scintillation cocktail and analyzed directly. Carbon dioxide was captured by one 15-ml Carbo-Sorb trap following the organic traps and 1 ml of Carbo-Sorb was mixed with Ecolite scintillation cocktail prior to counting. For counting, 1 ml of nitrogen-purged media was mixed with 15 ml of Ecolite scintillation cocktail. All samples were counted for radioactivity using a Packard Tri-Carb 2100TR scintillation counter via standard ^{14}C protocol. Counts per minute were converted to disintegration per minute by using an efficiency plot for known ^{14}C quench standards and SIS (spectral index of the sample) numbers.

RESULTS AND DISCUSSION

Degradation of PCE by Fenton reactions and modified Fenton reactions. PCE degradation was investigated by Fenton reactions, at pH 3.5, and modified Fenton reactions, at pH 7. The results (Figure 1) showed that, at pH 3.5, in accordance with the literature (Leung et al., 1992), degradation was very rapid and within 3 hours PCE concentration decreased to less than 0.2 µg/l. In buffered media (pH 7), degradation was slower; however, more than 95% of the PCE was degraded within 5 hours. Additionally, DCAA was confirmed qualitatively as an abiotic intermediary product (data not shown).

Following the confirmation of PCE degradation and DCAA yield during buffered Fenton reactions, the reaction medium was amended with vitamins and minerals for use as a growth medium (Büyüksönmez et al., 1998). Hydrogen peroxide and ferrous-NTA solutions were added hourly for the first 5 hours of the experiments then the bottles were incubated up to 72 hours. The results shown in

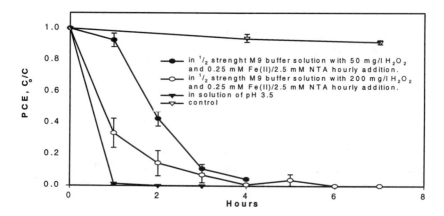

FIGURE 1. Degradation of PCE by modified Fenton reactions.

Figure 2, as compared to those in Figure 1, showed that addition of other media components lowered the PCE degradation by approximately 10%, perhaps due to mineral solution interference with hydrogen peroxide dissociation or vitamin-induced quenching of the reactive oxygen species (ROS). The addition of microorganisms (10^6 cells/ml) further decreased the PCE degradation by about another 10%, again explained by the quenching of ROS by microbial enzymes such as catalase. We also investigated microbial survival throughout the experiments (Figure 3). As expected, consecutive additions of hydrogen peroxide and ferrous solutions every hour resulted in rapid decline in cellular numbers. However, when reactant addition ceased, microorganisms numbers increased.

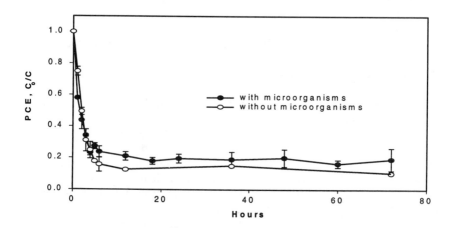

FIGURE 2. Degradation of PCE in simultaneous abiotic-biotic reactions.

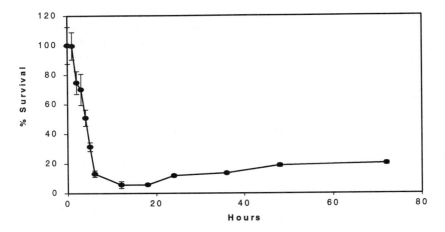

FIGURE 3. Microbial survival rates during simultaneous abiotic-biotic reactions.

Mineralization experiments. A set of experiments, similar to those above, was conducted with $^{14}C[1,2]$-PCE. The results, shown in Figure 4, revealed that the addition of microorganisms at a concentration of 10^6 cells/ml increased the overall extent of PCE mineralization about 15%, as compared to the sterile control. However, when cell concentration was increased to 10^7 cells/ml, PCE degradation was less than that seen with 10^6 cells/ml, yet mineralization extent equal to the control was obtained. This suggested that microorganisms were still involved in the mineralization process, as the extent of mineralization to degradation ratio was still higher than that of the control.

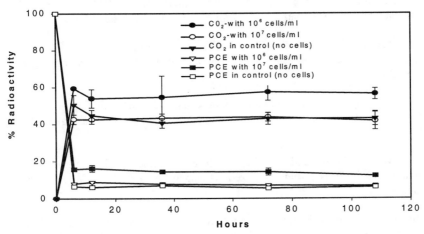

FIGURE 4. Radiolabeled CO2 and PCE during simultaneous abiotic-biotic reactions.

In this study, we showed the effects of simultaneous abiotic-biotic reactions on the mineralization of PCE. Certain factors, however, remain to be assessed prior to determining the overall efficacy of coexisting abiotic-biotic systems, including optimization of process variables and media formulations. Experimentation on these factors is currently in progress.

REFERENCES

Brown, R.A. and R.D. Norris. 1994. "The Evolution of a Technology: Hydrogen Peroxide in In-situ Bioremediation." In R.E. Hinchee, B.C. Alleman, R.E. Hoeppel, and R.N. Miller (Eds.), *Hydrocarbon Bioremediation*. Boca Raton, FL: Lewis Publishers, p.148.

Büyüksönmez, F., T.F. Hess, R.L. Crawford, and R.J. Watts. 1988."Toxicity of Modified Fenton's Reaction on *Xanthobacter flavus* FB71." *Appl. Environ. Microbiol.*, submitted

Carberry, J.B. and T.M. Benzing. 1991. "Peroxide Pre-Oxidation of Recalcitrant Toxic Waste to Enhance Biodegradation." *Water Sci. Technol.* 23:367-376.

Greenberg, A.E., L.S. Clesceri, and A.D. Eaton. 1992. *Standard Methods for the Examination of Water and Wastewater, 18th ed.* Washington, DC: American Public Health Association.

Heinze, U. and H.J. Rehm. 1998. "Biodegradation of Dichloroacetic Acid by Entrapped and Adsorptive Immobilized *Xanthobacter autotrophicus* GJ10." *Appl. Microbiol. Biotechnol.* 40:158-164.

Lee, S.H. and J.B. Carberry. 1992. "Biodegradation of PCP Enhanced by Chemical Oxidation Pretreatment." *Water Environment Res.* 64:682-690.

Leung, S.W., R.J. Watts, and G.C. Miller. 1992. "Degradation of Perchloro-ethylene by Fenton's Reagent: Speciation and Pathway." *J. Environ. Qual.* 21:377-381.

Scott, J.P. and D.F. Ollis. "Integration of Chemical and Biological Oxidation Processes for Water Treatment: Review and Recommendations." *Environ. Prog.* 14(1995):88-103.

Tyre, B.W., R.J. Watts, and G.C. Miller. 1991. "Treatment of Four Biorefractory Contaminants in Soils Using Catalyzed Hydrogen Peroxide." *J. Environ. Qual.* 20:832-838.

Watts, R.J., A.P. Jones, P.H. Chen, and A. Kenny. 1997. "Mineral-Catalyzed Fenton-Like Oxidation of Sorbed Chlorobenzenes." *Wat. Environ. Res.* 69:269-275.

Watts, R.J. and S.E. Dilly. 1996. "Evaluation of Iron Catalysts for the Fenton-Like Remediation of Diesel-Contaminated Soils." *J. Haz. Mat.* 51:209-224.

USING AN INTEGRATED APPROACH TO REMEDIATE A TRICHLOROETHYLENE-CONTAMINATED AQUIFER

Cliff Wright (Gannett Fleming, Inc., Madison, Wisconsin)
D. Robert Gan (Gannett Fleming, Inc., Princeton, New Jersey)

ABSTRACT: A field-scale project that has been in progress since September 1995 uses an integrated approach to remediate an aquifer contaminated with trichloroethylene (TCE) by applying institutional controls to eliminate the risks to human health, soil vapor extraction (SVE) to remove the source, and natural attenuation to remediate the groundwater. Residences in the area have been connected to municipal water. SVE exhaust gas data document that approximately 18 kg of TCE have been removed from the unsaturated soils in the source area, and soil gas monitoring data indicate that TCE concentrations have decreased over 99 percent from their initial concentrations within the radius of influence (ROI) of the remediation system. The removal of TCE from the unsaturated soils by SVE is beginning to improve groundwater quality through natural attenuation. Quarterly groundwater monitoring results indicate that TCE concentrations measured in the groundwater monitoring well nearest the source area, for example, have decreased 99 percent since the SVE system began operating.

INTRODUCTION

This paper presents a case study of remediation of TCE-contaminated soil and groundwater at an industrial site in the north-central U.S. When the contamination was first identified in 1993, the plume extended off site and affected several private water supply wells about 2,000 ft (600 m) downgradient. Based on the nature and extent of the contamination and the hydrogeology of the site, an integrated approach was developed: The risks to human health from TCE contamination were eliminated by connecting residences in the area to municipal water, while the source area soils and groundwater plume were cost-effectively remediated through SVE and natural attenuation, respectively.

The site is underlain by a permeable unit of well graded, fine-to-coarse sand, with some gravel, from the ground surface to about 95 ft (29.0 m) below ground surface (bgs). The sand is underlain by shale bedrock. Water levels in on-site monitoring wells have ranged from 10 to 20 ft (3.0 to 6.1 m) bgs. The groundwater discharges into a large river, which is about 3,000 ft (900 m) from the site, at an average linear flow velocity ranging between 40 and 330 ft/yr (12.2 and 101 m/yr).

SVE PILOT TEST RESULTS AND FULL-SCALE SYSTEM DESIGN

A pilot test vent well, VW-1, and five soil gas monitoring points, MP-1 through MP-5, were installed in August 1995, when the depth to water was approximately 16.5 ft (5.0 m) bgs. The vent well was constructed of 4-inch (10-cm) diameter PVC, screened from 5 to 15 ft (1.5 to 4.6 m) bgs. Figure 1 shows the

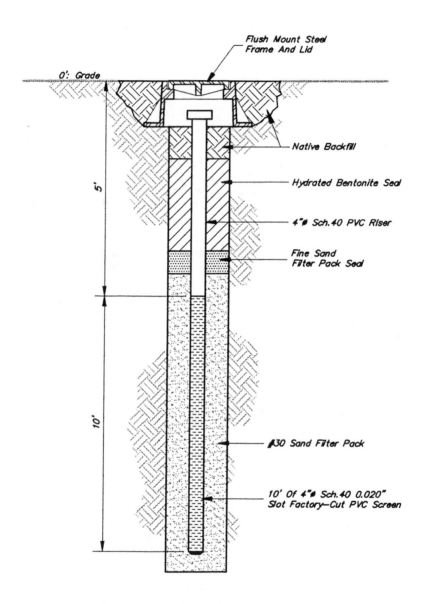

FIGURE 1. Pilot–Test Well Construction Detail.

construction details for the vent well. The monitoring points were constructed of 1-inch (2.5-cm) diameter PVC, screened from 9.5 to 10 ft (2.9 to 3.0 m) bgs.

The pilot tests were conducted during September and October 1995 using a portable SVE unit equipped with a vacuum blower to withdraw soil gas from the vent well while changes in the subsurface vacuum were measured at the five monitoring points. Exhaust gas samples were periodically collected and screened using a portable gas chromatograph (GC) in order to estimate total volatile organic compound (VOC) emissions. In addition, soil gas samples were collected from the monitoring points before, during, and after the tests and screened using the portable GC.

Results of the tests indicated that the ROI of VW-1 was 85 ft (25.9 m), which is comparable to ROIs reported by others at similar sites. The estimated permeability of the soil was 2.5×10^{-7} cm^2, which is within the expected range for sand. VOC concentrations in the exhaust and soil gas samples dropped off rapidly during the 23-day pilot test. For example, measured total VOC concentrations in the soil gas at the three monitoring points within the ROI of VW-1 decreased 97, 96, and 85 percent.

Based on the results of the pilot tests, two additional vent wells, VW-2 and VW-3, were installed in January 1996 to complete the full-scale SVE system. Figure 2 shows the SVE well layout and the estimated ROI of each vent well. Isoconcentration contours from an April 1993 VOC soil gas survey are also included, illustrating the extent of unsaturated soil contamination in the source area. Note that instead of installing a fixed blower in a building with underground piping to the vent wells, a portable regenerative SVE vacuum blower was delivered to the site by February 1996, reducing the design/construction time to a minimum and saving the client money.

RESULTS AND DISCUSSION

The full-scale SVE system was operated from February through August 1996 and January through December 1997. The portable blower was attached to each vent well in succession using a flexible hose. Vapors were extracted from each well for 14 to 28 days, after which the blower was turned off, disconnected from one well, moved and connected to the next well, and restarted.

SVE exhaust gas data document that approximately 36 kg of VOC contaminants, including about 18 kg of TCE, were removed from the unsaturated soils in the source area between September 1995 and December 1997. Soil gas monitoring data indicate that VOC contaminant concentrations decreased 74 to 90 percent from their initial concentrations, while TCE concentrations decreased over 99 percent from their initial concentrations within the ROI of the SVE system.

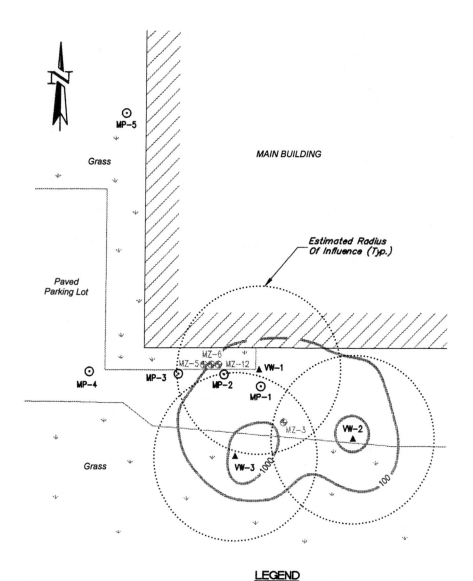

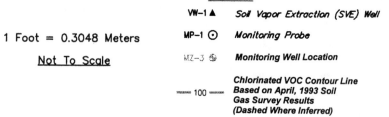

FIGURE 2. Site plan and SVE well layout

The removal of TCE from the unsaturated soils is beginning to improve groundwater quality through natural attenuation. Quarterly groundwater monitoring results indicated that TCE concentrations measured in the groundwater monitoring well nearest the source area, for example, have decreased 99 percent since the SVE system began operating. Figure 3 presents TCE isopleth maps of the plume in October 1993 and November 1997. A comparison of the two maps shows that TCE concentrations have decreased from 930 to 1.25 µg/L in MZ-3 and from 270 to 203 µg/L in MZ-8. These data illustrate that:

- The area with the highest concentrations of TCE, the centroid of the plume, has migrated downgradient of the source area near MZ-3 to the area between MZ-8 and PW-8.
- TCE concentrations in the centroid of the plume are decreasing due to natural attenuation.
- The TCE concentrations in the groundwater reaching the river will not affect surface water quality.

A long-term monitoring program has been implemented to document how the natural attenuation process is improving groundwater quality and to assess whether a significant amount of residual TCE remains in the capillary fringe where the SVE system cannot reach it.

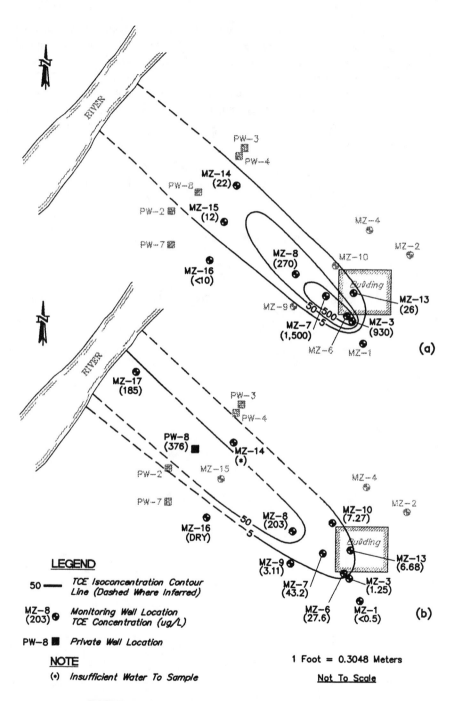

FIGURE 3. Groundwater monitoring TCE isopleth
maps. (a) 10/5/93 data. (b) 11/3/97 data.

IN SITU SEQUENTIAL TREATMENT OF A MIXED ORGANIC PLUME USING GRANULAR IRON, O₂ AND CO₂ SPARGING

Mary Morkin (University of Waterloo, Waterloo, Ontario, Canada)
Dr.J.Barker (University of Waterloo, Waterloo, Ontario, Canada)
Dr.R.Devlin (University of Waterloo, Waterloo, Ontario, Canada)
Michaye McMaster (Beak International Inc., Guelph, Ontario, Canada)

ABSTRACT: Groundwater plumes often contain a mix of contaminants that cannot easily be remediated in situ using a single technology. The purpose of this research is to evaluate an in situ treatment sequence for the control of a mixed organic plume (chlorinated ethenes and petroleum hydrocarbons) using two proven technologies within a Funnel-and-Gate. Contaminated groundwater is funneled into a gate, 3 m wide, 4.5 m long and 6 m deep where treatment occurs. The initial gate segment consists of granular iron, for the reductive dechlorination of the chlorinated ethenes. The second segment promotes aerobic biodegradation of petroleum hydrocarbons and any remaining lesser-chlorinated compounds, stimulated by oxygen (O_2) and carbon dioxide (CO_2) additions via an in situ sparge system (CO_2 is used to decrease high pH values produced from reactions in the iron wall). When average influent concentrations for cis 1,2 dichloroethene (*cis* 1,2 DCE) and vinyl chloride (VC) were 220 mg/L and 46 mg/L respectively, the granular iron wall removed >91% of the total mass. When the influent concentrations decreased to 26 mg/L and 19 mg/L for *cis* 1,2 DCE and VC respectively, >99% removal within the iron wall was attained. The biosparge zone supported aerobic biodegradation of VC and *cis* 1,2 DCE and volatilized these compounds to some extent. Based on data fitting with a mixed bioreactor model, mass removal of *cis* 1,2 DCE appeared to be predominantly by biodegradation, while the dominant removal process for VC was volatilization. Complete attenuation was not achieved, although concentration reductions were substantial. Laboratory microcosm results supported biodegradation of *cis* 1,2 DCE and VC but found that VC was readily biotransformed to levels below detection whereas *cis* 1,2 DCE was only partially biotransformed.

INTRODUCTION

It is not uncommon for groundwater plumes to contain a mix of contaminants such as chlorinated ethenes and petroleum hydrocarbons. Although these contaminants may be treatable with existing in situ technologies, they are not normally treated using the same technology. The objective of this research was to treat a mixed organic plume using two in situ treatment technologies in series within a Funnel-and-Gate (Starr et al., 1994). The contaminants of interest included trichloroethene (TCE), dichloroethene isomers (DCE) and vinyl chloride (VC), as well as benzene, toluene, ethylbenzene and the xylene isomers, (BTEX). A segment of the shallow, unconfined aquifer at Site 1, Alameda Point (formerly

Alameda Naval Air Station), Alameda, California was found to contain a mixed organic plume with the above mentioned contaminants and was selected as the site for the field demonstration. The in situ technologies installed within the Funnel-and-Gate included: 1) a permeable reactive barrier composed of granular iron for the abiotic reductive dechlorination of chlorinated ethenes (Gillham and O'Hannesin, 1994; Gillham, 1996) and 2) aerobic biodegradation of BTEX and any remaining 1 or 2 chloroaliphatic compounds, stimulated by O_2 and CO_2 additions via an in situ sparge system. O_2 was added to increase the dissolved oxygen levels to ~20 mg/L from <1 mg/L. CO_2 was added to reduce the high pH levels (8-11) produced from reactions in the iron wall to levels ranging from 7-8 (Figure 1).

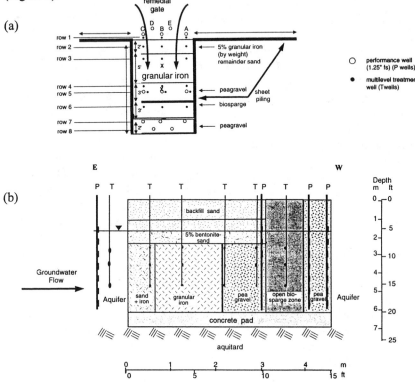

Figure 1. (a) Locations of the groundwater monitoring wells and (b) a cross section of the treatment gate

FIELD SITE DESCRIPTION

The field demonstration is located on the north western tip of Alameda Island, adjacent to San Francisco Bay. It is situated on sandy artificial fill on top of Holocene Bay mud, (estuarine deposits composed of silts and clays) which was the sea floor prior to the emplacement of the artificial fill. The Bay Mud unit is approximately 4.6 to 6 m thick in the western portion of the field site. The artificial fill unit is about 7 m thick and comprises silty sand to sand.

Beginning in the 1940's, cleansing solvents and waste petroleum hydrocarbons were disposed in unlined waste pits excavated in the artificial fill unit. The contaminant plume studied here is thought to have originated from these waste pits. To accurately position the Funnel-and-Gate to intercept the core of the plume, hydrogeologic investigations were completed to evaluate in three-dimensions the subsurface conditions as well as the contaminant distribution. Continuous soil cores were collected to evaluate lithologic variations (Einarson, 1995) and depth discrete water samples were obtained using the Waterloo Drive-Point Profiler to delineate contaminant distributions (Pitkin et al, 1995). It was found that the groundwater contained up to 218 mg/L of *cis* 1,2 DCE, 16 mg/L of VC, and <1 mg/L of the other chlorinated ethenes. Toluene concentrations as high as 4 mg/L were detected and concentrations of benzene, ethylbenzene and the xylenes were all <1 mg/L. The highest concentrations of the chlorinated ethenes and BTEX occurred together and were located within the upper portion of the artificial sand unit, 3.7 m below ground surface (bgs) (unconfined groundwater was located approximately 1.2 to 2.1 m bgs).

Treatment System. Contaminated groundwater was funneled into a gate 3 m wide, 4.5 m long and 6 m deep where treatment. The funnels were keyed 3 m into the aquitard and a 0.6 m cement floor was laid down to prevent aquifer and aquitard material from entering the gate and to prevent the introduction or loss of contaminants from the gate via vertical flow or diffusion. The first section in the gate was a 0.6 m permeable zone packed with a mixture of sand and granular iron, approximately 3-5% by weight of the iron. The next zone consisted of 1.5 m of 100% granular iron followed by 0.9 m of pea gravel (the pea gravel was used to separate the anaerobic from the aerobic treatment technology). Next was the 0.9 m biosparge zone partially packed with bacterial growth support material followed by a final pea gravel zone, 0.6 m thick. For purposes of experimentation, the Funnel-and-Gate was operated under controlled groundwater flow conditions. Two groundwater extraction wells were placed in the downgradient pea gravel section (Figure 1), and the downgradient sheet piles were left in place to seal the gate. Groundwater was then drawn through the gate at controlled rates by operating the extraction wells at 0.34 m^3/day. The pumps were turned on early February 1997 and the system continues to be being monitored.

RESULTS

Granular Iron Zone. Figure 2 illustrates the behaviour of the *cis* 1,2 DCE and VC throughout the remedial gate on June 16, July 31, September 30 1997 and January 12 1998. These data represent concentrations in fully screened wells from the south side of the gate (believed to contain a highly concentrated "core" of the plume). Profiles for TCE, 1,1 DCE and *trans* 1,2 DCE are not shown but in general the influent concentrations were low (0.01-0.60 mg/L but from April -

June 1997, TCE concentrations ranged from 26-74 mg/L) and the compounds were treated to <0.01 mg/L within the iron wall. A decrease in the influent concentrations of *cis* 1,2 DCE and VC through reductive dechlorination is illustrated from these two figures.

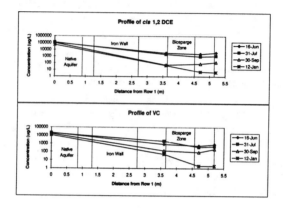

Figure 2. Profiles for *cis* 1,2 DCE and VC through the treatment gate

Table 1 summarizes the percent mass removal based on the total solute masses passing Row 1 (influent masses) and Row 4 (downgradient of the iron wall) over a six month period from April to September 1997.

TABLE 1: Percent mass removal from remediation within the iron wall

Contaminant	Row 1 (mg)	Row 4 (mg)	% Mass Removal
TCE	107,500	278	99.7
1,1 DCE	3595	137	96.1
cis 1,2 DCE	1,347,000	116,500	91.3
trans 1,2 DCE	2997	66	97.8
VC	571,900	37,180	93.5

Note: Data collected from multilevel piezometers
         January 1998 data is not included

Although the percent removal due to reductive dechlorination in the granular iron wall was high, 100% attenuation was not observed especially when the influent concentrations for *cis* 1,2 DCE and VC increased to ~220 mg/L and 64 mg/L respectively. Two possible explanations were considered.

1) The flow-through thickness of the iron wall was too short: With a groundwater flow velocity of 6 cm/day, the residence time in the iron wall is approximately 25 days. Assuming the packing of the iron in the column was representative of that in the wall, and using the half life ($t_{1/2}$) calculated from the bench scale treatability study ($t_{1/2}$ for *cis* 1,2 DCE = 12.1 hours for an influent concentration of 32 mg/L and 11.2 hours for a VC concentration of 26 mg/L), a residence time in the iron wall of between 7-8 days should have been sufficient to

treat 220 mg/L of *cis* 1,2 DCE and 46 mg/L of VC. Thus the iron wall thickness does not appear to be the limiting factor. However, apparent half lives calculated from field data (multilevel monitoring wells) were 70 hours and 26 hours for *cis* 1,2 DCE and VC respectively, which are substantially larger than those determined in the treatability study. Normalized half lives (normalized to 1 m^2 of iron surface per mL of solution) for *cis* 1,2 DCE calculated from lab and field data were 46.46 hours and 376.32 hours respectively. As these normalized half lives based on lab and field data are still significantly different, it may be that the available surface area of the iron affects the half lives.

2) Surface saturation of the iron: The dechlorination reaction is considered to be surface related, hence as the solute concentrations increase, the possibility of surface saturation becomes a concern. As the surface becomes saturated, the pseudo-first order rate constants would be expected to decrease, thereby increasing the chances of breakthrough. Column experiments are currently being conducted to examine the passivation of the iron's reactive sites as the concentration of *cis* 1,2 DCE is increased up to ~220 mg/L (preliminary results are pending).

During the early part of the field demonstration (February - April 1997), none of the BTEX compounds were detected in the groundwater downgradient of the iron wall. However, since these compounds are not reactive on granular iron and appreciable concentrations (4-5 mg/L Toluene, <1 mg/l of BEX) were detected in Row 1, BTEX compounds were probably retarded in the iron wall due to sorption. In late April 1997, benzene and toluene began to break through in the concentration range 0.02-0.10 mg/L for benzene and 0.01-0.06 µg/L for toluene. However, benzene and toluene were rarely detected as far as Row 6, in the biosparge zone. The attenuation of these compounds is due to some combination of aerobic biodegradation and volatilization. Insufficient data were available to resolve the relative contributions of these two mechanisms to the overall rate of mass removal.

Biosparge Zone. Because none of the BTEX compounds were detected in the biosparge zone above their LOQ (B=0.025 mg/L, T=0.016 mg/L, E=0.031 mg/L, X=0.021 mg/L), aerobic degradation was applied to *cis* 1,2 DCE and VC, the only organic contaminants to reach the biosparge zone. Laboratory microcosms, using site groundwater and sediment, supported biodegradation of *cis* 1,2 DCE and VC finding that VC was readily biotransformed to levels below detection whereas *cis* 1,2 DCE was only partially biotransformed. Modeling mass removal at the field site was undertaken by representing the biosparge zone as a large, completely mixed bioreactor, and accounting for instantaneous partitioning of the organic contaminants into the sparge bubbles, as well as storage in the headspace. Mass removal via biodegradation was described by first order kinetics. The modeling suggested that of the mass removed in the biosparge zone (66% for both *cis* 1,2 DCE and VC), *cis* 1,2 DCE was approximately 65% biodegraded and 35% was lost due to volatilization. For VC, 70% of the mass was removed via volatilization and only 30% was due to biodegradation.

CONCLUSIONS

Based on the high percent mass removals (Table 1), the performance of the treatment gate was judged excellent. Remediation of *cis* 1,2 DCE and VC within the granular iron was at times incomplete, which was probably related to the temporally variable and localized high influent concentrations. However, when the influent concentrations were reduced, >99% removal was observed suggesting that the iron wall at Alameda may have an upper concentration limit that it can successfully treat. Further research into whether the breakthrough was caused by an under designed wall (i.e. lab t½ were too short) or if it relates to surface saturation phenomenon is underway.

With respect to BTEX, an accurate assessment of its aerobic biodegradation could not be inferred because compounds detected in the biosparge zone were at or below their LOQ. The field demonstration was not of sufficient duration to allow the BTEX compounds to break through the iron wall. However, of the total mass removed in the biosparge zone (66%), aerobic biodegradation was responsible for 65% mass removal for *cis* 1,2 DCE but only 30% for VC. Further monitoring may find that mass removal could be improved once a larger active biomass is established in the biosparge zone.

In general, this field demonstration has shown that treatment of mixed contaminant plumes by sequential use of granular iron with a biosparge zone is a viable option for controlling groundwater contamination.

ACKNOWLEDGMENT

Funding for this work was provided wholly or in part by the United States Department of Defense under Grant No. DACA 39-93-1-001, to Rice University for the Advanced Applied Technology Demonstration Facility for Environmental Technology Program (AATDF). This work does not necessarily reflect the position of these organizations and no official endorsement should be inferred.

REFERENCES

Einarson, M.D. 1995. "Enviro-Core®-A New Direct-Push Technology for Collecting Continuous Soil Cores." *9th National Outdoor Action Conference, Las Vegas, Nevada,* NGWA.

Gillham, R.W. 1996. "In Situ Treatment of Groundwater: Metal-Enhanced Degradation of Chlorinated Organic Contaminants." *Advances in Groundwater Pollution Control and Remediation,* M.M. Aral (ed), Kluwer Academic Publishers, Netherlands, pp 249-274.

Gillham, R.W., and S.F. O'Hannesin. 1994. "Enhanced Degradation of Halogenated Aliphatics by Zero-Valent Iron." *Ground Water.* 32(6): 958-967.

Starr, R.C., and J.A. Cherry. 1994. "In Situ Remediation of Contaminated Groundwater: The Funnel-and-Gate System." *Ground Water.* 32(3): 465-476.

Pitkin, S., R.A. Ingleton, and J.A. Cherry. 1995. "Use of a Drive Point Sampling Device for Detailed Characterization of a PCE Plume in a Sand Aquifer at a Dry Cleaning Facility." in Proceedings of the *8th National Outdoor Action Conference and Exposition on Aquifer Remediation, Ground Water Monitoring, and Geophysical Methods.* NGWA. 395-412.

EVALUATING THE EFFECTIVENESS OF OZONATION AND
COMBINED OZONATION/BIOREMEDIATION TECHNOLOGIES

James R. Lute (Fluor Daniel GTI, Trenton, New Jersey, USA)
George J. Skladany (Fluor Daniel GTI, Trenton, New Jersey, USA)
Christopher H. Nelson (Fluor Daniel GTI, Golden, Colorado, USA)

ABSTRACT: Bench-scale experiments were performed to evaluate the effectiveness of treating PAH- and PCP-contaminated soils with ozone or combined ozonation/biological treatment processes. In less than 24 hours of continuous exposure to a 5-6% ozone stream, total and carcinogenic PAH levels decreased by 98% and 97% under the soil slurry conditions employed. With soil columns, total and carcinogenic PAH levels were reduced by 50% and 50% in 40 hours under similar treatment conditions. Microbial viability was essentially eliminated using continuous ozonation. By limiting ozonation to discrete daily periods, conditions were established which minimized the destructive effects of ozone on the resident microbial population. Such conditions were found to be site specific, and the contribution of chemical oxidation and biological destruction under combined ozone/biological treatment conditions are under further investigation.

INTRODUCTION

Polynuclear aromatic hydrocarbons (PAHs) and pentachlorophenol (PCP) can be chemically or biologically treated *in situ* to various degrees. Ozonation alone is effective at destroying these compounds, although concern exists over its cost and potential adverse impact on subsurface microbial viability. Bioremediation alone is relatively inexpensive to implement, but may be ineffective or require impractical periods of time in order to effectively treat high concentrations of PAHs and/or PCP. Successful simultaneous exploitation of the strengths of ozonation and biological treatment technologies may provide the most efficient and cost-effective *in situ* remediation process.

CONTINUOUS OZONATION SLURRY EXPERIMENTS

Each slurry system consists of a five liter round-bottom glass flask containing four liters of a 20% soil:water mixture. Slurries were mechanically mixed. Gas flow (5-6% ozone in the reactive flask and nitrogen or oxygen in control flasks) to each reactor was 300 mL/min. At each sampling time, slurry samples were taken by aspiration from the reactor flasks. The samples were centrifuged, and the resulting soil and water fractions were analyzed separately. Figure 1 shows typical reactive flask influent and effluent ozone concentration trends as a function of time. Initial effluent ozone levels are low as the ozone reacts simultaneously with any reactive materials present (measured as Total Organic Carbon) as well as the specific contaminants of concern (Figures 2, 3 and 4). As materials are oxidized over time, the effluent gas ozone concentration

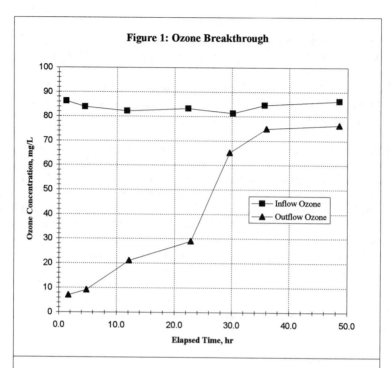

Figure 1: Ozone Breakthrough

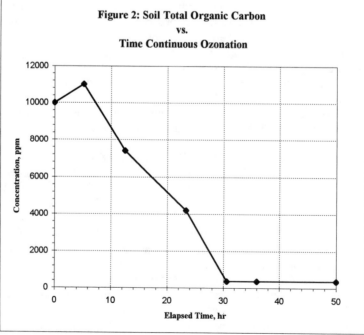

Figure 2: Soil Total Organic Carbon
vs.
Time Continuous Ozonation

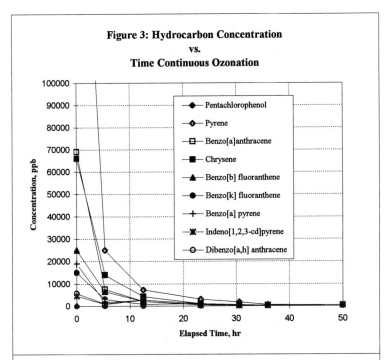

Figure 3: Hydrocarbon Concentration
vs.
Time Continuous Ozonation

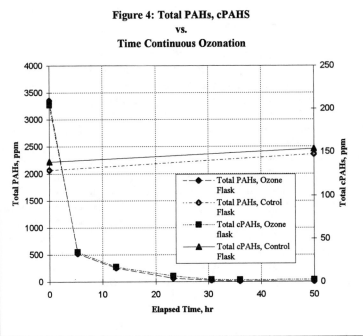

Figure 4: Total PAHs, cPAHS
vs.
Time Continuous Ozonation

increases to a maximum level. In this study, total soil PAH levels were reduced by 99%. While more recalcitrant PAHs such as dibenzofuran and fluoranthene took longer to degrade, their concentrations were also reduced to below detection limits (300 ug/L). With the corresponding nitrogen control flask, total soil PAH concentrations were essentially unchanged. PCP levels were reduced 99% in the ozonated flask and 50% in the control flask. These results are typical of several independent studies conducted with PAH and/or PCP contaminated soils.

MICROBIAL DIE-OFF STUDIES

The sterilizing effect of continuous ozonation on the indigenous soil microorganisms was monitored using standard spread plate and modified Most Probable Number (MPN) tube methods. The quick and deleterious effect of ozonation was most evident in the soil slurry systems described above. Figure 5 shows a significant three order of magnitude reduction (>99%) in population density within the first hour of exposure, although a significant viable microbial pop-ulation still is detectable. Four hours of continuous ozonation reduced the indi-genous population numbers to below detection limits (<10 Colony Forming Units per milliliter [CFU/mL]).

The sterilizing effect of ozone was not quite as rapid or complete when soil columns (2" diameter. by 6" in height; approximately 450 gm) were treated with ozone flowing from the bottom to the top (Figure 6). A column using soils from one PAH site showed three orders of magnitude reduction in population density within four hours of exposure, and remained at this level even after five additional hours of exposure. A column using soil from a second PAH site showed a reduction to undetectable levels (eight orders of magnitude reduction) after only four hours of exposure.

COMBINED OZONATION/BIOREMEDIATION TREATMENT

A conservative ozonation period and frequency of 15 minutes every other weekday was chosen to treat a column packed with PAH-contaminated soil. Air was passed continuously through the column between ozonation periods. Micro-bial concentrations were monitored using plate count methods. Gross community structure was monitored using phospholipid fatty acid analysis (PLFA). A robust microbial population remained viable throughout the ten week period, and showed a one to two orders of magnitude increase in density over time (Figure 6). The PLFA analyses (Figure 7) showed that a relatively diverse microbial community was sustained throughout the elapsed time of the study.

Figure 8 shows that even with a significantly reduced total quantity of ozone applied to the system, total and carcinogenic PAH levels were reduced by more than 70%. The PAH reduction due to ozone vs. biological treatment could not be determined from the experimental design employed. Additional experi-ments to differentiate the contribution made by chemical oxidation from that of biodegradation, as well as process optimization, are being conducted.

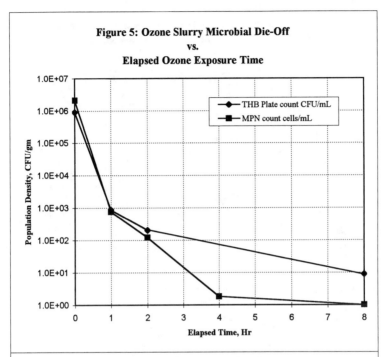

Figure 5: Ozone Slurry Microbial Die-Off
vs.
Elapsed Ozone Exposure Time

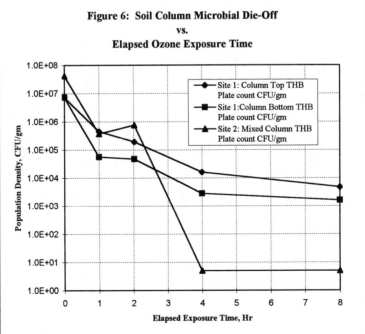

Figure 6: Soil Column Microbial Die-Off
vs.
Elapsed Ozone Exposure Time

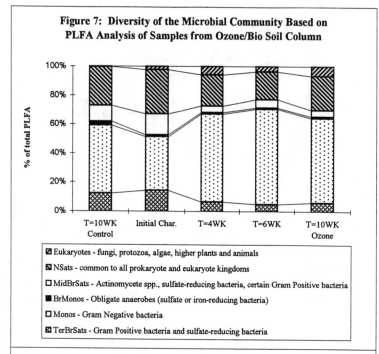

Figure 7: Diversity of the Microbial Community Based on PLFA Analysis of Samples from Ozone/Bio Soil Column

☑ Eukaryotes - fungi, protozoa, algae, higher plants and animals

☒ NSats - common to all prokaryote and eukaryote kingdoms

☐ MidBrSats - Actinomycete spp., sulfate-reducing bacteria, certain Gram Positive bacteria

■ BrMonos - Obligate anaerobes (sulfate or iron-reducing bacteria)

☐ Monos - Gram Negative bacteria

☒ TerBrSats - Gram Positive bacteria and sulfate-reducing bacteria

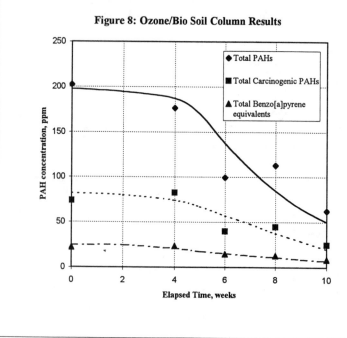

Figure 8: Ozone/Bio Soil Column Results

COMBINING AND SEQUENCING TECHNOLOGIES TO TREAT CHLORINATED VOCS

Richard A. Brown, Fluor Daniel GTI, Trenton, NJ, USA
Ray Lees, Fluor Daniel GTI, Marlton, NJ, USA

INTRODUCTION

There has been a continuing development and improvement of technologies used to treat chlorinated solvents. Over the last few years air sparging, dual phase extraction, and chemical oxidation have been developed as treatment technologies for dealing with chlorinated solvent contamination. While these technologies have been demonstrated as treating chlorinated solvent contamination, all have limitations with respect to cost and performance, as is typical of any technology.

It is generally held that all technologies eventually become transport limited. This occurs when either access to or removal of the contaminant is no longer based on advective flow but is diffusion controlled. This limits the cost effectiveness of the technology. Typically the limitations of cost and performance are magnified when technologies are used solely.

A novel approach to improving performance is to combine technologies that are based on different transport mechanisms to minimize overall performance limitations. For example air sparging can be followed by oxidation, or dual phase extraction can be cycled with air sparging. Experience has shown that when using combinations of technologies, "the whole is often greater than the sum of the parts." One of the reasons that combinations of technologies may work better than the single technologies is that each of the technologies employ a different transport mechanism, and thus reach a different limitation, which can then "cancel each other out."

There are several approaches to combining technologies. Technologies can be used sequentially, varied aerially, used for different aspects of a site, or used cyclically. Choosing the right combination of technologies is a function of a number of factors: the remedial goals that need to be achieved, the constraints of cost and time, and engineering compatibility.

REMEDIATION OF CHLORINATED SOLVENT CONTAMINATION

There is a high degree of variability in chlorinated solvent sites. This variability is a function of how the solvent was lost, what it was, and, importantly, the geological context into which it was lost. There are two basic means by which chlorinated solvents are lost - as neat liquids or as solutions. Loss of the neat solvent occurs due to the storage, use or transport of the solvent itself. Loss of solutions occurs when parts that are degreased with the solvent are subsequently washed to remove the residual solvent. When lost as neat liquids, chlorinated solvents penetrate the soil, become entrained in soils and, if of sufficient quantity can accumulate into DNAPL layers. The soil contamination and DNAPLs can impact groundwater.

When present, DNAPLs are extremely difficult to treat. When lost as a solution (wash waters) the contamination is primarily a groundwater issue.

Chlorinated solvents are typically, volatile, have moderate to low solubility, and have limited reactivity both biologically and chemically. These properties help determine the appropriate technologies. These general properties, however vary with specific compounds. The generalization can, however be modified by classes of compounds. As shown in Table 1, there are three classes of common chlorinated compounds. Table 1 lists the common solvents and characterizes them with respect to volatility, solubility and reactivity. Table 2 lists the appropriate technologies for the individual chlorinated solvents, classified as extraction (based on solubility), volatilization (based on volatility), and reactivity (chemical and biological).

The geological context (permeability and saturation) is an important factor in the selection of technologies. There is a continuum of treatability for the different geological contexts. As shown in Figure 1 the most treatable context is high permeability and low saturation; the least, saturated clays.

In a generalized view, chlorinated contamination has several aspects. As pictured in Figure 2 these include the source area where most of the mass is present as soil contamination or as NAPLs, the groundwater plume where active processes attenuate the dissolved contamination, and the compliance point which defines/limits the impact. Treatment involves a number of these aspects. Generally the goal of remediation is to attain an endpoint defined at a compliance point. This may be achieved creating a barrier, by increasing the process of removal of contaminants from the groundwater plume, or by reducing the source area. To accomplish these different aspects effectively may necessitate the use of combinations of technologies.

The technologies appropriate for the different aspects do vary. The following organizes the technologies listed in Table 2 by source, plume and barrier technologies.

Source technologies: SVE, Air Sparging, NAPL recovery, oxidation
Plume technologies: Air Sparging, Oxidation, Pump and Treat, Intrinsic bioremediation, Bioremediation
Barrier technologies: Air Sparging, Iron Walls, Oxidation, Bio-barriers

COMBINING TECHNOLOGIES

There are several approaches that can be used for combining technologies. Technologies can be used sequentially, varied aerially, used for different aspects of contamination, or used cyclically. Sequential applications involve a life cycle approach. Processes are phased in and out as aspects of the site are remediated. One common sequence is to do NAPL recovery before SVE. A second type of sequence is to switch from active systems to passive or intrinsic systems. A third would be to switch the technology to address residual problems, such as ozonating after air sparging. Areal variability involves application of different systems for different areas of the plume. For example a barrier wall can be placed on the downgradient property boundary to prevent off site migration while a treatment system is operated upgradient. Designing different processes for different aspects of the site contamination involves differentiating the contamination as a function of mass and/or

TABLE 1. Classes of chlorinated solvents.

Chlorinated Solvent	Solubility, mg/L	Volatility, mm Hg	Reactivity - C(chemical), B(biological), L(low)
CHLOROMETHANES			
Carbon Tetrachloride CCl_4	1160	56	L
Chloroform $CHCl_3$	9300	160	L
Methylene Chloride (dichloromethane /DCM) CH_2Cl_2	20000	500	B
Chloromethane CH_3Cl	40000	3800	B
CHLOROETHANES			
1,1,1Trichloroethane CCl_3-CH_3	4400	100	L
1,1,2,2 Tetrachloroethane $CHCl_2$-$CHCl_2$	2900	5	L
1,1,2 Trichloroethane CH_2Cl- CH_2Cl	4500	19	L
1,1 Dichloroethane $CHCl_2$-CH_3	5500	180	L
1,2 Dichloroethane CH_2Cl-CH_2Cl	9200	40	L
CHLOROETHENES			
Tetrachloroethene (PCE) $CCl_2=CCl_2$	150	14	C
Trichloroethene $CCl_2=CHCl$	1100	60	C
cis, trans 1,2 Dichloroethene $CHCl=CHCl$	800	200	C, B
Vinyl Chloride $CHCl=CH_2$	1100	2500	C, B

TABLE 2. Remedial options for chlorinated solvents.

CHLORINATED SOLVENT	Extraction	Volatilitization	Reaction
CHLOROMETHANES			
Carbon Tetrachloride CCl_4	Dual Phase	SVE Air Sparging	Intrinsic Bio
Chloroform $CHCl_3$	Pump & Treat Dual Phase	SVE Air Sparging	Intrinsic Bio
Methylene Chloride (dichloromethane - DCM) CH_2Cl_2	Pump & Treat Dual Phase	SVE Air Sparging	Anaerobic, Cometabolism Intrinsic
Chloromethane CH_3Cl	Pump & Treat Dual Phase	SVE Air Sparging	Anaerobic, Cometabolism Intrinsic
CHLOROETHANES			
1,1,1Trichloroethane CCl_3-CH_3	Pump & Treat Dual Phase	SVE Air Sparging	Intrinsic Bio
1,1,2,2 Tetrachloroethane $CHCl_2$-$CHCl_2$	Pump & Treat Dual Phase	SVE Air Sparging	Intrinsic Bio
1,1,2 Trichloroethane CH_2Cl-CH_2Cl	Pump & Treat Dual Phase	SVE Air Sparging	Intrinsic Bio
1,1 Dichloroethane $CHCl_2$-CH_3	Pump & Treat Dual Phase	SVE Air Sparging	Anaerobic, Cometabolism Intrinsic
1,2 Dichloroethane CH_2Cl- CH_2Cl	Pump & Treat Dual Phase	SVE Air Sparging	Aerobic/Anaerobic, Cometabolism Intrinsic
CHLOROETHENES			
Tetrachloroethene (PCE) CCl_2=CCl_2	Dual Phase	SVE Air Sparging	Oxidation Intrinsic bio
Trichloroethene CCl_2=$CHCl$	Dual Phase	SVE Air Sparging	Oxidation Intrinsic bio
cis, trans 1,2 Dichloroethene $CHCl$=$CHCl$	Dual Phase	SVE Air Sparging	Oxidation Intrinsic bio Aeroc/Anaerobic, Cometabolism
Vinyl Chloride $CHCl$=CH_2	Dual Phase	SVE Air Sparging	Oxidation Intrinsic bio Aerobic/Anaerobic, Cometabolism

Figure 1: Treatability of Chlorinated Solvents as a Function of Geological Context

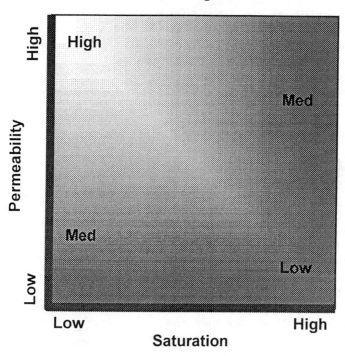

FIGURE 2. Aspects of remediation.

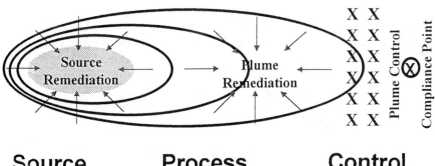

impact. There are three general aspects to be dealt with - the source area, the on-site plume and the offsite plume. Finally cycling technology can be beneficial. Cycling sparging and dual phase extraction can maximize removal. Cycling anaerobic and aerobic conditions can degrade recalcitrant organics.

DESIGNING FOR TECHNOLOGY COMBINATIONS

There are many applicable technologies for treating sites contaminated with chlorinated volatile organic compounds (CVOCs). The effectiveness of these technologies however, is dependent on the regulatory framework, site conditions, contaminants and ultimately how the technology is applied to achieve closure. The remedial design must be both flexible, to allow quick modifications during the project life-cycle, and easily buildable. Remediation technologies can work together to complement their effectiveness in mitigating the risk at the site. Unfortunately, little thought is typically given toward designing flexibility into the remedial action at the site.

The major steps in successful application of multiple technologies are (1) generating an accurate design basis for site conditions, (2) understanding the principals of the technologies in question, and (3) cost-effective application of the technology(s) to reach closure.

An accurate design basis is the first step in a flexible design. The design basis consists of defining the influent constituent loading, effluent cleanup goals and permit requirements (regulatory constraints). It is important as well as to understand the fate and transport issues, including mass balance distribution and the potential receptors existing at the site.

The second stage is to choose the right technology(s) for the site. It is important to understand the general principals of the applicable technologies. Is it a mobility based (extraction) or reactivity based system? What are its limitations? Will it work well with other technologies?

The third stage is proper application. After selection of the method(s) of treatment, the treatment system must be defined. These are the aspects of the project that, when given careful consideration, will minimize the cost, add to the attractiveness of the treatment system, fit the operational needs of the process and plant personnel, and offer flexibility for incorporating design modifications for future needs. The treatment system design helps convert the functional needs of the project from concept into reality.

Engineering the remedial action is an important step prior to implementation. Engineering drawings will allow the proposed remedial solution to be communicated to the regulators and public community. Any remedial system can be broken-up into three parts:

- Delivery/withdrawal components
- Transmission line components
- Treatment plant layout

The delivery/withdrawal component of the remedial action consists of the pumps, blowers, and well (recovery/ injection) network. The transmission line conveys the recovered vapor and/or liquids from the well network back to the treatment plant.

Designing the optimum plant layout for current and future needs requires considerable effort, particularly for the medium and larger sized remedial systems. The sizing and arranging of facilities, evaluation of alternatives, and planning for future expansion and upgrading require thoughtful consideration. Although process units, because of space and cost considerations, are generally arranged first, the design engineer should be aware that the layout must integrate the functions of all components.

The most important control documents at the beginning stages of the design are the Process Flow Diagram (PFD) and Piping and Instrumentation Diagram (P&ID). The PFD provides a process-flow sequence for the unit operations involved in the remedial system. This drawing provides for a mass-balance around the treatment system and allows for emission concentrations to be accurately quantified. It is important to accurately represent the mass balance evolution at the site. For example, a PFD should be constructed for the minimum, average, and maximum influent VOC concentrations over the project life span. This will allow the designer to forecast unit operation treatment performance for different operational phases. In addition, predetermined set-points on when to switch out technologies or modify current technology can be planned.

The P&ID would be generated to correspond with the phased-PFDs. The P&ID, which details the operational sequence and relationship between the various unit operations is the key drawing allowing the designer to convey important operational settings. Some key engineering principals that would apply to the P&ID are as follows:

- Pump/Blower Flexibility: Using variable speed drive motors will allow a high degree of fluid-flow range versus the standard fix-drive motors.

- Compatibility: The transmission lines need to be compatible for the various operating pressure and vacuum regimes. For example, an air sparge deliver line operation at 25 psig may be later converted to recover soil vapors and/or groundwater with applied vacuums up to 27 inches of mercury.

- Utilities: For medium to larger systems, a 3-phase electrical service will allow more flexibility in the design of the motors within the treatment system. However, a step-down transformer will be necessary to single-phase motors and low-voltage equipment.

CONCLUSION

There are a number of technologies that are applicable to chlorinated solvents. None of these are equally efficient in dealing with all aspects of site contamination. Combining technologies can lead to greater remedial efficiency. This is because all technologies eventually become diffusion limited.   Combining technologies can overcome these limitations. In combination, technologies can be used sequentially, varied aerially, used for different aspects of contamination, or used cyclically.

Combining technologies requires a more rigorous engineering than is common for single systems. The remedial design must be both flexible, to allow quick modifications during the project life-cycle, and easily buildable. When carefully designed, combined technologies will minimize cost, add to the effectiveness of the treatment system, fit the operational needs of the process and plant personnel, and offer flexibility for future needs.

PHYSICAL REMEDIATION OF A TRICHLOROETHYLENE CONTAMINATED INDUSTRIAL SITE

Paolo Muzzin (Ambiente, Italy)
Stefano Tenenti (Ambiente, Italy)
Maurizio Buzzelli (Ambiente, Italy)

ABSTRACT: A full scale remediation project was carried out in a synthetic fabric factory in Italy. Two different technologies for the unsaturated and the saturated zone, contaminated by trichloroethylene (TCE), were applied; both technologies were integrated with the TCE recovery plant of the factory. A Soil Vapor Extraction (SVE) system was used to extract up to 800 Nm^3/h of soil vapors; three hydraulic barriers were used to pump up to 50 m^3/h of water and send it to an air stripping plant. The vapor streams from the SVE system and the air-stripping off-gases were sent to the TCE recovery system. The recovered TCE was sent back in the production cycle. All the system was supervised by the control room of the factory.

INTRODUCTION

During the past 20 years many diffused TCE spills occurred in a synthetic fiber factory in Italy and TCE contaminated both saturated and unsaturated zones; three main contaminated areas were identified: below the aboveground storage tanks, below the production plants and below the TCE recovery section.

In the unsaturated zone the contamination plume was about 50 m wide, 200 m long and 10 m deep, in the saturated zone the plume was about 500 m wide and at least 25 m deep. TCE concentrations were as follows: up to 9000 µg/L in the unsaturated zone, close to the storage tanks; up to 205 mg/L in the saturated zone.In a spring situated outside the facility boundaries and at the base of the above mentioned hill, the TCE concentration values ranged between 4 and 2700 µg/L, very often exceeding limits set by the Italian Law.

A soil vapor extraction (SVE) technology was chosen to remediate the unsaturated zone whilst a Pump and Treat (P&T) system was considered as safety action for the saturated zone.The remediation plants were designed and completed so as to be integrated in the existing factory facilities and production processes.

PRELIMINARY ASSUMPTIONS

Site description. The industrial area is located on alluvial deposits (from the Quaternary period) close to a river approximately 100 m downgradient of the site.

The aquifer is unconfined and consists of fine to medium-grained sand and gravel, with discontinuous beds of silty sand.

Water table elevation ranged from 10 to 12 m bgs; transmissivity is 9.9 x 10^{-4} and hydraulic gradient is 0.025.

Project assumptions and constraints. The main assumptions and constraints were:

•TCE release was interrupted

•contamination extension, both in the saturated and unsaturated zone, was the one detected by previous Client's consultant

•availability of a concrete tank (free volume of 600 m^3) in a favourable position;

•probable availability of an activated carbon TCE contaminated air treatment plant with a remaining treatment capacity of 5,000 m^3/h.

•geological and hydrogeological data were given by Client and needed to be verified by further investigation

•Client's request to realize four water extraction wells in the TCE storage tanks area

•Client's need to increase the water supply

Remediation goals. The remediation goals, in terms of concentration limits to be achieved, were:

•50 mg/kg for the unsaturated zone (Regione Toscana Regulation)

•30μg/L (MCL) for the saturated zone (DPR 236/88-Drinking Water Standard)

•1 mg/L for effluent discharges into surface waters (L.319/76)

DESCRIPTION OF THE REMEDIATION SYSTEMS

SVE system. One of the most important parameters considered in the the Soil Vapor Extraction system dimensioning was soil permeability which, determined on the basis of a field pilot test, allowed the calculation of the ROI (approx.15 m).

The average depth of the unsaturated zone was about 11 m and the plume extended for at least 10,000 m^2 under the aboveground storage tanks and under the TCE recovery section. The data gathered led to the placement and the installation of 24 vapor phase extraction wells, in order to treat all the contaminated soil volume. The wells were 11m deep and screened for 10.5 m.

The SVE wells were then connected together and to a vacuum pump station whose additional purpose was to send the extracted vapor to the existing central gas treatment system (activated carbon) of the factory.

The vapor extraction rate was about 800 m^3/h.

The area on which the SVE system insisted was covered with HDPE sheets in order to limit the TCE exhalations in the working areas, and to enhance the ROI and the performance of the extraction wells.

Hydraulic barriers and air stripping plant. The Pump & Treat system was constituted by three hydraulic barriers of four wells each and by an air stripping system for the treatment of the contaminated groundwater; the vapor stream from this plant was sent to the activated carbon treatment system, already present in the factory.

The dimensioning of the hydraulic barriers has been carried out on the basis of the data gathered during field tests (particularly, two pumping tests near two high concentration areas), of data supplied by the Client, of calculations executed with traditional methods and of simulations carried out with a computer model.

Two of the three barriers were placed along the factory boundaries, in order to create two capture zones of approximately 150 m each along the river; the third barrier, placed near the storage tanks, had also the purpose to protect the groundwater in case of new spills in the most hazardous area of the factory (presence of TCE storage tanks, pipelines, TCE loading/unloading station, etc.).

However, the hydraulic barriers along the river were overdimensioned and deeper than the river water level (25 m).

The total groundwater extraction rate was about 40-50 m^3/h and the average concentration was around 20-30 mg/L (with occasional concentrations of 50-60 mg/L).

Many hydraulic discontinuities were observed in the aquifer at very short lateral distances, in particular high differences of the piezometric level (up to 3 m) between adjacent wells (30 m), and the presence of a paleoaquifer with permeability parameters much higher than the surrounding aquifer.

The TCE concentration value changed with time and this could be correlated with natural fluctuations of the groundwater level caused by rainfall regime and with new spills.

There was no apparent correlation between static level, pumping rates and concentrations of TCE, in that the above mentioned had the same order of magnitude in all the wells. All was caused by evident lateral discontinuities of alluvial deposits present in the area were there was the presence of substantial geologic and hydrogeologic variations.

The possible treatment alternatives for the contaminated water were represented by diffused aeration in a tank and stripping tower. The obtainable levels of decontamination, under certain conditions, could be of the same order of magnitude. There was the possibility of using an existing system in order to treat the air/gas recovered. The system had a capacity of treating roughly 4,000 m^3/h. There was also the possibility of using an existing tank (not equipped for our intention) which could although be easily modified in accordance to our schemes. After considerable technical and economical considerations on the application of both systems the aeration tank was chosen.

For the sizing of the aeration tank, and in particular taking in consideration substantial variations in operating conditions (concentrations, flow rates and temperature) allowing to achieve national environmental standards, mathematical models were developed in order to achieve the desired goal. In support of the model and in order to verify levels of confidence, laboratory tests were carried out to back all considerations made which resulted in a positive confirmation. Given the particular nature of TCE it was not considered propitious to implement an open aeration tank but it was chosen to use a triple stage tank with diffused aeration with a stripping gas recovery system. The adoption of a triple stage aeration tank allowed to avoid building preferential currents in the tank and to have two operating stages in almost countercurrent (first and third), an advantage of more uniform effluent characteristics and of a more global efficiency of the system as a whole.

For the aeration porous tube diffusers were used which had the advantage of allowing an optimal air distribution in the water to be treated and increasing the A/V ratio (interface area between gas and water/unit of volume of water).

The necessary flow rate of air was produced by three blowers having air filters in order to prevent impurities (dust, oil etc.) and possible clogging of the porous tubes.

The stripping tank was covered and three hoods were added (one per stage) to capture the off gas and send it through a dedicated line to the gas treatment facility within the factory. This was all done with the intention of preventing TCE vapor from diffusing within and outside the factory.

In reference to the main system characteristics, considering a 3 m water level, the volume of treatment of each single stage was roughly 95 m^3, which comes out to 300 m^3.

The air used in the stripping tank was 3,600 Nm^3/h (minimum value) divided between the three stages, therefore 1,200 Nm^3/h per stage (about 40 $Nm^3/h/m^2$).

The maximum flow rate of the treated water was between 60 and 70 m^3/h, with a residence time of 4-5 hours.

Management Considerations. The normal funcioning of the wells which form the hydraulic barrier was managed by the on/off level indicator of the single submerged pumps as indicated by the engineering plans and such to guarantee the requested radius of influence. The control room of the plant supervised the operation af all the wells.

The monitoring plan was a weekly monitoring of the following parameters: dynamic level of the aquifer in the water extraction wells, piezometric level in the monitoring wells, flow rate of the single wells (cumulative and instant) and concentration of TCE in the single wells.

In steady state conditions the functioning of each well was decided also on the basis of the weekly monitoring. The control room of the plant also supervised the SVE and air stripping systems.

The monitoring plan for SVE system had also a weekly occurrence. The paramenters measured were: registration of the flow rates, and sampling extracted gas. Also the monitoring plan for the stripping system had a weekly occurrance and monitored: concentrations of TCE in water in inflow and outflow, sampling and analyses of gas and vapor sent to the treatment system within the plant, flow rates and pressure.

CONCLUSIONS

System performance. During the test-run and monitoring period (4 months) TCE concentrations after the diffused air stripping treatment dropped to 0.01 mg/L and a TCE recovery rate of 2.5 Kg/day from the hydraulic barriers was achieved. The spring contamination was drastically reduced.

Quite higher was the global effectiveness of the SVE system: depending on the extraction lines, it was possible to have TCE recovery rates ranging from 12 to 35 Kg/day, i.e. values between 1.6 and 4.3 Kg/day per well.

Recovered TCE was then reused in the production processes.

Project total cost. The realization of the remediation system, starting from the basic design to the test-run, required 9 months.

The cost was about 3 billion Italian Lira, approx. 1,700,000 U.S.Dollars, which appears quite high, but it's to be considered that the system was designed and constructed in accordance with the high standards of the production plant.

AUTHOR INDEX

This index contains names, affiliations, and book/page citations for all authors who contributed to the six books published in connection with the First International Conference on Remediation of Chlorinated and Recalcitrant Compounds, held in Monterey, California, in May 1998. Ordering information is provided on the back cover of this book.

The citations reference the six books as follows:

1(1): Wickramanayake, G.B., and R.E. Hinchee (Eds.). 1998. *Risk, Resource, and Regulatory Issues: Remediation of Chlorinated and Recalcitrant Compounds*. Battelle Press, Columbus, OH. 322 pp.

1(2): Wickramanayake, G.B., and R.E. Hinchee (Eds.). 1998. *Nonaqueous-Phase Liquids: Remediation of Chlorinated and Recalcitrant Compounds*. Battelle Press, Columbus, OH. 256 pp.

1(3): Wickramanayake, G.B., and R.E. Hinchee (Eds.). 1998. *Natural Attenuation: Chlorinated and Recalcitrant Compounds*. Battelle Press, Columbus, OH. 380 pp.

1(4): Wickramanayake, G.B., and R.E. Hinchee (Eds.). 1998. *Bioremediation and Phytoremediation: Chlorinated and Recalcitrant Compounds*. Battelle Press, Columbus, OH. 302 pp.

1(5): Wickramanayake, G.B., and R.E. Hinchee (Eds.). 1998. *Physical, Chemical, and Thermal Technologies: Remediation of Chlorinated and Recalcitrant Compounds*. Battelle Press, Columbus, OH. 512 pp.

1(6): Wickramanayake, G.B., and R.E. Hinchee (Eds.). 1998. *Designing and Applying Treatment Technologies: Remediation of Chlorinated and Recalcitrant Compounds*. Battelle Press, Columbus, OH. 348 pp.

KEYWORD INDEX

This index contains keyword terms assigned to the articles in the six books published in connection with the First International Conference on Remediation of Chlorinated and Recalcitrant Compounds, held in Monterey, California, in May 1998. Ordering information is provided on the back cover of this book.

In assigning the terms that appear in this index, no attempt was made to reference all subjects addressed. Instead, terms were assigned to each article to reflect the primary topics covered by that article. Authors' suggestions were taken into consideration and expanded or revised as necessary to produce a cohesive topic listing. The citations reference the six books as follows:

1(1): Wickramanayake, G.B., and R.E. Hinchee (Eds.). 1998. *Risk, Resource, and Regulatory Issues: Remediation of Chlorinated and Recalcitrant Compounds*. Battelle Press, Columbus, OH. 322 pp.

1(2): Wickramanayake, G.B., and R.E. Hinchee (Eds.). 1998. *Nonaqueous-Phase Liquids: Remediation of Chlorinated and Recalcitrant Compounds*. Battelle Press, Columbus, OH. 256 pp.

1(3): Wickramanayake, G.B., and R.E. Hinchee (Eds.). 1998. *Natural Attenuation: Chlorinated and Recalcitrant Compounds*. Battelle Press, Columbus, OH. 380 pp.

1(4): Wickramanayake, G.B., and R.E. Hinchee (Eds.). 1998. *Bioremediation and Phytoremediation: Chlorinated and Recalcitrant Compounds*. Battelle Press, Columbus, OH. 302 pp.

1(5): Wickramanayake, G.B., and R.E. Hinchee (Eds.). 1998. *Physical, Chemical, and Thermal Technologies: Remediation of Chlorinated and Recalcitrant Compounds*. Battelle Press, Columbus, OH. 512 pp.

1(6): Wickramanayake, G.B., and R.E. Hinchee (Eds.). 1998. *Designing and Applying Treatment Technologies: Remediation of Chlorinated and Recalcitrant Compounds*. Battelle Press, Columbus, OH. 348 pp.

A

accelerated cleanup (groundwater) **1(1):**253
activated charcoal (filter) **1(5):**187
adsorption **1(4):**91, **1(5):**237
air sparging **1(1):**205, **1(4):**51, **1(5):**247, 265, 279, 285, 293
air stripping **1(5):**187, 271, 205, 247, 415, 467, **1(6):**309
analytical methods **1(1):**13, **1(2):**125, **1(3):**111
aquifer modification **1(5):**193
arsenic **1(4):**257

B

barium **1(3):**33
barometric pumping **1(5):**161
base catalyzed degradation **1(6):**195
bedrock **1(2):**7, **1(5):**205
fractured **1(2):**7, **1(5):**205, **1(6):**85
beneficial use (groundwater) **1(1):**151
benzene, toluene, ethylbenzene, and xylenes (BTEX) **1(3):**333, **1(4):**215
bioaugmentation **1(4):**149